Ecological Management of Pine Forests

Ecological Management of Pine Forests

Special Issue Editors

Rodolfo Picchio

Roberto Mercurio

MDPI • Basel • Beijing • Wuhan • Barcelona • Belgrade

Special Issue Editors
Rodolfo Picchio
University of Tuscia
Italy

Roberto Mercurio
Italian Society of Forest Restoration (SIRF)
Italy

Editorial Office
MDPI
St. Alban-Anlage 66
4052 Basel, Switzerland

This is a reprint of articles from the Special Issue published online in the open access journal *Forests* (ISSN 1999-4907) from 2017 to 2018 (available at: https://www.mdpi.com/journal/forests/special_issues/ecological_pine)

For citation purposes, cite each article independently as indicated on the article page online and as indicated below:

LastName, A.A.; LastName, B.B.; LastName, C.C. Article Title. *Journal Name* **Year**, *Article Number*, Page Range.

ISBN 978-3-03897-425-3 (Pbk)
ISBN 978-3-03897-426-0 (PDF)

Cover image courtesy of Burak Kantarci

Contents

About the Special Issue Editors

Rodolfo Picchio obtained a Forest Science degree and a PhD degree in Forest Mechanization from Tuscia University in Viterbo. From 2004 to 2014, he worked as a university researcher and since 2014 has been an Associate Professor in Forest Logging and Wood Technologies. Also since 2004, he has lead the 'Forest Utilization Work Group' at Tuscia University (DAFNE). From 2015 to 2017, he served as the president of the Master program of Forest Science "Conservation and Restoration of the Forest Environment and Soil Defense" at Tuscia University (DAFNE). He has served on the board of the PhD programs of "Science and Technologies for Forest and Environmental Management" (2005–2013) and "Engineering for Energy and Environment" (since 2018) at Tuscia University. His fields of expertise include forest logging, forest mechanization, precision forestry, wood technologies, and the environmental impacts of wood harvesting operations. He has worked as the Italian representative in two EU COST Actions, on silviculture and biomass. Furthermore, he has co-authored 55 peer-reviewed papers in international journals indexed by WoS and Scopus, 15 technical books, and over 150 papers in technical magazines. He also works on the editorial boards of three international scientific journals covering forestry and forest engineering subjects. He is a member of two scientific societies: the Italian Society of Forest Restoration and the Italian Society of Silviculture and Forest Ecology.

Roberto Mercurio graduated from the Forestry Sciences program at the University of Florence (Italy). He worked as a senior researcher from 1979 to 1993 and from 1986 to 1992 as Appointed Research Director of Biology Lab at the Experimental Institute for Silviculture. He was employed as an Associate Professor of Forest Management and Silviculture at the Mediterranean University (Reggio Calabria) from 1993, and as a Full Professor from 2001 until the end of 2011. He has taught also at Marche Polytechnic University (Ancona) and Tuscia University (Viterbo). He has served on the board of the Ph.D. program of "Science and Technologies for Forest and Environmental Management" at Tuscia University (1993–2013). From 2004 to 2009, he was the Director of the Department of Agricultural and Forest Systems Management (GESAF) at the Mediterranean University (Reggio Calabria). Furthermore, he is a member of several scientific societies: the Italian Academy of Forest Sciences, the Italian Society of Forest Restoration, and the Academy of Letters Arts and Sciences "F. Petrarca". His areas of expertise include silviculture, forest management, and forest restoration.

Preface to "Ecological Management of Pine Forests"

Pine forests are widespread natural forest communities, and pine trees are one of the most-used species for reforestation. Therefore, these stands have large economical, ecological, and social importance, mainly due to their resilience against the effects of climatic changes. Still, this resilience needs to be further improved.

Their management attracts the interest of forest engineers, forest owners, and public opinion. As a result, management choices (preservative silvicultural treatment for pine forest or re-naturalization aimed to change forest composition and structure) may sometimes fuel arguments between different opinions in modern society.

This Special Issue contains 15 original papers, some of them reporting the outcomes of long-time experiences, reviewed by international experts in botanic, ecology, silviculture, forest restoration, and forest logging of natural and artificial pine forests, in different environments of boreal and austral hemispheres.

This Special Issue aims to increase the knowledge concerning the ecological management of pine forests, with practical implications at scientific and technical levels.

We are pleased to share these works with the scientific community, forest engineers, private owners, and public managing authorities, in the hopes that this edition will provide a cognitive base to improve pine forests ecological management and reduce arguments between different interests and opinions.

As Guest Editors, we want to express our enormous gratitude to MDPI, who agreed to publish this book; to the staff of Forests and MPDI, for their kindness and professional support; to all the reviewers, for improving original texts and clearing little slips; and to all the authors, for providing their papers with professionalism and scientific rigor. Thanks to the support of these people, this interesting and scrupulous overview of the ecological management of pine forests was made possible.

Guest Editor contributions: the two coeditors equally contributed to organizing the Special Issue, to the editorial work, and to writing this editorial.

Funding: No external funding was obtained for editing this Special Issue and for writing the preface.

Conflicts of Interest: The authors declare no conflict of interest.

This book collects a representative sample of the most recent papers on the subject, which come from many different countries and cover a variety of subjects, confirming the wide scope covered by the ecological management of pine forests.

Rodolfo Picchio, Roberto Mercurio

Special Issue Editors

forests

MDPI

Article

Dispersal Patterns of Pine Wilt Disease in the Early Stage of Its Invasion in South Korea

Won Il Choi [1,†], Hye Jung Song [2,†], Dong Soo Kim [3], Dae-Sung Lee [4], Cha-Young Lee [2], Youngwoo Nam [2], Joon-Bum Kim [3] and Young-Seuk Park [4,5,*]

1. Warm Temperate and Subtropical Forest Research Center of National Institute of Forest Science, Jeju 63582, Korea; wchoi71@korea.kr
2. Division of Forest Insect Pests and Diseases, National Institute of Forest Science, Seoul 02455, Korea; shj4621@naver.com (H.J.S.); ishursain06@korea.kr (C.-Y.L.); orangmania99@korea.kr (Y.N.)
3. Southern Forest Resources Research Center of National Institute of Forest Science, Jinju 52817, Korea; skimds@korea.kr (D.S.K.); jbkim1434@hanmail.net (J.-B.K.)
4. Department of Biology, Kyung Hee University, Seoul 02447, Korea; dleotjd520@naver.com
5. Department of Life and Nanopharmaceutical Sciences, Kyung Hee University, Seoul 02447, Korea

* Correspondence: parkys@khu.ac.kr; Tel.: +82-02-961-0946

† These authors contributed equally to this work.

Received: 26 September 2017; Accepted: 25 October 2017; Published: 30 October 2017

Abstract: We characterized the dispersal patterns of pine wilt disease (PWD) in the early stage of its invasion in the South Korea, and estimated the influence of environmental factors on the dispersal of PWD. Data were obtained in 10 regions with at least five consecutive years of data for 10 years from 1994 to 2005. The dispersal patterns of PWD were categorized into four types: type 1 is a jumping type of dispersal, forming new patches; type 2 infestations are ones without any expansion of patch size; and types 3 and 4, respectively, show uni-directional or multi-directional dispersal outward from an existing patch. Dispersal patterns changed during different phases of the pathogen's invasion history: type 1 was the most frequent in the early invasion stage. Annual dispersal distance showed regional variations. Human population density had a positive correlation with the dispersal distance of PWD, indicating that anthropogenic factors can contribute to the dispersal of PWD. Our results suggested that dispersal through jumping from areas occupied by PWD was the main dispersal route in the early stage of invasion and that after this phase, the existing colonies expanded and merged. These results supported the existence of stratified dispersal patterns of PWD.

Keywords: pine wood nematode; forest disease; spatial dispersal; patch; invasion history; invasive species; stratified dispersal

1. Introduction

Invasive species are among the most severe disturbances affecting ecosystems due to their potentially devastating effects on natural communities [1]. The invasion process can be categorized into four steps: introduction, establishment, expansion, and naturalization [2]. After the successful establishment of an invasive species in a new area, it will continue to occupy the new habitat by expanding its range up to its ecological limits [2]. Thus, the best management strategy to control invasive species is to prevent their establishment in the initial stage of an invasion. If they are established, the next best strategy is to slow down their dispersal speed to minimize their impact on ecosystems. A USA national program to manage the gypsy moth (*Lymantria dispar*) called "slow the spread" is an example of the successful use of this approach [3]. The dispersal speed of gypsy moth was reduced through the eradication of isolated colonies that had formed beyond the population dispersal front of the infested zone [4].

Understanding the dispersal characteristics of a particular invasive species is essential to determining a suitable strategy to slow down its dispersal. Generally, invertebrates display a stratified dispersal based on the combination of long distance dispersal (LDD) and short distance dispersal (SDD) [5]. There are many examples of stratified dispersal including gypsy moth (*L. dispar*) [6], Argentine ant (*Linepithema humile*) [7], and mountain pine beetle (*Dendroctonus ponderosae*) [8].

Pine wilt disease (PWD), caused by the pine wood nematode (PWN), *Bursaphelenchus xylophilus* Nickle, a species native to North America, is one of the most serious threats to pine trees in Asia and Europe [9]. The PWN has invaded many Asian and European countries, including Japan in 1905 [10], China in 1982 [11], Korea in 1988 [2], Portugal in 1999 and Spain in 2008 [12]. The PWN is a tree-parasitic nematode and its length ranges from 0.6 mm to 1.0 mm [2]. It develops to a female or male adult through several stages including egg and four juvenile stages. The third juveniles enter the body of vectors such as *Monochamus alternatus* through the tracheal system. The PWN takes five days to complete a generation at 25 °C [2]. The nematode forms novel mutualistic relationships with native *Monochamus* species, vectors of PWN, allowing the nematode to quickly and safely move between pine trees and increasing the number of freshly killed trees available for the reproduction of *Monochamus* beetles [13]. The area damaged by PWN was 72 ha in 1988 and increased to 7811 ha in 2005 [2], indicating that the PWN has been in the range expansion phase. The dispersal distance of PWD depends on the dispersal capacity of *Monochamus* beetles. Several approaches have been used to measure the dispersal capacity of PWD, including analysis of field occurrence data [2], and flight-mill experiments [14]. Therefore, it is important to identify dispersal patterns of PWD as well as major environmental factors influencing the dispersal speed for the management of PWD [15].

The first occurrence of PWD in the South Korea was reported at Mt. Geumjong in Busan in 1988 [2] although it is suspected that PWD invaded Korea before 1986 [16]. Until 1997, distribution of PWN was restricted to the Busan area. Before the first detection of PWN on Korean white pine (*Pinus koraiensis* Siebold et Zuccarini) near Seoul in 2006, Japanese red pine (*Pinus densiflora* Siebold et Zuccarini), and Japanese black pine (*Pinus thunbergii* Parlatore) were known to be host plants for pine wilt nematode. The Japanese pine sawyer, *Monochamus alternates* Hope, is the PWN vector for these two pine species, whereas *Monochamus saltuarius* Gebler is the vector for Korean white pine.

This study aims to characterize the dispersal patterns of PWD at both the stand (patch) and regional levels in the early stage of PWD invasion in South Korea. In addition, we identified the potential factors such as human population density, which affect the dispersal speed of PWD in the different study regions.

2. Materials and Methods

2.1. PWD Occurrence Data

PWD occurrence data were obtained from the 'Pine Wilt Disease Geographic Information Management System' operated by the Korean Forest Service. The analog maps of PWD occurrence data in 53 regions (i.e., cities, counties, or districts) from 1994 to 2007 were digitalized using ArcView 3.2 and ArcGIS 9.1 (ESRI, Redlands, USA) [17,18]. The dead pine trees were detected by ground survey to define the occurrence area of PWD. The dead trees inside the boundary of the PWD occurrence area in the previous year were considered as trees infected by PWD, whereas the dead trees outside of the boundary were examined for the infection status from the pine wood nematode based on wood samples collected from the trees by nematode experts. PWD occurrence data consisted of polygons without information on the number of trees infected by PWD. To estimate regional dispersal speed, data were selected in 10 regions where the occurrence data were recorded consecutively for at least 5 years (Figure 1). Among them, data from four regions (Gumi, Jinju, Tongyeong and Haman) were from 2001 to 2006, while those in the other six regions (Geoje, Gimhae, Miryang, Busan, Yangsan and Ulsan) were from 2001 to 2005.

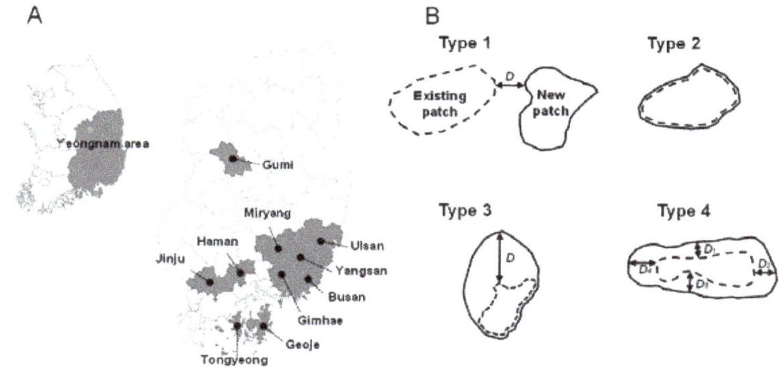

Figure 1. (**A**) Survey areas and (**B**) schematic diagrams for four dispersal types of pine wilt disease (PWD). D is the annual dispersal distance of PWD. Type 1: a jumping dispersal from a patch forming a new patch; type 2: without any dispersal; types 3 and 4 are respectively unidirectional and multi-directional dispersals expanding from an existing patch. Dispersal distance of type 4 is a mean of four Ds D1–D4.

To evaluate the factors affecting dispersal speed, we investigated two anthropogenic factors (road density and human population density) and a forestry factor (proportion of coniferous forests in each administrative region). Human population density was obtained from the Korean Statistical Information Service (http://www.kosis.go.kr), while road density was estimated from the GIS database as a ratio of the road area to total land area in each study region. The proportion of coniferous forest was estimated as a ratio to the total forest area in each region on the 1/25,000 scale map provided by the Korean Forest Service (http://www.forest.go.kr). We considered coniferous forests to be those areas with at least 25% coverage area of coniferous species.

2.2. Dispersal Pattern of Patches

PWD occurrence areas were recorded in polygons in each year. We considered polygons with a direct connection to another polygon as single patches. A patch cluster was defined as a group of patches with less than 2 km of uninfected areas between patches. To classify the dispersal pattern of PWD, annual changes in PWD patch shape and dispersal distances within and between the patches were analyzed at the regional level with reference to invasive history. We divided the regions into two groups based on the invasion history: initial stage of invasion if the regions were infested by PWD within the last four years and later stage of invasion if the regions had an invasion history greater than four years. We examined dispersal patterns for all patches between two consecutive years, and obtained approximately 10 different types with different dispersal distance at different directions. To simplify the analyses, we categorized dispersal patterns into four different types based on the shape of the patches. Type 1 is a jumping type of dispersal in which a new patch is formed, consisting of a colony of pine trees infected by the pine wood nematode; type 2 is a patch that does not expand; while types 3 and 4 of dispersal show uni-directional and multi-directional expansion outward from an existing patch (Figure 1). Dispersal distance from patches was estimated according to the dispersal pattern. The dispersal distance of type 1 events was estimated to be the distance between the nearest existing patch and a newly formed patch in two consecutive years (Figure 1), that of type 2 was considered to be 0 m (Figure 1), that of type 3 was estimated to be the nearest linear distance between the lines of polygons from two consecutive years (Figure 1), and the distance of type 4 was the mean value of dispersal distance of each of the four directional distances between lines of polygons over two consecutive years (Figure 1). We used the term "distance between patch clusters" for the distance between the centers of patch clusters.

2.3. Statistical Analysis

The relationships between dispersal distance and its relative frequency at the regional level were estimated using the exponential decay function [19], showing the best fitting curve based on the CurveExpert (ver. 1.4) [20]:

$$f(x) = N_0 e^{-bx}$$

where $f(x)$ is a frequency at the dispersal distance (m) x, N_0 is a frequency at the first smallest dispersal distance (i.e., initial value), and b is the decay constant.

Linear regression analysis was conducted to estimate the relationship between the dispersal distance of PWD and its related environmental factors at the regional level.

3. Results

3.1. Characteristics of Dispersal Patterns

Of the four dispersal patterns, type 1 was the most dominant (52% in the data) (Figure 2), followed by types 2, 4 and 3, accounting for 27%, 12% and 9% of dispersal, respectively. In particular, type 1 was dominant in the early stage of the invasion process, and its frequency decreased from the fifth year after the invasion (Figure 2). Dispersal distances between patches were 1.00 and 1.03 km at the initial and later stages, respectively (Figure 2), while dispersal distances between patch clusters were 1.15 and 0.74 km at the initial and later stages, respectively (Figure 2).

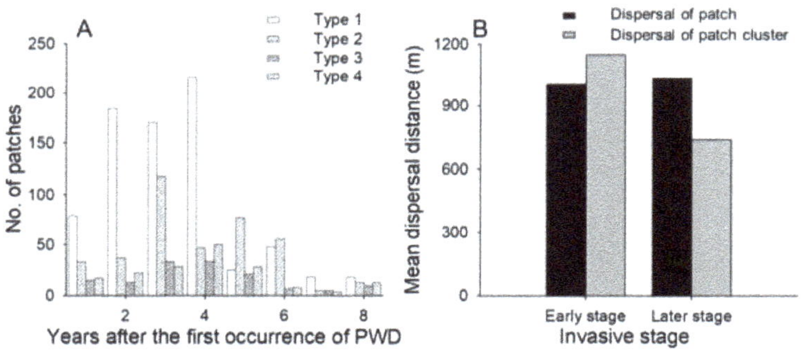

Figure 2. (**A**) Changes in the proportion of dispersal types according to the number of years after the first occurrence of PWD; (**B**) Annual dispersal distance at different invasive stages. Early stage: within four years after the invasion; later stage: more than four years after the invasion. Dispersal distance of a patch was measured as D in Figure 1.

3.2. Annual Dispersal Distance at Patch and Regional Levels

Most study regions displayed a high frequency of short annual dispersal distance of 0.37 km at the patch level, while some regions including Geoje, Busan, Ulsan, and Tongyeong displayed a relatively high frequency of long dispersal distance (Figure 3). Overall, 88.8% of annual dispersal distances were less than 1.0 km considering all regions together (Figure 3). Only 2.5% of annual dispersal exceeded 2.0 km, and the maximum annual dispersal distance was 7.71 km in Geoje (Figure 4).

The relationships between the annual dispersal distance at the patch level and its relative frequency in the ten regions were well fitted by the exponential decay function ($R^2 > 0.99$) with regional variations (Figure 3). The slopes of the function were highest in Miryang and lowest in Ulsan. Dispersal greater than 3 km was observed in Geoje, Gumi and Tongyeong.

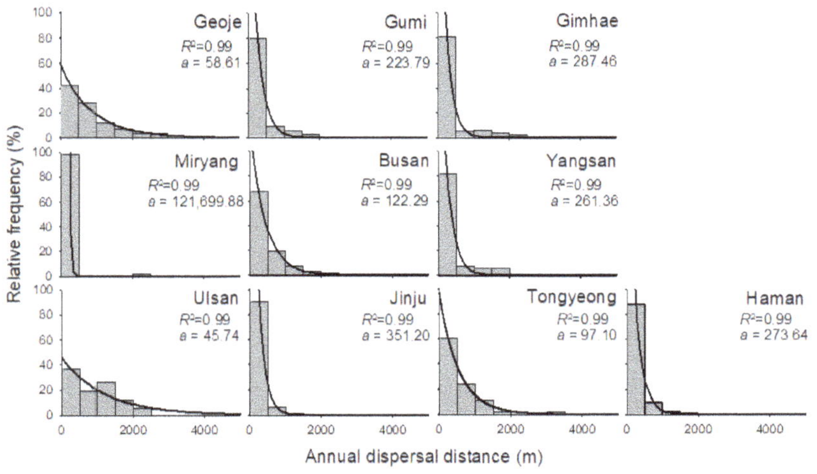

Figure 3. Frequency of annual dispersal distances in ten different study regions: Geoje, Gumi, Gimhea, Miryang, Busan, Yangsan, Ulsan, Jinju, Tongyeong, and Haman. The lines were fitted by the exponential decay function.

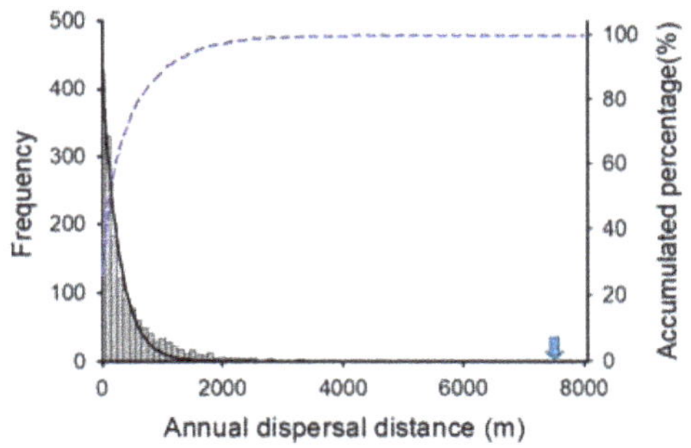

Figure 4. Frequency of annual dispersal distance with pooled data from the ten study regions and its accumulated percentage (dashed line). The solid line was fitted by the exponential decay function. An arrow indicates a maximum value (7710 m) which is not visualized because of a low frequency value.

Mean annual dispersal distances at the regional level varied by region and year. The longest and shortest mean annual dispersal distances at the regional level were 1.48 km in Geoje and 0.01 km in Miryang from 2001 to 2002. Deviations of mean annual dispersal distance between regions (0.71–0.86 km) were larger than deviations between years (0.35–0.61 km).

3.3. Influence of Socio-Environmental Factors

The human population density, road density, and proportion of coniferous forests all varied by region. Human population density was highest in Busan with 45.7 person/ha and lowest in Miryang with 1.4 person/ha. Road density was highest in Gimhae with 15.9% of land covered and lowest in Geoje with 4.0% of land covered. The proportion of coniferous forests was highest in Geoje with 42.5% and lowest in Gimhae with 23.8%. Annual dispersal distances increased as a function of the human population density (F = 11.02; df = 1, 7; $p < 0.013$; R^2 = 0.61) (Figure 5A). However, they were not affected by road density (F = 0.050; df = 1, 8; $p > 0.05$) (Figure 5B) or proportion of coniferous forests (F = 0.13; df = 1, 8; $p > 0.05$) (Figure 5C).

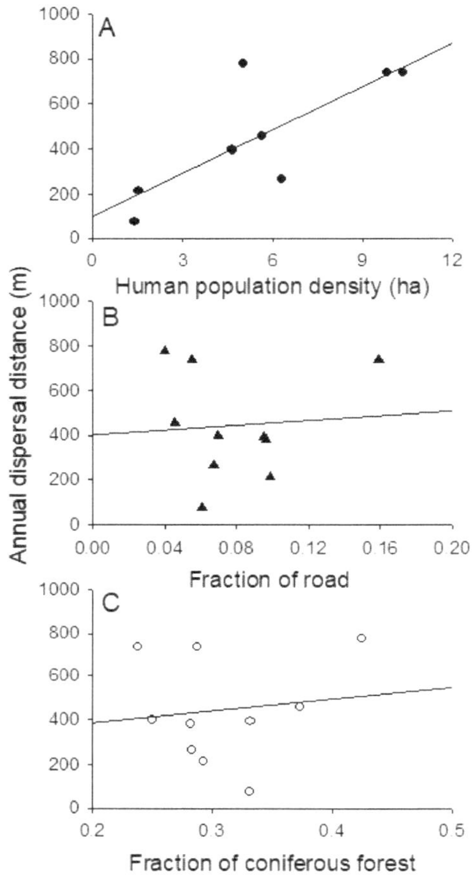

Figure 5. Relationships between annual dispersal distance and (**A**) human population density; (**B**) fraction of coniferous forest to total area and (**C**) fraction of road density to total land cover area.

4. Discussion

4.1. Characteristics of Dispersal Patterns

Our results showed that the frequency of dispersal patterns changed from a predominance of type 1 (displaying a jumping dispersal forming a new patch) at the early stage of the invasion process to types 3 or 4 (showing a uni- or multi-directional dispersal expanding from an existing patch) at the later stage of the invasion. This phenomenon can be attributed to the foraging behavior of the vector infected by pine wood nematode (PWN). Generally, cerambycid beetles, including *Monochamus* species, require weakened or freshly dead trees for oviposition. At the initial invasion stage, beetles infected by PWN search for such trees since available oviposition sites are probably limited, causing them to forage over wider areas to find suitable trees. During this foraging behavior, pine trees are infected by the maturation feeding of the adult beetles, and the number of pine trees infected by PWD increases. Beetle female fecundity is maximized if they can find suitable oviposition hosts near where they emerge, but in the early stages of invasion, suitable hosts so close at hand are often not available.

The foraging behaviors of *Monochamus* species, including *M. alternatus*, have been intensively studied because these species are the major vectors of PWD in many countries: *M. alternatus* in Japan [21–23], M. galloprovincialis in Europe [14,24,25], and M. carolinensis in America [26]. Flight distances of beetles heavily infected by PWN were shorter than those of lightly infected ones [26]. The beetles forage over broader areas to search for weakened or freshly dead pine trees at the initial stage of the PWN invasion. These foraging behaviors could cause the different spatial distributions of pine trees infected by PWN. Moreover, the proportion of pine trees with delayed symptoms may be different at different stages of the invasion. The proportion of trees with delayed symptoms was higher at the initial stage of the invasion than at the later stage, a phenomenon that might be caused by the lower phoretic pressure of PWN.

Kwon et al. [27] suggested that all pine trees close to trees infected by PWD should be cut to control asymptomatic trees infected by PWD with an optimum clear cut radius of 20 m at the initial stage of the invasion and 10 m at the later stage. Our results partially support those of Kwon et al. [27] with respect to the foraging behavior of *Monochamus* beetles, but the exact radius needed for sanitary clear cutting should be re-examined.

4.2. Annual Dispersal Distance at Patch and Regional Levels

The combination of long distance dispersal (LDD) and short distance dispersal (SDD) results in stratified dispersal, including the establishment of new colonies far from the moving population front, growth of individual colonies, and colony coalescence contributing to the advance of the population front [4]. In this study, the annual dispersal distance of *M. alternatus* was 0.37 km and ranged from 0 to 7.71 km. However, careful interpretation on these dispersal distances is required, for several reasons. The first of these is that the estimation of the dispersal distance on the basis of field observation is spatial scale-dependent. Choi and Park [2] showed that the dispersal speed of PWD in Korea changed from 1.1 km/year to 13.8 km/year according to the invasive stage in the nationwide scale. Their dispersal distances were estimated on a relatively large scale based on administrative regions without considering dispersal within the administrative regions, while the dispersal distance in our study was estimated at both patch and regional scales. A second reason is that the dispersal distances include SDD, LDD, and human-mediated dispersal distance. The relative frequency of SDD is higher than those of LDD and human-mediated dispersal distance because 89% of dispersal distances were less than 1 km/year. David et al. [14] reported that the mean dispersal distance of *M. galloprovincialis*, a main European vector of PWD, was 16 km over the lifetime of the beetle on the basis of flight mill experiments. A modeling study showed that the maximum dispersal distance was 464 m, whereas a lifetime dispersal distance of *M. galloprovincialis* was observed in the range of 107 to 122 m in the field [25]. Similarly, the dispersal capacity of *M. alternatus* ranged from 7.1 to 37.8 m in a *P. thunbergii* forest about 50 m in width [23] and the maximum dispersal distance in the forest was 54.9 m and 58.5 m for male and female beetles, respectively [22], suggesting that SDDs of *Monochamus* were a few hundred meters in the field and further dispersal distances might be underestimated. Therefore, the dispersal capacity of *M. alternatus* in Korea probably ranges from a few hundred meters to a few kilometers with a high probability that any single dispersal would be less than 1 km. Moreover, SDD might be the most dominant dispersal of the beetle and LDD and human-mediated dispersal could be the less frequent. This suggests that preventing longer dispersal by mediating human activity is the best strategy to slow down the dispersal speed of PWD.

The last reason to be cautious about dispersal distances is the potential inconsistency of data collection across studies. All data were collected by each local government of Korea, and the boundary criteria for patches could vary. Moreover, an exact criterion of boundary was not defined during data collection [18]. These factors have the potential to lead to differences in the estimation of occurrence areas in each local government and could induce errors in the estimation of dispersal distance of patches and between patches. In spite of these uncertainties, however, our results showed the dispersal characteristics of PWD in the field condition in spite of potential errors in estimating exact dispersal distances.

4.3. Influence of Socio-Environmental Factors on Dispersal

Interestingly, the dispersal speed of PWD was high in areas with high human population density, suggesting that human-mediated dispersal accelerates the dispersal speed of PWD. The role of human-mediated dispersal in the spread of PWD was also reported in China [28,29]. Human-mediated dispersal plays an important role in the invasion process of many organisms: the emerald ash borer, *Agrilus planipennis*, in the USA [30], the yellow-legged hornet, *Vespa velutina* in France [31] and 17 invasive plants in China [32]. Human activities such as logging and trade which require wooden packaging material can increase the risk for the accidental transportation of infected materials by PWN and may be responsible for the rapid spread of the nematode [28]. According to the Korea Forest Service [33], the human-mediated movement of wooden materials infected by PWN was the main cause of new occurrences of PWD in eight administrative regions where the first invasion of PWD was reported in 2013. Besides the movement of wooden material infected by PWN, the accidental movement of beetles vectoring PWD by vehicle may be one of the causes for the rapid dispersal of PWD in Korea. Similarly, adults of walnut husk fly, *Rhagoletis completa*, were found to be transported passively by vehicles in Europe [34].

In spite of the possibility that beetle movement was facilitated by vehicles, road density was not found to be correlated with the dispersal speed of PWD in our analysis. This might be because our analysis focused on movements within administrative regions on a relatively small scale. In China, long-distance human-mediated dispersal by wooden material infected by PWD was observed, ranging from 111 to 339 km [28], suggesting that such longer dispersal distance by human-mediated dispersal was not included in our data. In addition, our data reflected only the road density rather than traffic density. Finally, our data were not based on individual trees, but based on the area for the occurrence of PWD. Therefore, estimating the influence of road density on the dispersal speed of PWD was limited by the nature of our data. The influence of the proportion of coniferous forest was not probably observed for similar reasons. Similarly, the dispersal speed of pine needle gall midge (*Thecodiplosis japonensis*), an invasive pest on pine trees in Korea, was dependent on the density of pine forests in the landscape, with higher speed associated with lower density of pine forest [2]. In highly coniferous forests, insect vectors are more sedentary because favorable feeding conditions are locally abundant.

5. Conclusions

The dispersal patterns of PWD were categorized into four types based on the dispersal rate and directions. The dispersal patterns were changed by invasive history from type 1 (i.e., jumping type of dispersal, forming new patches) in the initial stage of invasion to type 4 (multi-directional dispersal outward from an existing patch) in the later stage, suggesting that dispersal through jumping from areas occupied by PWD was the main dispersal route in the early stage of invasion, and human activity plays an important role in the long-term dispersal of PWD. These stratified dispersal patterns could be implemented in the dispersal spatial models in future studies.

Acknowledgments: This study was supported by the National Institute of Forest Science and R&D Program for Forest Science Technology (Project No. 017042B10-1723-CA01)provided by Korea Forest Service (Korea Forestry Promotion Institute).

Author Contributions: W.I.C. and Y.-S.P. conceived and designed the study; D.S.K., C-Y.L. and J.-B.K. contributed to collect data; W.I.C., H.J.S., D.-S.L., Y.N. and Y.-S. P. analyzed the data; W.I.C, H.J.S., Y.N. and Y.-S.P. wrote the paper.

Conflicts of Interest: The authors declare no conflict of interest.

References

1. Mack, R.N.; Simberloff, D.; Lonsdale, W.M.; Evans, H.; Clout, M.; Bazzaz, F.A. Biotic Invasions: Causes, epidemiology, global consequences, and control. *Ecol. Appl.* **2000**, *10*, 689–710. [CrossRef]
2. Choi, W.I.; Park, Y.-S. Dispersal patterns of exotic forest pests in South Korea. *Insect Sci.* **2012**, *19*, 535–548. [CrossRef]
3. Sharov, A.A.; Leonard, D.; Liebhold, A.M.; Roberts, E.A.; Dickerson, W. "SLOW THE SPREAD" a national program to contain the gypsy moth. *J. For.* **2002**, *100*, 30–35.
4. Sharov, A.A.; Liebhold, A.M. Model of slowing the spread of gypsy moth (Lepidoptera: Lymantriidae) with a barrier zone. *Ecol. Appl.* **1998**, *8*, 1170–1179. [CrossRef]
5. Hengeveld, E. *Dynamics of Biological Invasions*; Chapman and Hall: London, UK, 1989; 160p.
6. Tobin, P.C.; Blackburn, L.M. Long-distance dispersal of the gypsy moth (Lepidoptera: Lymantriidae) facilitated its initial invasion of Wisconsin. *Environ. Entomol.* **2008**, *37*, 87–93. [CrossRef]
7. Suarez, A.V.; Holway, D.A.; Case, T.J. Patterns of spread in biological invasions dominated by long-distance jump dispersal: Insights from Argentine ants. *Proc. Natl. Acad. Sci. USA* **2001**, *98*, 1095–1100. [CrossRef] [PubMed]
8. Chen, H.; Walton, A. Mountain pine beetle dispersal: Spatiotemporal patterns and role in the spread and expansion of the present outbreak. *Ecosphere* **2011**, *2*, 1–17. [CrossRef]
9. Jung, C.S.; Koh, S.-H.; Nam, Y.; Ahn, J.J.; Choi, W.I. A forecasting model for predicting the spring emergence of *Monochamus saltuarius* (Coleoptera: Cerambycidae) on Korean white pine, *Pinus koraiensis*. *J. Econ. Entomol.* **2015**, *108*, 1830–1836. [CrossRef] [PubMed]
10. Togashi, K.; Shigesada, N. Spread of the pinewood nematode vectored by the Japanese pine sawyer: Modeling and analytical approaches. *Popul. Ecol.* **2006**, *48*, 271–283. [CrossRef]
11. Wang, Z.; Zhang, Y.; Shi, J.; Luo, Y.-Q.; Ren, L.-L.; Shi, Y.-M. Effect of pine wood nematode invasion on pine community functions in the Pinghu region, Zhejiang Province, Eastern China. *For. Sci. Pract.* **2013**, *15*, 302–309. [CrossRef]
12. Abelleira, A.: Picoaga, A.; Mansilla, J.P.; Aguin, O. Detection of *Bursaphelenchus xylophilus*, causal agent of pine wilt disease on *Pinus pinaster* in Northwestern Spain. *Plant Dis.* **2011**, *95*, 776. [CrossRef]
13. Akbulut, S.; Stamps, W.T. Insect vectors of the pinewood nematode: A review of the biology and ecology of *Monochamus* species. *For. Pathol.* **2011**, *42*, 89–99. [CrossRef]
14. David, G.; Giffard, B.; Piou, D.; Jactel, H. Dispersal capacity of *Monochamus galloprovincialis*, the European vector of the pine wood nematode, on flight mills. *J. Appl. Entomol.* **2014**, *138*, 566–576. [CrossRef]
15. Park, Y.-S.; Chung, Y.-J.; Moon, Y.-S. Hazard ratings of pine forests to a pine wilt disease at two spatial scales (individual trees and stands) using self-organizing map and random forest. *Ecol. Inform.* **2013**, *13*, 40–46. [CrossRef]
16. Enda, N. The status of pine wilting disease caused by *Bursaphelenchus xylophilus* (Steiner et Buhrer) Nickle and its control in Korea. *J. Korean For. Soc.* **1989**, *78*, 248–253.
17. Environmental Systems Research Institute (ESRI). *ArcGIS 9.1*, ESRI: Redlands, CA, USA, 2005.
18. Kim, D.Y. Development of Pine Wilt Disease Spatial Information Management System Using Web-GIS. Master's Thesis, The Graduate School, Kyungil University, Gyeongsan, Korea, 2008. (In Korean).
19. SAS Institute. *SAS/STATVR 9.2 User's Guide*; SAS Institute: Cary, NC, USA, 2010.
20. Hyames, D. *CurveExpert Version 1.4*, A Comprehensive Curve Fitting System for Windows; Daniel G. Hyams: Hixson, TN, USA, 2009.
21. Shibata, E. Spatial distribution pattern of the Japanese pine sawyer, *Monochamus alternatus* Hope (Coleoptera: Cearambycidae), on dead pine trees. *Appl. Entomol. Zool.* **1984**, *19*, 361–366. [CrossRef]
22. Shibata, E. Dispersal movement of the adult Japanese pine sawyer, *Monochamus alternates* Hope (Coleoptera: Cerambycidae) in a young pine forest. *Appl. Entomol. Zool.* **1986**, *21*, 184–186. [CrossRef]
23. Togashi, K. A field experiment on dispersal of newly emerged adults of *Monochamus alternatus* (Coleoptera: Cerambycidae). *Appl. Entomol. Zool.* **1990**, *32*, 184–186.
24. Etxebeste, I.; Sanchez-Husillos, E.; Álvarez, G.; Mas i Gisbert, H.; Pajares, J. Dispersal of *Monochamus galloprovincialis* (Col.: Cerambycidae) as recorded by mark–release–recapture using pheromone traps. *J. Appl. Entomol.* **2016**, *140*, 485–499. [CrossRef]

25. Torres-Vila, L.M.; Zugasti, C.; De-Juan, J.M.; Oliva, M.J.; Montero, C.; Mendiola, F.J.; Conejo, Y.; Sánchez, Á.; Fernández, F.; Ponce, F.; et al. Mark-recapture of *Monochamus galloprovincialis* with semio chemical baited traps: Population density, attraction distance, flight behavior and mass trapping efficiency. *Forestry* **2015**, *88*, 224–236. [CrossRef]
26. Humphry, S.J.; Linit, M.J. Effect of pinewood nematode density on tethered flight of *Monochamus carolinensis* (Coleoptera: Cerambycidae). *Environ. Entomol.* **1989**, *18*, 670–673. [CrossRef]
27. Kwon, T.S.; Shin, J.H.; Lim, J.H.; Kim, Y.K.; Lee, E.J. Management of pine wilt disease in Korea through preventative silvicultural control. *For. Ecol. Manag.* **2011**, *261*, 562–569. [CrossRef]
28. Robinet, C.; Roques, A.; Pan, H.; Fang, G.; Ye, J.; Zhang, Y.; Sun, J. Role of human-mediated dispersal in the spread of the pinewood nematode in China. *PLoS ONE* **2009**, *4*, e4646. [CrossRef] [PubMed]
29. Robinet, C.; Van Opstal, N.; Baker, R.; Roques, A. Applying a spread model to identify the entry points from which the pinewood nematode, the vector of pine wilt disease, would spread most rapidly across Europe. *Biol. Invasions* **2011**, *13*, 2981–2995. [CrossRef]
30. Prasad, A.M.; Iverson, L.R.; Peters, M.P.; Bossenbroek, J.M.; Matthews, S.N.; Sydnor, T.D.; Schwartz, M.W. Modeling the invasive emerald ash borer risk of spread using a spatially explicit cellular model. *Landsc. Ecol.* **2010**, *25*, 353–369. [CrossRef]
31. Robinet, C.; Suppo, C.; Darrouzet, E. Rapid spread of the invasive yellow-legged hornet in France: The role of human-mediated dispersal and the effects of control measures. *J. Appl. Ecol.* **2017**, *54*, 205–215. [CrossRef]
32. Horvitz, N.; Wang, R.; Wan, F.-H.; Nathan, R. Pervasive human-mediated large-scale invasion: Analysis of spread patterns and their underlying mechanisms in 17 of China's worst invasive plants. *J. Ecol.* **2017**, *105*, 85–94. [CrossRef]
33. Korea Forest Service (KFS). *Statistical Yearbook of Forestry*; Korea Forest Service: Daejeon, Korea, 2014; 473p. (In Korean with English summary)
34. Verheggen, F.; Verhaeghe, A.; Glordanengo, P.; Tassus, X.; Escobar-Gutiérrez, A. Walnut husk fly, *Rhagoletis complete* (Diptera: Tephritidae), invades Europe: Invasion potential and control strategies. *Appl. Entomol. Zool.* **2017**, *52*, 1–7. [CrossRef]

forests

MDPI

Article

Post-Fire Regeneration and Diversity Response to Burn Severity in *Pinus halepensis* Mill. Forests

Sonsoles González-De Vega, Jorge de las Heras and Daniel Moya *

Escuela Técnica Superior Ingenieros Agrónomos y Montes, Universidad de Castilla-La Mancha, Campus Universitario, 02071 Albacete, Spain; Sonsoles.GonzalezVega@alu.uclm.es (S.G.-D.V.); jorge.heras@uclm.es (J.d.l.H.)

* Correspondence: daniel.moya@uclm.es; Tel.: +34-967-599-200

Received: 20 April 2018; Accepted: 24 May 2018; Published: 26 May 2018

Abstract: In recent decades, fire regimes have been modified by various factors such as changes in land use, global change or forest management policies. The vulnerability of Mediterranean terrestrial ecosystems is increasing due to more severe and frequent droughts. This study aimed to determine the plant response of ecosystems during the short-term post-fire period by relating alpha diversity, floristic richness and tree recruitment dynamics to burn severity 5 years after a wildfire. Our results conclude that in the short term, *Pinus halepensis* Mill. stands in southeastern Spain quickly recovered alpha diversity values, mainly in areas burned with low severity. We observed that moderate and high severities affected the ecosystem more significantly, showing higher values for the Shannon Index but lower for the Simpson index. Pine recruitment was higher in burned areas, and we found the highest number of Aleppo pine seedlings under a moderate burn severity. Post-fire regeneration functional groups (obligate seeders and resprouters) were promoted under moderate and high burn severity, increasing their abundance. Annual species (mainly herbs) colonized burned areas, persisting with higher presence under moderate burn severity. Restoration tools should be focused on reducing fire severity, mainly in areas at high risk of desertification, and promoting resistance, vulnerability and resilience of these ecosystems.

Keywords: burn severity; fire ecology; ecosystem recovery; natural regeneration; Aleppo pine recruitment; post-fire restoration

1. Introduction

Forest fires have been described as natural processes inherent to terrestrial biomes as they modify ecosystems to shape the current general global distribution [1,2]. For this reason, forest fires and climate are key processes that condition vegetal dynamics, biome development, soil formation, and hydrological and erosive cycles [3]. The first signs of natural forest fires date back to the Carboniferous Era, and human activity managing fire was evidenced in the early Holocene in the Mediterranean Basin, changing the natural fire regime [4].

A fire regime is defined as the characteristic of wildfire activity that prevails in a given area, mainly determined by fuel consumption patterns, intensity and severity, fire frequency, patch size and seasonality [5]. The fire regime in the Mediterranean Basin has been affected by several factors, mainly by changes in land use, global warming, prolonged drought period, changes in forest policies, and fauna or invasive plant species [6].

Wildfires of the Mediterranean Basin are a dramatic hazard for dense populations but also for terrestrial ecosystems, since alteration of fire patterns modifies vegetation resilience to fire, and predictions point to a general rise in fire risk [7]. The main changes include increases in large forest fires, which promote the size of burned areas affected by high fire severity [8,9]. In this way, monitoring and data assessment of ecosystem response to disturbances is a research priority, and Mediterranean terrestrial ecosystems serve as exemplary natural laboratories in which to study impacts and drivers [10].

The predominant plant communities in the Mediterranean Basin are characterized by their rate of natural recovery to reach a previous equilibrium following disturbance, termed engineering resilience [11]. The two most common regeneration strategies after fire are resprouters (buds in soil or protected by tissues) or obligate seeders (seeds from unburned areas, buried in the soil bank or enclosed in aerial banks) [12,13]. In some cases, seed germination can be initiated or improved by heat, the presence of burnt wood or ash as scarifying agents [14,15]. High temperatures can also trigger the opening of serotinous cones or fruits to release seeds after a fire [16].

In the Mediterranean Basin, modal natural fire frequency ranges from 25 to 50 years, which promotes a high alpha diversity after disturbance, but shorter-lived species are rapidly excluded [17]. However, fire exclusion reduces fire frequency, and promotes fire severity, which could induce changes in post-fire recovery patterns related to burn severity [18]. To avoid confounding recovery patterns related to other fire regime components, monitoring of ecosystems in the same post-fire succession stage, but burned on different dates, is required [10]. However, the large spatial scales and time periods needed to establish a fire history in an ecosystem lead to pseudo-replication [19] and have been proven to be unavoidable in our research on burn severity and α-diversity.

There are several studies on *Pinus halepensis* Mill. (Aleppo pine) forests in the western Mediterranean Basin and on the effects that fire severity has on plant recovery and pine recruitment, diversity and time to recover ecosystem services (e.g., soil protection, vulnerability to new fire, nutrient cycling, etc.) [13,16,18,20–22]. However, the importance of burn severity and the short-term ecosystem changes are important for management decisions about potential intervention in burned areas [18,23,24]. Aleppo pine is one of the most widely distributed pine species throughout low altitudinal areas of the Mediterranean Basin; this is a tree species that has adapted to fire and drought [25,26]. This pine species is a pioneering and obligate seeder, shows precocious reproduction and a dual-life strategy, bearing partially non-serotinous cones releasing seeds and storing a canopy seed bank in serotinous cones to be released after fire [27,28]. However, the amount of serotinous cones increases with higher fire recurrence [29]. Burn severity, defined as the loss or change in organic matter both aboveground and belowground [28], influences the dynamics of the structure and vegetal composition after a fire [30]. In this way, the ecosystem's ability to return to the pre-fire state, fulfilling similar functions and services as those prior to the disturbance, could be affected [11].

The objective of this research was to assess the effect of burn severity in the short term after fire on vegetation dynamics and community structure, including pine recruitment, plant regeneration and diversity performances following a diachronic approach. For this purpose, we carried out floristic inventories by recording the recruitment of Aleppo pine and the main companion species in four different burn severity classes in stands in the south-eastern part of the Iberian Peninsula with a semiarid climate. Our goal was to improve knowledge to support tools for post-fire planning and decision making for the stabilization, rehabilitation and restoration of burned areas depending on the fire damage.

2. Materials and Methods

2.1. Study Area

A wildfire from 1–6 July 2012 burned almost 6500 ha of the forests in "Sierra de los Donceles", close to Hellin (the south-eastern area of the province of Albacete in Spain (ignition point: 30 S 611,807 E, 4,251,765 N)) (Figure 1) where no recurrent fires occurred for more than 80 years. During the wildfire, the maximum temperatures ranged from 31 to 37 °C (minimum average temperature was 16 °C), mean humidity ranged from 30 to 80% (although was less than 20% on the first day) and the mean wind speed was 1–3 m s^{-1}, reaching gusts of 7–9 m s^{-1}. The fire perimeter and weather conditions were provided by the Forest Services of the Regional Government of Castilla-La Mancha.

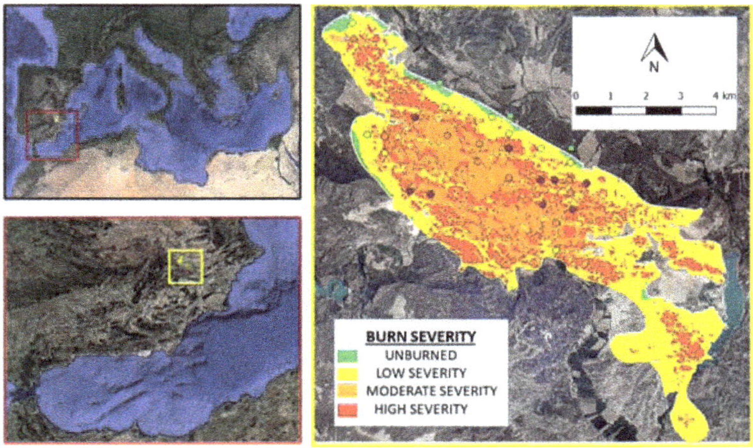

Figure 1. Sierra de Los Donceles wildfire, close to Hellin (province of Albacete, Spain). Burn severity levels were: unburned (uncolored), low severity (yellow), moderate severity (orange) and high severity (red) obtained from [35]. The sampling plots were represented according burn severity: unburned in green, low severity in yellow, moderate severity in orange and high severity in red.

The study area is a semiarid Mediterranean climate, BSk, according to the Köppen-Geiger classification [31], which is in the upper meso-Mediterranean bioclimatic belt. The average precipitation and annual temperature values are 278.50 mm and 15.85 °C, respectively (based on the 25-year data period (1990–2014) provided by the National Meteorological Agency). The dry period extends from June to September, during which relative humidity is less than 50%. According to Garcia-Morote et al. [32], the total precipitation values recorded in the study area from early 2014 to late 2016 were slightly higher than the average values.

According to the Spanish Soil Map and following the USDA Soil Classification [33], the studied soils were classified as Aridisols (Lithic Haplocalcids), and as calcic soils with long periods (>90 days) of water not available for plants. According to Garcia-Morote et al. [32], they are alkaline soils with a clay content above 30% with a high cation exchange capacity (CEC) and low available P. The predominant landform is composed of dolomitic limestones forming rocky slopes that range from 500 to 700 m a.s.l.

The potential vegetation series belonged to *Rhamno lycioidis-Querceto cocciferae* sigmetum, but *Pinus halepensis* was the main tree species. The most common plant community in the area before fires occurred was *Pinus halepensis* Mill. (Aleppo pine) forests, accompanied by several scrub species and steppic grasslands with *Macrochloa tenacissima* (L.) Kunth (Alpha grass), *Quercus coccifera* L. (Kermes oak) and *Pistacia lentiscus* L. (Mastic tree).

The study site and burn severity were evaluated by using remote sensing and digital cartography provided by the Geographic Information National Centre, specifically, Digital Terrain Models (pixel size of 5 m, matrix ASCII file format, geodesic reference system ETRS89), and orthophotos (National Aviation Plan Orthophotography mosaics; ECW format, geodetic reference system ETRS89). The distribution units were set according to MTN50 (Spanish National Topographic Map, 1:50,000)) to create a map with GIS software (QGIS, GNU General Public License, June 1991, 2.8.2-Wien). Remote-sensing imagery from pre- and post-fire data (Enhanced Thematic Mapper Plus (ETM+) sensor on Landsat 7), after atmospheric correction undertaken by using the dark body subtraction technique and gap-filled, was used to create a burn severity map by calculating the dNBR (delta normalized burn ratio; [34]) by using pre- and post-fire images (23 June 2012 and 9 August 2012, respectively). We implemented a supervised classification by using real-ground data to identify and discriminate the categories of burn severity to be discriminated [35]. Following the Spanish technical guide for burned forest management [36], a systematic scheme overlapped the burned area (500 × 500 m grid cell) to validate the burn severity obtained. In 30 plots of the grid, we evaluated the burn severity in the field (November 2012) by recording ash color, soil affected and amount of burned biomass (in both the crown of trees and shrubs) [37].

2.2. Field Sampling

Following Gomez-Sanchez et al. [35], for our study area, we established four levels of burn severity (SEV hereafter): unburned for dNBR < 0.300 (UB), low severity for the values of dNBR ranging from 0.300 to 0.486 (Low), moderate severity for the dNBR threshold from 0.487 to 0.567 (Mod) and high severity for those higher than 0.568 (High).

We analyzed the response and dynamics of the plant community depending on the time after fire (TIME hereafter). A diachronic approach was followed, including three data collection campaigns (late spring in 2013, 2015 and 2017) to characterize 1, 3 and 5 years post-fire (short-term period after fire). To acquire information about the mature plant community, three control plots were sampled in undisturbed and mature stands. In the burned areas, we set nine circular plots (5-m radius; 78.5 m^2) in each burn severity class, by using a Monterra GPS receiver (Garmin International, Inc., Olathe, KS, USA). A total of thirty plots were established and monitored in the spring of 2013, 2015 and 2017.

The selected plots had a similar orientation, slope and percentage of bare soil to reduce variability. Regarding vegetation, both structure and composition were evaluated by recording the coverage and plant abundance using a nested sampling technique [38]. Floristic inventory of all species in the plot followed the Braun-Blanquet approach, modified by Van der Maarel [39], and by the line transect method for plant surveys by recording species and the length intersecting the line (all on a perpendicular plane). Using the georeferenced center of the plot, three transects (3-m long) were arranged, with one oriented northward and forming 120° angles. We also established three square subplots (1 m^2) in each transect to record and monitor pine seedlings. The plant survey and quantification of land cover type were carried out by the point intercept method because it is simple, unbiased, low-cost and suitable for our study area [40].

2.3. Plant Diversity and Regeneration

To characterize the diversity within a community, we used three indices related to α-diversity. We regarded the number of plant species or floristic richness (FR, number of species) and heterogeneity, including the measure of species abundance, i.e., the Shannon index (H′, nondimensional) [41] and the measure of species dominance, i.e., the Simpson index (D, nondimensional) [42].

FR was calculated as the number of plant species recorded on the sampling lines in each plot, including additional sampling in the entire plot to account for all those present in each one. H′ was measured as the relative abundance of the different species (1), and D evaluated the probability that two individuals chosen at random belonged to the same species (2).

$$H' = -\sum_{i=1}^{S} p_i \ln p_i \quad (1)$$

where p_i = coverage of each species (%)

$$D = 1 - \frac{\sum n(n-1)}{N(N-1)} \quad (2)$$

where n = % coverage of each species and N = % total of the present organisms.

To characterize the Aleppo pine recruitment rate, we counted and recorded all the pine seedlings inside the three-square subplots (1 × 1 m, 1 m^2). Pine recruitment was calculated as the mean pine seedling density (RECRUIT, pine m^{-2}).

We classified the species identified in the field sampling according to their adaptive strategy to fire following the plant trait database for Mediterranean Basin species (BROT): non-adaptive species (R−S−); obligate seeders (R−S+); obligate resprouters (R+S−); facultative seeders (R+S+) [43]. To obtain life-form compositions, plant species were classified according to the life-form categories: trees (TREE), shrubs (SHRUB), dwarf shrubs (DWARF-S) and herbs (HERB) [44,45].

2.4. Statistical Analysis

Generalized linear models (GLMs) were used to evaluate the effects of categorical factors (TIME and SEV) and their interactions on the studied variables. Squared correlation coefficients and empirical *p*-values were calculated by one-way analysis of variance (ANOVA). We selected a critical *p*-value below 0.05 for the significant results. We also used the adjusted R-squared statistic, which is more suitable for comparing models than the R-squared value with different numbers of independent variables, and then the Durbin-Watson statistic to test residuals to determine whether there was any significant correlation based on the order in which they occurred in the data files. Log-transformed or rank-transformed data were used for those variables that did not meet the assumptions of normality of residuals and homogeneity of variance. The method currently used to discriminate among means is Tukey's honestly significant difference (HSD) procedure.

Principal component analysis (PCA, varimax rotation method) was performed with all variables to obtain the structure of the dependence and correlation between variables and factors, at two levels (time after fire and burn severity). The PCA created a standardized correlation matrix covariance list-wise and reduced the variables used by employing linear combinations and accounted for most of the variability contained in the original data. Statistical analyses were performed using Statgraphics Centurion 18.1.03 (Statgraphics Technologies, Inc., Virginia, USA) [46] and the R software 3.2.1 (R Development Core Team) [47].

3. Results

In a first step, we verified that the control plots displayed no variations in the sampling period for structure and plant characteristics since the ANOVA for FR, H′, D and RECRUIT showed no significant differences. By running a GLM to analyze the result of the variables TIME, SEV and interactions were meaningful (Table 1). We found RECRUIT and FR significantly related to SE. H′ was significantly influenced by TIME and the interaction TIME*SEV, whereas D was significantly linked to TIME and SEV.

Table 1. General Linear Models relating α-diversity variables: H′ (Shannon Index, nondimensional), D (Simpson Index, nondimensional), RECRUIT (recruitment, pine seedlings m^{-2}) and FR (floristic richness, number of species).

	p-Value	F-Ratio	Adj R^2	Durbin Watson	Factor	F-Ratio	*p*-Value
H′	<0.01	18.78	27.66	1.20	TIME TIME*SEV	28.96 6.42	<0.01 0.01
D	<0.01	29.39	37.91	1.77	TIME SEV	17.45 37.39	<0.01 <0.01
RECRUIT	0.01	6.42	5.56	2.18	SEV	6.42	0.01
FR	0.02	5.26	4.34	1.51	SEV	5.26	0.02

Factors studied were years after fire (TIME), burn severity level (SEV) and the interaction of both (TIME*SEV).

The ANOVA showed a significant influence of TIME and SEV on the studied variables (Figure S1). Independent of TIME, the values of variables studied in the control plots were not different, which implied that mature stands showed the same characteristics in the short term after the fire. One year after the fire, H′ and D were lower in burned plots than in unburned areas, whereas FR and RECRUIT were higher in burned areas. Three of the studied parameters (H′, RECRUIT and FR) increased three years after the fire, with all three SEV classes having higher values than UB, whereas D showed lower values. Five years after, H′ and FR (in all three burn severity classes) decreased to reach similar values to those found in mature stands. However, RECRUIT was significantly higher in the Low and Mod classes, but D showed lower values in Mod and High than in UB.

With the GLM, we ran ANOVAs for each variable according to the significant parameters. According to H′ (Figure 2), we observed that 1 year after the fire, H′ presented lower values in the burned plots than in the control plots, and roughly similar values among the severity values 3 years after the fire, H′ values increased, which were higher than the control plots with increases noted for SEV, and with higher values according to increasing burn severity. However, H′ showed a maximum value 3 years after the fire, whereas the values decreased 5 years later, with no differences between unburned plots and those burned with low severity.

According to the analysis of the Simpson Index (Figure 3), we found that within 1 and 3 years after the fire, the D values were more similar than 5 years after the fire, which presented slightly higher values. However, when considering severity in the short term, the D values of the burned plots were always below the values presented by the control plots, and severity had a significant effect and resulted in the highest D values under low severities.

When analyzing the results obtained from the analysis without the control plots, we found that the plots affected by low severities presented slightly higher values than those affected by moderate severities, with two significantly different homogeneous groups appeared, while the plots affected by high severities presented medium values that belonged to both groups.

According to the ANOVA carried out to analyze RECRUIT (Figure 4) related to SEV, we observed that RECRUIT was lower in the UB plots than in burned areas, with the exception of High SEV, and the highest RECRUIT was found in Mod SEV (Figure 4a). We ran an ANOVA with only data from burned areas (Figure 4c) and found no significant differences for Low and High SEV, but values were lower than those observed for Mod. Regarding FR (Figure 4b), the study showed that in the short term, the values of this parameter were always higher than for the control plots, and these values were significantly similar in severity terms, although the plots affected by moderate severities had slightly higher values. When we ran a one-way ANOVA with no control plots, we observed no significant differences in the FR values according to severity.

When we carried out a descriptive analysis to represent the presence of species according to their adaptive strategy, we represented the mean values related to TIME and SEV (Table 2). We found that 1 year after the fire, the most representative adaptive strategy, independent of the burn severity level, was R−S+ (with almost 50% of the species present). During our study period, R−S+ maintained high percentages of species' presence (always >40%) with the highest values for the Mod and High severities. However, these values slightly decreased with time. We found no changes in the percentage of the presence of R+S−, showing approximately 20% regardless of burn severity. Finally, the percentages of R+S+ were close to 20–25%, which corresponded to an increase 5 years after the fire, with similar values to the mature stands (30%).

According to growth forms and after another descriptive analysis, we observed that 1 year after the fire (in 2013), the most representative growth forms were HERB because these species had the highest presence percentages. For SEV, HERB showed a higher presence in Low SEV, decreasing according to increasing SEV. However, 3 years after the fire, the HERB percentage decreased and was not related to SEV because we found similar values in all the burned plots, with the predominant growth shown for DWARF-S. Five years after the fire, the presence of HERB and Dwarf-S was poor in the burned plots, with SHRUB as the most common growth form. This stage showed similar percentages for each growth form as the mature and unburned stands. TREE values had the lowest percentage values, which were similar when comparing TIME and SEV.

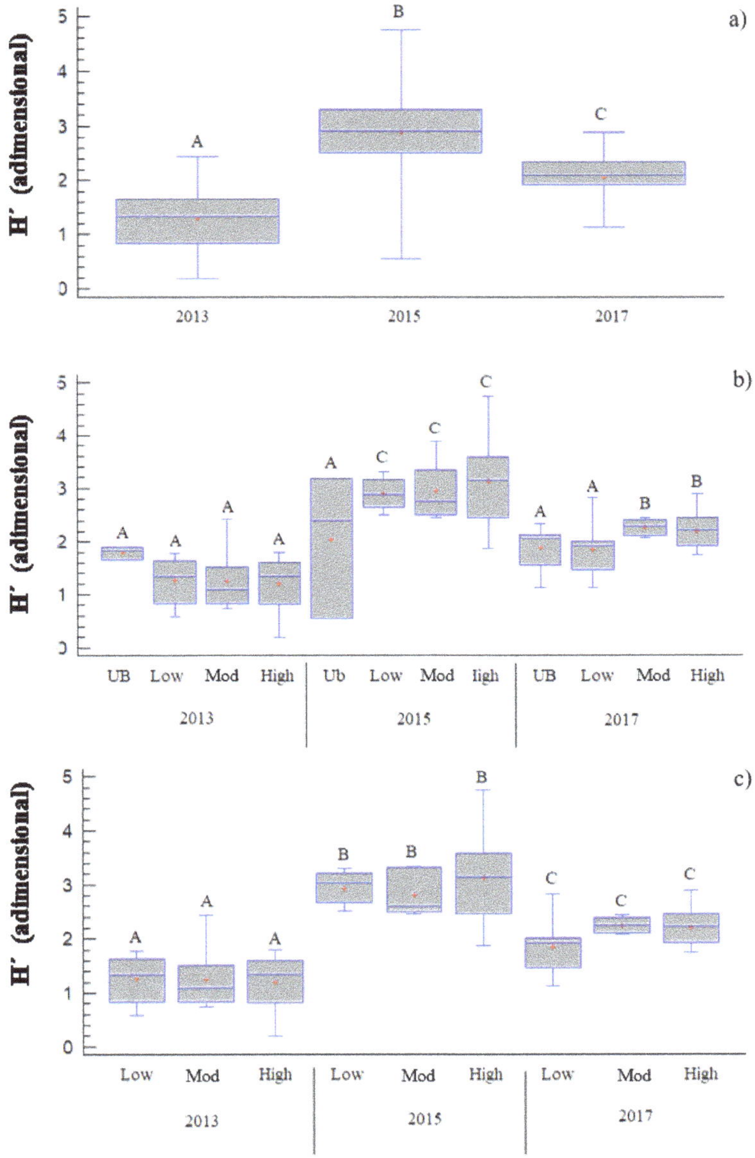

Figure 2. Box and Whisker plot of the one-way ANOVA relating the Shannon Index (H′, nondimensional) to (**a**) time after fire (TIME, years) and (**b**) the interaction of time after fire and burn severity classes (SEV, UB = unburned; Low = low burn severity; Mod = moderate burn severity; High = high burn severity). In addition, we checked the significance between (**c**) time after fire and burn severity classes, but we excluded the unburned plots. The different capital letters indicate significant differences between means of groups (ANOVA according to Tukey's HSD method).

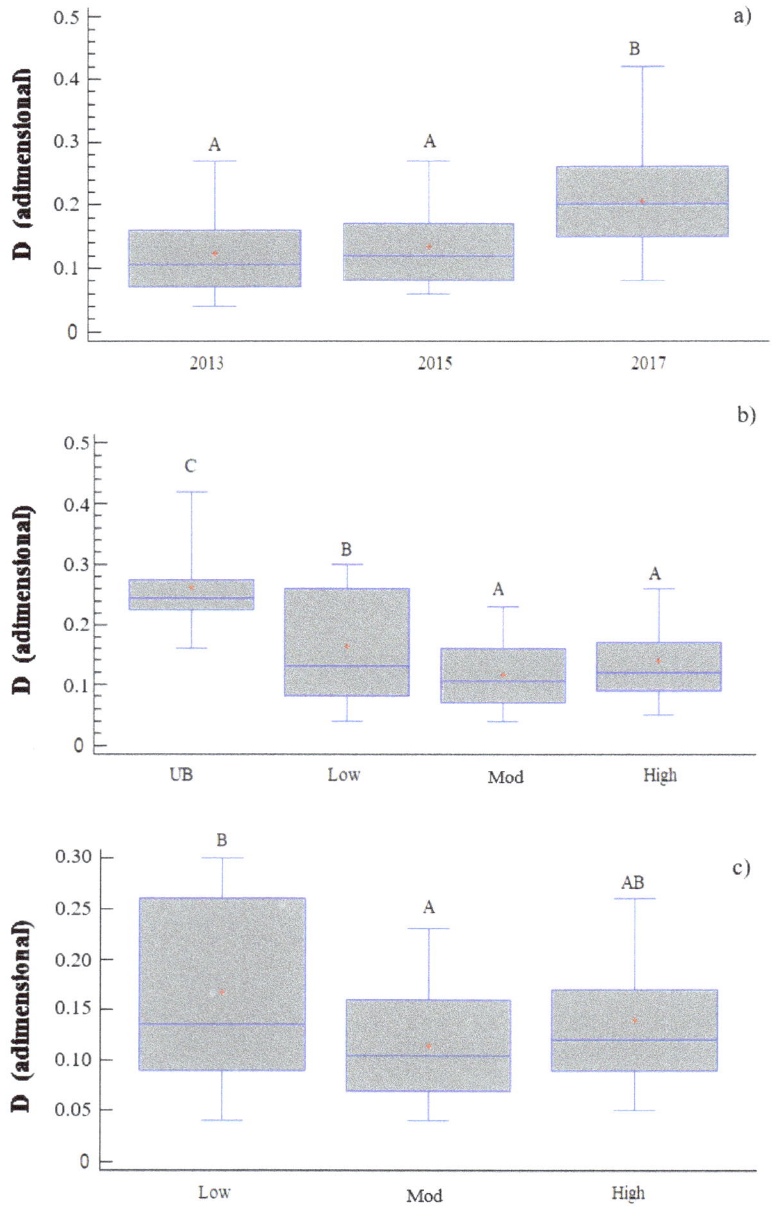

Figure 3. Box and Whisker plot of the one-way ANOVA relating the Simpson Index (D, nondimensional) to (**a**) time after fire (TIME, years) and (**b**) burn severity classes (SEV, UB = unburned; Low = low burn severity; Mod = moderate burn severity; High = high burn severity). In addition, (**c**) burn severity classes and the Simpson Index were related after excluding the unburned plots. The different capital letters indicate significant differences between the means of groups (ANOVA according Tukey's HSD method).

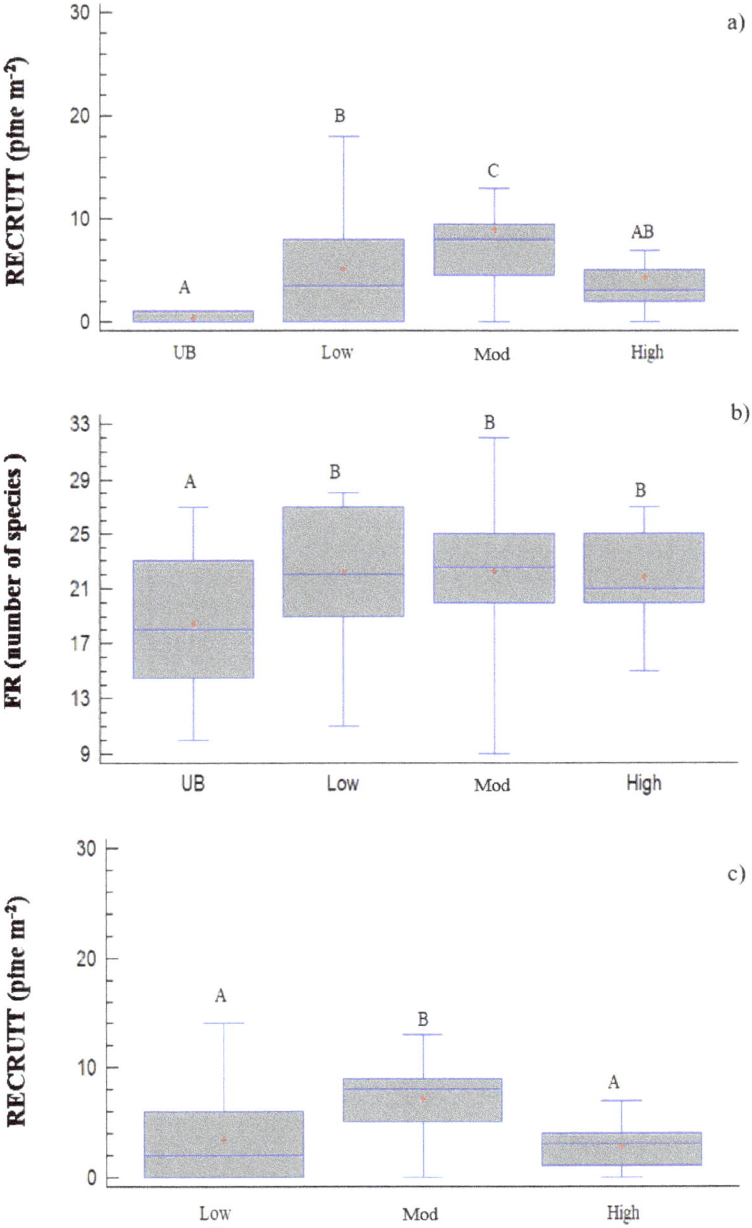

Figure 4. Box and Whisker plot of the one-way ANOVA relating the burn severity classes (SEV, UB = unburned; Low = low burn severity; Mod = moderate burn severity; High = high burn severity) to (**a**) the Aleppo pine recruitment rate, calculated as the mean pine seedling density (RECRUIT, pine m^{-2}) and (**b**) floristic richness (FR, number of species). In addition, (**c**) burn severity classes and floristic richness were related when the unburned plots were excluded. The different capital letters indicate significant differences between means of groups (ANOVA according Tukey's HSD method).

Table 2. Descriptive analysis including the mean values of reproductive strategies (non-adapted species (R−S−), obligate seeders (R−S+), obligate resprouters (R+S−) and facultative seeders (R+S+)) and growth form categories (trees (TREE), shrubs (SHRUB), dwarf shrubs (DWARF-S) and herbs (HERB) related to both time after fire (1, 3 and 5 years after fire, i.e. sampling carried out in 2013, 2015 and 2017 respectively) and the burn severity classes (UB = unburned; Low = low burn severity; Mod = moderate burn severity; High = high burn severity).

		Regeneration Strategy				Growth Form			
		R−S−	R−S+	R+S−	R+S+	TREE	SHRUB	DWARF-S	HERB
2013	UB	0.00 ± 0.00	50.00 ± 5.77	16.11 ± 4.55	33.89 ± 3.89	8.89 ± 1.11	23.89 ± 4.55	39.44 ± 5.30	27.78 ± 9.09
	Low	0.00 ± 0.00	54.54 ± 7.06	16.08 ± 2.29	23.54 ± 1.33	11.56 ± 1.35	16.07 ± 2.35	19.37 ± 1.44	50.66 ± 1.80
	Mid	1.04 ± 0.68	54.10 ± 2.40	18.03 ± 2.25	26.83 ± 1.73	12.12 ± 2.03	25.49 ± 1.66	16.23 ± 1.88	46.15 ± 3.79
	High	0.35 ± 0.35	56.39 ± 1.81	18.30 ± 1.28	25.91 ± 2.04	11.82 ± 1.63	26.74 ± 1.89	19.01 ± 2.74	42.44 ± 3.03
2015	UB	2.38 ± 2.38	36.96 ± 6.19	22.67 ± 4.85	37.99 ± 3.96	11.08 ± 1.97	20.19 ± 2.56	38.92 ± 4.15	29.81 ± 4.79
	Low	0.41 ± 0.41	46.34 ± 6.13	14.40 ± 1.86	32.62 ± 2.27	9.52 ± 1.44	16.00 ± 2.34	33.54 ± 2.53	38.71 ± 2.99
	Mid	0.00 ± 0.00	54.88 ± 2.35	14.88 ± 1.86	30.24 ± 2.70	8.54 ± 0.94	24.01 ± 1.97	27.13 ± 1.73	40.33 ± 2.20
	High	0.82 ± 0.52	53.25 ± 2.32	19.45 ± 1.89	26.49 ± 2.95	9.44 ± 1.23	24.01 ± 1.29	29.72 ± 1.72	36.83 ± 2.04
2017	UB	0.00 ± 0.00	46.85 ± 8.59	17.60 ± 5.07	38.33 ± 9.22	11.44 ± 3.45	22.66 ± 3.45	35.62 ± 7.10	30.28 ± 7.52
	Low	0.41 ± 0.41	44.08 ± 6.38	17.55 ± 2.22	33.10 ± 2.89	12.97 ± 1.76	20.40 ± 2.70	28.78 ± 2.33	34.38 ± 2.10
	Mid	0.00 ± 0.00	46.38 ± 3.73	18.61 ± 2.99	35.01 ± 3.09	10.49 ± 2.15	28.73 ± 1.99	23.51 ± 2.49	37.28 ± 5.07
	High	0.34 ± 0.34	45.13 ± 1.53	20.84 ± 2.29	33.69 ± 3.30	11.05 ± 1.35	26.10 ± 2.10	28.08 ± 2.66	34.77 ± 3.09

A PCA (Figure 5), which included the results for the burned area, reduced the variables to a five-component solution (eigenvalues > 1) and explained more than 73.84% of variability but this was simplified to a two-component solution which included 39.17% of the variability in the original data. The first component (COMP1) accounted for 21.17%, while the second explained 18.00% (COMP2). The ecosystem responses to burn severity (SEV) were clustered along the COMP1 axis, while that related to time after the fire (YEAR) was clustered on the COMP2 axis.

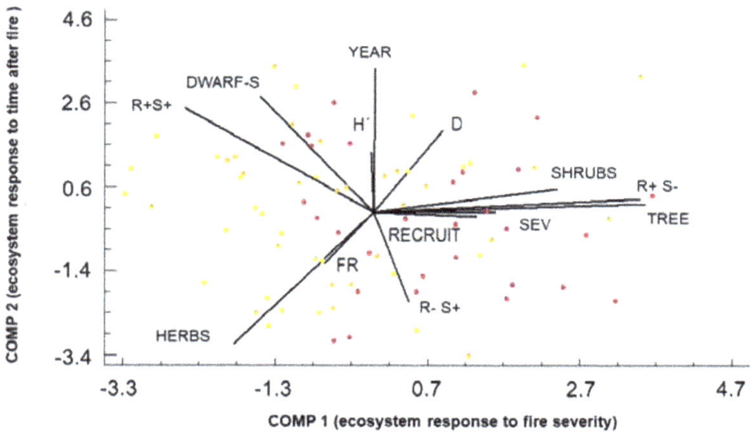

Figure 5. The PCA that included the recorded variables in the burned area: H′ (Shannon Diversity Index, nondimensional), FR (Floristic Richness, number of species), SEV (burn severity, quantitative levels), TIME (period of time since the wildfire event to the sampling date, years), adaptive strategies (R−S+, R+S− and R+S+) and growth form (trees (TREE), shrubs (SHRUB), dwarf shrubs (DWARF-S) and herbs (HERB)). Dots represent the standardized values for the short-term period after the wildfire event (1, 3 and 5 years after) in light orange (low burn severity), dark orange (moderate burn severity) and red (high burn severity). COMP1: first component (ecosystem response to burn severity) accounts for 21.17% of the variance; COMP2: second component (ecosystem response to time after fire) explains 18.00% of the variance (both accounted for 39.17% of the total variability contained in the original data when added together).

4. Discussion

Changes in fire regimes and the effects on plant communities, mainly in terms of plant (alpha) diversity and recovery time of Mediterranean plant communities, are a major trend to manage natural and seminatural landscapes in ecosystem conservation [48]. Thus, our approach demonstrated a relationship between burn severity, time after fire passage and the recovery/regeneration of the plant community in *P. halepensis* stands in the SE Iberian Peninsula.

The results indicated that the α-diversity values were lower in the burned area than in the control plots, mainly in the areas burned by high fire severity, which reduced seedbanks and increased colonizer species cover [49,50]. Lopez-Poma and Bautista [22] found different effects of fire severity, depending on soil damage, which we linked to species and adaptive strategies (related to growth forms). We also found higher and increasing floristic richness in burned areas, but the highest values were recorded for the Moderate severity, which also had the highest RECRUIT value that would encourage the plant community toward a more mature successional stage in the short term. On the contrary, H' increased according to the time after fire and burn severity, showing a maximum value under the high burn severity in 2017. To avoid an interaction of the unburned plots, we found marked differences in the specific diversity values for years 5 after the fire, with values remaining above those of the more severely burned plots. This indicated that there is a sustainable fire frequency (more than 25 years) since the high alpha diversity decreases rapidly in the short term [51] although a high burn severity could induce changes in post-fire recovery patterns [17,22].

Post-fire succession chrono-sequences in the Mediterranean Basin suggest that post-fire succession is dominated by herbaceous taxa in the short term [52], according to the initial floristic composition model of Egler [20], which concluded that herbaceous plants rapidly appeared and colonized bare ground after fire. In our assessment, the highest cumulative number of species was found 3 years after the fire due to the presence of herbaceous species, heliophiles and pioneer species, established due to low interspecific competition. These colonizer and pioneer species disappeared 3 years after the fire, probably due to successional dynamics, including successive species removal related to intra- and inter-specific competition [53].

The plant community included no fire-intolerant plant species, with post-fire obligate seeders predominating, mainly those comprising persistent seed banks enclosed in soil (*Cistus*) or in the canopy (*Pinus*). Nonetheless, some species (HERBS in early years) may come from unburned patches or outside the fire perimeter (metapopulation dynamics) [54]. The facultative seeders showed their highest percentages under the moderate burn severity and obligate resprouters were not apparently affected by burn severity, similar to other Mediterranean communities [55]. According to time after fire, heliophiles and pioneer species were replaced with shrubs, with increasing tree cover, but obligate resprouters were not affected by burn severity. We found no clear evidence that variation in burn severity modulates the seeders to resprouters abundance ratio, which contrasts with other findings [23,56,57]. However, RECRUIT was promoted by fire, with lower values for Low and High severities than for Moderate severity, and where the conditions had been related to the killing of seeds enclosed in cones [58], competition of germinated seedlings [59] or diminished regeneration capacity [21].

5. Conclusions

Adaptive strategies developed by plants evolved in Mediterranean climates provide resilience to regenerate after disturbances such as wildfires (time variation according to natural productivity and sustainable fire regime) [60]. However, it is important to acquire knowledge about the ecological effects of fire severity and the influence on the time extension for the immaturity risk [61] to apply a proper adaptive forest management under changing scenarios.

A diachronic study was used to improve the assessment of fire effects in Mediterranean forests, mainly in the short term. In Aleppo pine forests in semiarid stands, alpha diversity quickly recovers after fire, but is higher in the plots burned according to moderate and high burn severity in terms of species complexity and abundance. Floristic richness and pine recruitment are promoted by moderate burn severity. When considering the adaptive strategies and growth forms, we conclude that 5 years after the fire, plant composition apparently follows the mature successional stage and shows similar values to closer mature stands, but moderate and high severity will induce higher risk of fire due to the remaining percentages of herbs and dwarf shrubs.

Therefore, fire prevention should not focus only on prevention of fire risk but also on promoting the frequency and severity to similar levels as in a sustainable fire regime [28]. Once fire occurs, it is important to bear in mind to include in post-fire management and restoration actions that moderately burned areas will have a high natural potential recovery, but those under high severity could show a lack of pine recruitment and dominance of species.

Supplementary Materials: The following are available online at http://www.mdpi.com/1999-4907/9//299/s1, Figure S1: Descriptive comparison procedure to determine patterns of (**a**) Shannon index (H′, nondimensional), (**b**) Simpson index (D, nondimensional), (**c**) Aleppo pine recruitment rate (RECRUIT, pine seedlings m^{-2}) and (**d**) Floristic richness index (FR, number of species) related to time after fire (TIME, years after fire (one (2013), three (2015) and five (2017))) and burn severity classes (SEV, UB = Unburned (green line)).

Author Contributions: All authors participated and significantly contributed to the editing of the manuscript. J.D.l.H. supervised and coordinated the experiment and improved the experimental design. D.M. and S.G.-D.V. oversaw the experimental design, conducted the field experiments and analyzed the results, ran statistical analyses, and compiled the Tables and Figures. All authors contributed to manuscript writing and figure drafting.

Funding: This research was funded by the Excma. Diputación Albacete with a grant for research initiation (DIPU4-AB2015), the Spanish National Institute for Agricultural and Food Research and Technology (INIA), i.e., the GEPRIF project (RTA2014-00011-C06-05) and the Universidad de Castilla-La Mancha with support to the Forest Ecology Research Group.

Acknowledgments: We acknowledge the Forest Services of the Regional Government of Castilla-La Mancha (Junta de Comunidades de Castilla-La Mancha) for field support, mainly Elena Gomez. We also thank Helen Warburton for the professional English editing and suggestions.

Conflicts of Interest: The authors declare no conflict of interest.

References

1. Bond, W.J.; Keeley, J.E. Fire as a global 'herbivore': The ecology and evolution of flammable ecosystems. *Trends Ecol. Evol.* **2005**, *20*, 387–394. [CrossRef] [PubMed]
2. Pausas, J.G.; Keeley, J.E. A burning story: The role of fire in the history of life. *BioScience* **2009**, *59*, 593–601. [CrossRef]
3. Cerdá, A.; Mataix-Solera, J. *Efectos de los Incendios Forestales Sobre los Suelos en España: El Estado de la Cuestión visto por los Científicos Españoles*; FUEGORED, Cátedra Divulgación de la Ciencia (Universitat de Valencia): Valencia, Spain, 2009; 529p.
4. Scott, A.C. The Pre-Quaternary history of fire. *Palaeogeogr. Palaeoclimatol. Palaeoecol.* **2000**, *164*, 281–329. [CrossRef]
5. Pausas, J.G.; Keeley, J.E. Abrupt climate-independent fire regime changes. *Ecosystems* **2014**, *17*, 1109–1120. [CrossRef]
6. Dale, V.H.; Joyce, L.A.; McNulty, S.; Neilson, R.P.; Ayres, M.P.; Flannigan, M.D.; Hanson, P.J. Climate change and forest disturbances. *BioScience* **2001**, *51*, 723–734. [CrossRef]
7. Doblas-Miranda, E.; Martínez-Vilalta, J.; Lloret, F.; Álvarez, A.; Ávila, A.; Bonet, F.J.; Brotons, L.; Castro, J.; Curiel Yuste, J.; Díaz, M.; et al. Reassessing global change research priorities in Mediterranean terrestrial ecosystems: How far have we come and where do we go from here? *Glob. Ecol. Biog.* **2015**, *24*, 25–43. [CrossRef]
8. Moriondo, M.; Good, P.; Durao, R.; Bindi, M.; Giannakopoulos, C.; Corte-Real, J. Potential impact of climate change on fire risk in the Mediterranean area. *Clim. Res.* **2006**, *31*, 85–95. [CrossRef]
9. San-Miguel-Ayanz, J.; Schulte, E.; Schmuck, G.; Camia, A.; Strobl, P.; Liberta, G.; McInerney, D. Comprehensive monitoring of wildfires in Europe: The European forest fire information system (EFFIS). *Approaches to Managing Disaster-Assessing Hazards, Emergencies and Disaster Impacts*. Intech Open. 2012. Available online: https://www.intechopen.com/books/approaches-to-managing-disaster-assessing-hazards-emergencies-and-disaster-impacts/comprehensive-monitoring-of-wildfires-in-europe-the-european-forest-fire-information-system-effis- (accessed on 25 May 2018).

10. Doblas-Miranda, E.; Alonso, R.; Arnan, X.; Bermejo, V.; Brotons, L.; de las Heras, J.; Estiarte, M.; Hódar, J.A.; Llorens, P.; Lloret, F.; et al. A review of the combination among global change factors in forests, shrublands and pastures of the Mediterranean Region: Beyond drought effects. *Glob. Planet. Chang.* **2017**, *148*, 42–54. [CrossRef]
11. Newton, A.C.; Cantarello, E. Restoration of forest resilience: An achievable goal? *New For.* **2015**, *46*, 645–668. [CrossRef]
12. Clarke, P.J.; Lawes, M.J.; Midgley, J.J.; Lamont, B.B.; Ojeda, F.; Burrows, G.E.; Knox, K.J.E. Resprouting as a key functional trait: How buds, protection and resources drive persistence after fire. *New Phytol.* **2013**, *197*, 19–35. [CrossRef] [PubMed]
13. Tapias, R.; Gil, L.; Fuentes-Utrilla, P.; Pardos, J.A. Canopy seed banks in Mediterranean pines of South-Eastern Spain: A comparison between *Pinus halepensis* Mill.; *P. pinaster* Ait.; *P. nigra* Arn. and *P. pinea* L. *J. Ecol.* **2001**, *89*, 629–638. [CrossRef]
14. Reyes, O.; Trabaud, L. Germination behaviour of 14 Mediterranean species in relation to fire factors: Smoke and heat. *Plant Ecol.* **2009**, *202*, 113. [CrossRef]
15. Moreira, F.; Catry, F.X.; Rego, F.; Bacao, F. Size-dependent pattern of wildfire ignitions in Portugal: When do ignitions turn into big fires? *Landsc. Ecol.* **2010**, *25*, 1405–1417. [CrossRef]
16. Moya, D.; De las Heras, J.; López-Serrano, F.R.; Leone, V. Optimal intensity and age of management in young Aleppo pine stands for post-fire resilience. *For. Ecol. Manag.* **2008**, *255*, 3270–3280. [CrossRef]
17. González-De Vega, S.; De Las Heras, J.; Moya, D. Resilience of Mediterranean terrestrial ecosystems and fire severity in semiarid areas: Responses of Aleppo pine forests in the short, mid and long term. *Sci. Total Environ.* **2016**, *573*, 1171. [CrossRef] [PubMed]
18. Viana-Soto, A.; Aguado, I.; Martínez, S. Assessment of Post-Fire Vegetation Recovery Using Fire Severity and Geographical Data in the Mediterranean Region (Spain). *Environments* **2017**, *4*, 9. [CrossRef]
19. Schwilk, D.W; Keeley, J.E.; Bond, W.J. The intermediate disturbance hypothesis does not explain fire and diversity pattern in Fynbos. *Plant Ecol.* **1997**, *132*, 77–84. [CrossRef]
20. Verroios, G.; Georgiadis, T. Post-fire vegetation succession: The case of Aleppo pine (*Pinus halepensis* Miller) forests of Northern Achaia (Greece). *Fresenius Environ. Bull.* **2002**, *11*, 186–193.
21. García-Jiménez, R.; Palmero-Iniesta, M.; Espelta, J.M. Contrasting effects of fire severity on the regeneration of *Pinus halepensis* Mill. and resprouter species in recently thinned thickets. *Forests* **2017**, *8*, 55. [CrossRef]
22. López-Poma, R.; Bautista, S. Plant regeneration functional groups modulate the response to fire of soil enzyme activities in a Mediterranean shrubland. *Soil Biol. Biochem.* **2014**, *79*, 5–13. [CrossRef]
23. Pereira, P.; Francos, M.; Brevik, E.C.; Ubeda, X.; Bogunovic, I. Post-fire soil management. *Curr. Opin. Environ. Sci. Health* **2018**, *5*, 26–32. [CrossRef]
24. Pasho, E.; Camarero, J.J.; de Luis, M.; Vicente-Serrano, S.M. Impacts of drought at different time scales on forest growth across a wide climatic gradient in north-eastern Spain. *Agric. For. Meteor.* **2011**, *151*, 1800–1811. [CrossRef]
25. Quezel, P. Taxonomy and biogeography of Mediterranean pines (Pinus halepensis and P. brutia). In *Ecology, Biogeography and Management of Pinus halepensis and P. brutia Forest Ecosystems in the Mediterranean Basin*; Ne'eman, G., Trabaud, L., Eds.; Backhuys Publishers: Leiden, The Netherlands, 2000; pp. 1–12.
26. Sarris, D.; Christodoulakis, D.; Korner, C. Impact of recent climatic change on growth of low elevation eastern Mediterranean forest trees. *Clim. Chang.* **2011**, *106*, 203–223. [CrossRef]
27. Goubitz, S.; Nathan, R.; Roitemberg, D.; Shmida, A.; Ne'eman, G. Canopy seed bank structure in relation to: Fire, tree size and density. *Plant Ecol.* **2004**, *173*, 191–201. [CrossRef]
28. Keeley, J.E.; Bond, W.J.; Bradstock, R.A.; Pausas, J.G.; Rundel, P.W. *Fire in Mediterranean Ecosystems: Ecology, Evolution and Management*; Cambridge University Press: Cambridge, UK, 2012.
29. Hernández-Serrano, A.; Verdú, M.; González-Martínez, S.C.; Pausas, J.G. Fire structures pine serotiny at different scales. *Am. J. Bot.* **2013**, *100*, 2349–2356. [CrossRef] [PubMed]
30. DeRose, R.J.; Long, J.N. Resistance and resilience: A conceptual framework for silviculture. *For. Sci.* **2014**, *60*, 1205–1212. [CrossRef]
31. Kottek, M.; Grieser, J.; Rieser, C.; Beck, C.; Rudolf, B.; Rubel, F. World map of the Koppen-Geiger climate classification updated. *Meteorol. Z.* **2006**, *15*, 259–263. [CrossRef]

32. García-Morote, F.A.; Martínez-García, E.; Andrés-Abellán, M.; Caballero, E.R.; Miettinen, H.; López-Serrano, F.R. Direct seeding of *Pinus halepensis* Mill. for recovery of burned Semi-Arid Forests: Implications for post-fire management for improving natural regeneration. *Forests* **2017**, *8*, 353.
33. Soil Survey Staff. *Keys to Soil Taxonomy*, 12th ed.; USDA-Natural Resources Conservation Service: Washington, DC, USA, 2014.
34. Miller, J.D.; Thode, A.E. Quantifying burn severity in a heterogeneous landscape with a relative version of the delta normalized burn ratio (dNBR). *Remote Sens. Environ.* **2007**, *109*, 66–80. [CrossRef]
35. Gómez-Sánchez, E.; de las Heras, J.; Lucas-Borja, M.; Moya, D. Assessing fire severity in semi-arid environments: Application in Donceles 2012 wildfire (SE Spain). *Revista Teledetección* **2017**, *49*, 103–113. [CrossRef]
36. Alloza, J.A.; García, S.; Gimeno, T.; Baeza, M.J.; Vallejo, V.R. *Guía Técnica Para la Gestión de Montes Quemados*; Ministerio de Agricultura, Alimentación y Medio Ambiente: Madrid, Spain, 2014; 188p.
37. Ryan, K.; Noste, N. Evaluating prescribed fires. *Proceedings of the Symposium and Workshop on Wilderness Fire, Missoula, Montana, 15 November 1983*; USDA Forest Service Intermountain Forest and Range Experiment Station, General Technical Report INT 182; 1985; pp. 230–238. Available online: https://www.fs.fed.us/rm/publications/titles/int_gtr.html (accessed on 25 May 2018).
38. Stohlgren, T.J.; Falkner, M.; Schell, D.L. A Modified-Whittaker nested vegetation sampling method. *Plant Ecol.* **1995**, *117*, 113–121. [CrossRef]
39. Van der Maarel, E. Transformation of cover-abundance values for appropriate numerical treatment–Alternatives to the proposals by Podani. *J. Veg. Sci.* **2007**, *18*, 767–770.
40. Buckland, S.T.; Borchers, D.L.; Johnston, A.; Henrys, P.A.; Marques, T.A. Line transect methods for plant surveys. *Biometrics* **2007**, *63*, 989–998. [CrossRef] [PubMed]
41. Shannon, C. A mathematical theory of communication. *Bell. Syst. Tech. J.* **1948**, *27*, 379–423. [CrossRef]
42. Simpson, E.H. Measurement of diversity. *Nature* **1949**, *163*, 688. [CrossRef]
43. Paula, S.; Arianoutsou, M.; Kazanis, D.; Tavsanoglu, Ç.; Lloret, F.; Buhk, C.; Ojeda, F.; Luna, B.; Moreno, J.M.; Rodrigo, A.; et al. Fire-related traits for plant species of the Mediterranean Basin. *Ecology* **2009**, *90*, 1420. [CrossRef]
44. Tutin, T.G.; Heywood, V.H.; Burges, N.A.; Moore, D.M.; Valentine, D.H.; Walters, S.M.; Webb, D.A. *Flora Europaea*; Cambridge University Press: Cambridge, UK, 1964.
45. Castroviejo, S. Flora ibérica. Available online: http://www.floraiberica.es/ (accessed on 24 May 2018).
46. *Statgraphics Centurion 18.1.03*; Statpoint Technologies Inc.: Warrenton, VA, USA, 2018.
47. R Development Core Team. *R: A Language and Environment for Statistical Computing*; R Foundation for Statistical Computing: Vienna, Austria, 2015.
48. Kruger, F.J. Plant Community Diversity and Dynamics in Relation to Fire. In *Mediterranean-Type Ecosystems*; Kruger, F.J., Mitchell, D.T., Jarvis, J.U.M., Eds.; Ecological Studies (Analysis and Synthesis); Springer: Berlin/Heidelberg, Germany, 1983.
49. Ghermandi, L.; Gonzalez, S.; Lescano, M.N.; Oddi, F. Effects of fire severity on early recovery of Patagonian steppes. *Int. J. Wildland Fire* **2013**, *22*, 1055–1062. [CrossRef]
50. Zedler, P.H.; Gautier, C.R.; McMaster, G.S. Vegetation change in response to extreme events: The effect of a short interval between fires in California chaparral and coastal scrub. *Ecology* **1983**, *64*, 809–818. [CrossRef]
51. Cowling, R.M.; Rundel, P.W.; Lamont, B.B.; Arroyo, M.K.; Arianoutsou, M. Plant diversity in mediterranean-climate regions. *Trends Ecol. Evol.* **1996**, *11*, 362–366. [CrossRef]
52. Kazanis, D.; Arianoutsou, M. Vegetation Composition in a Post-Fire Successional Gradient of *Pinus halepensis* Forests in Attica, Greece. *Int. J. Wildland Fire* **1996**, *6*, 83–91. [CrossRef]
53. Bond, W.J.; Van Wilgen, B.W. Fire and Plants (population and community biology series 14). *Popul. Community Biol.* **1996**, *263*.
54. Pausas, J.G.; Keeley, J. Evolutionary ecology of resprouting and seeding in fire-prone ecosystems. *New Phytol.* **2014**, *204*, 55–65. [CrossRef] [PubMed]
55. Marais, K.E.; Pratt, R.B.; Jacobs, S.M.; Jacobsen, A.L.; Esler, K.J. Postfire regeneration of resprouting mountain fynbos shrubs: Differentiating obligate resprouters and facultative seeders. *Plant Ecol.* **2014**, *215*, 195–208. [CrossRef]
56. Pausas, J.G.; Bradstock, R.A.; Keith, D.A.; Keeley, J.E. Plant Functional traits in relation to fire in crown-fire ecosystems. *Ecology* **2004**, *85*, 1085–1100. [CrossRef]

57. Hollingsworth, T.N.; Johnstone, J.F.; Bernhardt, E.L.; Chapin, F.S., III. Fire severity filters regeneration traits to shape community assembly in Alaska's boreal forest. *PLoS ONE* **2013**, *8*, e56033. [CrossRef] [PubMed]
58. Moya, D.; De las Heras, J.; Salvatore, R.; Valero, E.; Leone, V. Fire intensity and serotiny: Response of germination and enzymatic activity in seeds of *Pinus halepensis* Mill. from southern Italy. *Ann. For. Sci.* **2013**, *70*, 49–59. [CrossRef]
59. Sagra, J.; Plaza-Álvarez, P.A.; Moya, D.; Lucas-Borja, M.E.; Alfaro Sánchez, R.; de las Heras Ibáñez, J.; Ferrandis, P. Predation influence on seedling recruitment in Mediterranean forests after prescribed burning. *Forests* **2017**, *8*, 243. [CrossRef]
60. Pausas, J.G.; Keeley, J.E.; Schwilk, D.W. Flammability as an ecological and evolutionary driver. *J. Ecol.* **2017**, *105*, 289–297. [CrossRef]
61. Ne'eman, G.; Goubitz, S.; Nathan, R. Reproductive traits of *Pinus halepensis* in the light of fire-A critical review. *Plant Ecol.* **2004**, *171*, 69–79. [CrossRef]

forests

MDPI

Article

Effects of Management Practices and Topography on Ectomycorrhizal Fungi of Maritime Pine during Seedling Recruitment

Arthur Guignabert [1,*], Florian Delerue [2,3], Maya Gonzalez [1], Laurent Augusto [1] and Mark R. Bakker [1]

1 UMR 1391 ISPA (Atmosphere Plant Soil Interactions), Bordeaux Sciences Agro, INRA, 33140 Villenave d'Ornon, France; maya.gonzalez@agro-bordeaux.fr (M.G.); laurent.augusto@inra.fr (L.A.); mark.bakker@inra.fr (M.R.B.)
2 Bordeaux INP, G&E, EA 4592, F-33600 Pessac, France; florian.delerue@bordeaux-inp.fr
3 Georesources & Environment, Bordeaux Montaigne University, EA 4592, F-33600 Pessac, France
* Correspondence: arthur.guignabert@inra.fr; Tel.: +33-5-5712-2523

Received: 13 April 2018; Accepted: 2 May 2018; Published: 3 May 2018

Abstract: Symbiosis with ectomycorrhizal (ECM) fungi can be important for regeneration success. In a context of increasing regeneration failures in the coastal forest of maritime pine in Southwest France, we tried to identity whether differences in ECM communities could partly explain the variation of regeneration success and how they are influenced by forest practices and stand characteristics. In particular, we focused on the effects of harvesting methods (comparing mature forest with seed-tree regeneration and clear-cuts) and topography (bottom-, mid-, and top positions). Five field trials (two in regeneration failure areas and three in successful areas) were used to sample 450 one-year-old seedlings. Assessments of ECM of seedling nutrient concentrations and of seedling growth based on exploration types were made. ECM root colonisation was similar in all harvesting treatments, suggesting that enough inoculum remained alive after logging. Harvesting-induced effects modifying soil properties and light availability respectively impacted ECM composition and seedling growth. Topography-induced variations in water and nutrient availability led to changes in ECM composition, but had little impact on seedling growth. Contact, short-distance, and long-distance exploration types improved the nutritional status of seedlings (Ca, K, and N), showing that mycorrhization could play an important role in seedling vitality. However, neither ECM root colonisation nor exploration types could be related to regeneration failures.

Keywords: Ectomycorrhizal fungi; *Pinus pinaster*; seedlings; natural regeneration; harvesting practices; topography; precipitation

1. Introduction

Maritime pine (*Pinus pinaster* Aït in Soland) is the dominant forestry tree species in south-western France, covering almost one million ha [1]. Most of the forest range in the interior land plains is renewed by planting, whereas natural regeneration is used in the coastal area, which represents 10% of this forest [1]. In addition to wood production, these coastal forests have a multifunctional role, including soil erosion protection, preservation of biodiversity, and tourism/public usage [2]. To accommodate these multiple objectives, the use of natural regeneration to renew forest stands limits soil disturbance (no soil ploughing, such as before plantations) and improves the conservation of genetic diversity and the capacity of forests to adapt to global change. However, in recent years, regeneration failures have increased in some areas, leading to significant economic losses [2,3]. The regeneration status is estimated by counting the number of seedlings two to three years after clearcutting: regeneration is considered successful if the number of seedlings per hectare is higher than 3000; failed if there are less than 1500 seedlings/ha; and semi-successful (i.e., it could be sufficient, but the stand should remain under surveillance) if the number of seedlings is between these two thresholds [2].

The period between germination and seedling establishment is often a crucial step in the process of natural regeneration [4,5], and is well-known to depend on soil, climate, and biotic factors [6–8]. This is particularly the case for forests that have soil seed banks with a short life lifespan (such as maritime pine) where regeneration occurs from the soil seed bank after clear-cutting. In areas where regeneration failures are observed, "security sowing" is generally applied to ensure regeneration [2]. One of the most important biotic factors involved in natural regeneration, but which is often overlooked, is the association between plants and ectomycorrhizae (ECM) fungi [9–11]. Throughout their extramatrical mycelial network, ECM fungi can represent up to 75% of the absorptive area and over 99% of the absorptive length [12]. This symbiotic relationship, obligatory for almost all conifer species, provides water and nutrients to the seedling in exchange for carbohydrates, thus improving seedling growth, nutrient concentrations, and the success of seedling recruitment [13,14]. It also improves seedling resistance against drought, pathogens, and heavy metals [15–17]. Considering the importance of ECM for plant nutrition and water acquisition in oligotrophic and dry systems [17–19], like the sandy-soil system of the coastal dune forest, this particular biotic interaction could be involved in the differences of natural regeneration observed in the region.

Hence, this study examined the ectomycorrhizal composition of pine seedlings one year after logging by using five field trials with contrasting natural regeneration success which are located throughout this region. It corresponds to an initial examination of the subject to determine whether this biotic interaction should be studied in more depth. ECM fungi were classified according to the typology of Agerer [20], which classified ECM into different exploration types (i.e., their ability to explore the soil) based on the extent of mycelium and the presence and differentiation of rhizomorphs. Shorter exploration types represent a lower carbon cost to the plant and are more prevalent in wet and/or nutrient-rich areas, whereas longer exploration types (i.e., fungi with rhizomorphs) are more prevalent in dry and/or nutrient-poor areas where seedlings need to increase their absorptive area [21–23].

ECM communities can be influenced by a large number of factors including soil properties such as pH, soil temperature, nutrient availability, soil moisture, and climatic factors [24–26]. Soil moisture is an important driver of fungal composition. In dry soils, a decrease of ECM colonisation can occur [27], as well as a shift in ECM species, leading to the appearance of ECM which are well-adapted to dry soils and less expensive in carbon for the host plant or which develop highly differentiated rhizomorphs [22,28]. Studies at regional scales have shown that ECM fungal composition could vary with precipitation and temperature [25,29,30].

ECM communities are also highly sensitive to forestry practices in relation to tree harvesting, which can strongly affect the development of new seedlings during the next forest rotation. Briefly, tree harvesting practices could: (i) represent a potential disturbance of topsoil layers where ECM fungi are abundant [31]; (ii) improve the colonisation of seedling roots by ECM fungi in the case of seed-tree cuts relative to clear-cut systems; and (iii) modify the overall aforementioned soil properties. More precisely, the effects of clear-cutting, reviewed by Jones et al. [32], are closely associated with the level of soil disturbance. When the forest floor is little disturbed during harvesting, clear cutting results in a decrease in ECM species richness and a change in species composition rather than a reduction in root colonization. Mycorrhizal inoculum can remain active in the soil for one to two years after harvesting [9] and does not appear to be limiting for seedling colonisation [32]. When soil is heavily disturbed (forest floor removal and/or soil compaction), harvesting also impacts ECM colonisation, reducing the available inoculum by breaking mycelial networks and modifying soil properties [11,33–35]. Secondly, the preservation of seed-trees in the stand after harvest, either by leaving individual trees [36] or by leaving patches of trees [37], can improve the fungal root colonization of seedlings and could enhance natural regeneration, especially under drought conditions [38,39]. Indeed, the pre-existing mycelial network established by the roots of surrounding adult trees permits the seedlings to connect to this network at a low carbon cost and this can improve seedling regeneration [14,40,41]. Partial cutting can promote the maintenance of active root tips, and seed-tree cuts can maintain an intermediate level of ECM colonisation between clear-cuts and forests [36,42]. Thirdly, the impacts of clear-cuts or seed-tree cuts on ECM communities are also due to the associated modifications of soil properties relative to hydric conditions, soil temperature, or nutrient availability [32].

In the face of observed regeneration failure, the more specific context of coastal dune forests was considered and regeneration practice other than standard clearcutting is currently being tested by preserving seed-trees in the stand. In addition, the topography induced by the dune ecosystem is a major characteristic of these coastal forests, and strongly impacts the distance to the water table which varies from 1–3 m at the bottom to 17 m at the top of the dunes [43]. Consequently, harvest practices (clear-cuts and seed-tree cuts vs mature stands) and topographical positions (bottom-, mid-, and top positions) were tested in this experiment as factors potentially affecting ECM communities and their relationship with seedling nutrition.

More precisely, we examined the variations in ECM colonization between sites in relation to regeneration status, harvesting practices, and environmental conditions (i.e., topography, meteorological conditions, and soil properties). We hypothesized that: (i) seedlings from the sites with regeneration failures could have a lower ectomycorrhizal colonisation and a lower diversity of ectomycorrhizal exploration types; (ii) ectomycorrhizal colonisation could be lower in clear-cuts than forests due to the loss of inoculum; and (iii) ECM fungi could modify root development, and could lead to better needle nutrient concentrations.

2. Materials and Methods

2.1. Study Area Description

To test our hypotheses, we established trials in five forests along the coast in SW France (Figure 1a). Two of the sites are within the area of chronic stand regeneration failure (hereafter referenced to as sites F1 and F2), while the three other sites are in areas with high values of regeneration success (S1, S2, and S3). Since regeneration failures are very heterogeneous within the same forest, germination and survival were monitored during the three years following the clear-cut to confirm the regeneration status of the sites (Table 1).

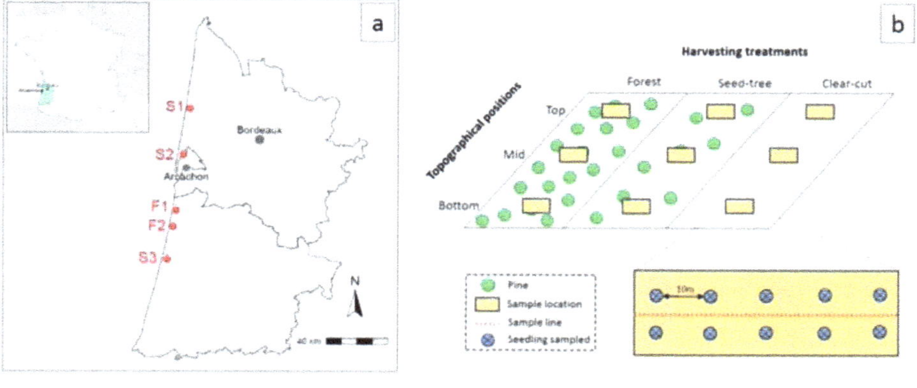

Figure 1. Location of the five study sites (**a**) and the seedling sampling design (**b**).

Table 1. Characteristics of the five study sites.

	S1	S2	F1	F2	S3
Regeneration status [a]					
Survival (%)	59.8	31.0	15.8	25.7	66.2
Seedlings/ha	3062	3875	1000	1188	8438
Location					
Forest	Lacanau	Lège-Cap-Ferret	Biscarrosse	Sainte-Eulalie-en-Born	Saint-Julien-en-Born
Latitude	45°1′43.2″ N	44°44′9.4″ N	44°24′37.1″ N	44°18′14″ N	44°6′8″ N
Longitude	1°10′44.1″ W	1°12′35.6″ W	1°14′3.4″ W	1°15′8.2″ W	1°17′27″ W
Meteo					
Annual P (mm)	709 (879)	608 (840)	659 (894)	718 (936)	880 (1007)
Summer P (mm)	191 (155)	139 (141)	127 (151)	147 (157)	174 (165)
Annual T (°C)	13.8 (13.7)	14.4 (14.1)	14.4 (14.1)	14.4 (14.1)	14.2 (14.0)
Summer T (°C)	19.5 (19.7)	20.2 (20.3)	20.2 (20.2)	20.2 (20.2)	19.9 (20.1)
Annual PET (mm)	759 (756)	780 (773)	778 (770)	778 (771)	774 (769)
Soil					
pH (H_2O) ± SE	4.46 ± 0.14^{a}	4.82 ± 0.21^{c}	4.57 ± 0.13^{ab}	4.68 ± 0.19^{bc}	4.44 ± 0.14^{a}
OM (%) ± SE	42.5 ± 13.7^{d}	24.8 ± 13.9^{ab}	28.2 ± 10.9^{bc}	36.8 ± 15.7^{cd}	14.3 ± 8.5^{a}

[a] Values in March 2018 (i.e., three years after logging), resulting from a three-year monitoring of germination and seedling survival under standard practice (i.e., clear-cut) on these five sites. Values for precipitations (P), temperature (T), and potential evapotranspiration (PET) represent values for the year 2015, with the average of the 2006–2016 period in brackets. Summer T and P correspond to values from July to September 2015. Meteorological data were calculated for each site, by linear interpolation of the three closest Météo-France weather stations. Organic matter (OM) and pH are means of 20 soil samples per site (10 in seed-trees and 10 in clear-cuts). Each sample was taken from an area of 0.125 m^2 and 5 cm deep, and represents a mixture of forest floor and upper mineral horizon. Different letters after mean values indicate significant differences between sites at $p < 0.05$ determined by ANOVA and a Tukey test.

The climate in the region is temperate oceanic. Annual average precipitations varied from 840 mm to 1007 mm and annual average temperature was around 14 °C for all sites (2006–2016, Météo-France data; Table 1). The year of germination and early growth of seedlings (i.e., 2015) was the driest year of the decade, with precipitations between 608 mm and 880 mm, and average annual temperatures around 14 °C (Météo-France data; Table 1). All sites are near a monospecific forest of *Pinus pinaster* (Aït in Soland), with some small individuals of oak (*Quercus robur* L., *Q. ilex* L., and *Q. suber* L.) within the stands or in the margins. Understorey was mainly composed of *Arbutus unedo* L., as well as small amounts of other shrubs (*Ulex Europaeus* L., *Cytisus scoparius* (L.) Link), grasses (*Holcus lanatus* L., *Deschampsia flexuosa* (L.) Trin.), and ericaceous plants (*Erica cinerea* L., *Erica scoparia* L.). All sites were chosen on westerly facing slopes at about 2.5 km from the ocean, with an average slope of 10°. Soils were young sandy soils (WRB (World Reference Base) classification: arenosols; USDA (United States Department of Agriculture) classification: entisols), developed from Aeolian deposits that occurred during the Holocene period [44]. These soils are mainly composed of coarse sands (96–97%), are slightly acidic (topsoil values of pH = 4.5–5.0; base saturation = 32–54%), have a low water holding capacity, and are extremely poor in nutrients [44]. Forest floor organic layer thickness varies between 0.5 and 4 cm (unpublished data [44]).

2.2. Tree Harvesting and Stand Regeneration Management Methods

Because tree harvesting and regeneration methods may impact ECM communities, the three following treatments were tested at each site: (1) control (i.e., no tree harvest and stand disturbance; hereafter referred to as "Forest" treatment); (2) seed-tree regeneration method (i.e., natural regeneration with seed-trees (70 trees ha^{-1}; "Seed-tree" treatment); and (3) clear-cut, which corresponds to the dominant harvesting and regeneration method in coastal dune forest ("Clear-cut" treatment). Logging was carried out mechanically between December 2014 and March 2015 in the seed-tree and clear-cut treatments. Other current practices were carried out before logging in these two treatments: understorey vegetation was mechanically removed to limit post-logging competition, and was combined with mechanical tillage to increase soil aeration and the availability of nutrients. This tillage mixed the forest floor organic layers with the mineral topsoil layer.

Light availability for each harvesting treatment was calculated from the diameter, age, and density of the trees, following equations of Porté et al. [45] and Berbigier and Bonnefond [46]. It was significantly different between harvesting treatments (Kruskal–Wallis rank sum test, $p = 0.002$), with the lowest values in forests (mean: 80 ± 2.2%), intermediate values in seed-trees (mean: 92 ± 0.7%), and maximal values in clear-cuts (mean: 100%) (Tukey test: a, b, and c, respectively).

2.3. Sampling of Seedlings

In this study, we chose to sample seedlings that had regenerated naturally rather than planting nursery grown seedlings, in order to remain as close as possible to natural regeneration conditions. Indeed, ECM communities between seedlings planted under nursery conditions and those that had regenerated naturally in the stand are different [14]. The nursery fungi suppress the initial colonization by native fungi, and do not reflect a difference in local inoculum [32]. Seedlings were selected and sampled from all five sites between the end of November 2015 and early February 2016. Seedlings almost only germinate in spring (between the end of March and the beginning of June), and were eight to 10 months old when sampled. At each site, for each harvesting treatment and for each topographical position, a central area of about 50 m × 8 m was identified for seedling sampling (=9 areas per site; Figure 1b). Within each area, two seedlings were selected every 10 m as close as possible to the line splitting the area in half, with one on each side of the line (Figure 1b). Non-browsed seedlings were selected (between 5 and 20 cm in height). Areas too close to other tree species (especially oaks) were avoided. Seedlings with their entire root system and a small amount of soil were gently removed using a shovel and then stored in plastic bags at 4 °C. In total, 450 seedlings were sampled (450 = 5 sites × 3 harvest treatments × 3 topographical positions × 10 seedlings).

2.4. Assessment of Seedling Dimensions and Ectomycorrhizal Status

Seedlings were cut at the root collar. Shoot stem diameter and total height were measured. Roots were then washed carefully over a plastic tray to remove soil adhering to the roots without disrupting the ECM material. Tap root length was recorded. Then, roots were divided into coarse roots (CR, diameter > 2 mm) and fine roots (FR, diameter < 2 mm). CR Lengths were measured with a caliper. For FR, we calculated mycorrhizal and non-mycorrhizal root length based on the line intersect method [47]. Root tips were observed using a binocular microscope, and ECM fungi were classified into the four main morphotypes reported by Agerer [20], depending on the exploration type of fungi. Contact exploration types have a smooth mantle, sometimes with a few hyphae. Short-distance exploration types correspond to ECM with many hyphae and no rhizomorphs. Medium-distance exploration types are represented by fungae with hyphae and rhizomorphs, which ramify and interconnect repeatedly. Long-distance exploration types include smooth ECM with few but highly differentiated rhizomorphs [20]. Non-mycorrhizal root tips were also counted. Shoots, CR, and FR samples were dried at 60 °C to obtain biomass values. ECM root colonization (%), specific root length (SRL, m g^{-1}), total seedling biomass (g), root:shoot (R:S, g g^{-1}), and height:diameter (H:D, mm mm^{-1}) ratios were calculated from these measurements. Measurements from the ten seedlings from each sample location (same harvest treatment and same topographical position) were averaged for statistical analyses.

2.5. Needle Nutrient Concentrations

Needles from the ten seedlings of each sample location were grouped into a composite sample ($n = 45$ composite samples) for mineral analysis (N, P, K, Mg, Ca). Nutrient concentrations were analysed after digestion in sulphuric acid and hydrogen peroxide. Nitrogen and phosphorus were determined colorimetrically with a Technicon auto analyser II. Potassium, calcium, and magnesium were determined with a Varian SpectrAA-20 flame atomic absorption spectrophotometer (Varian, Mulgrave, Australia).

Nutrient concentration values were compared to two sufficiency thresholds (i.e., where growth is medium to good) from a literature compilation by van den Burg [48]. These thresholds correspond to the mean values of several pine species, discerning values obtained from studies in sand culture from those in pot trials (Supplementary Materials, Table S1).

2.6. Data Treatments and Statistics

All statistical analyses were realised with R software version 3.4.1. [49].

2.6.1. ECM Status

ECM root colonization, expressed as a percentage, was logit-transformed following Warton and Hui [50] and total ECM root tips and number of root tips for each exploration type were expressed as number per meter of fine root length.

First, we scrutinized potential differences of ECM status between sites and according to meteorological differences between the sites. Analysis of variance was used to compare ECM status between sites, using data for all seedlings from all treatments within each site. Correlations between ECM variables and site characteristics (pH, OM, and precipitation variables) were prospected with Bravais-Pearson correlation tests.

To investigate variations of ECM status within the different sites according to harvesting treatments and topography, we used linear mixed modelling with harvesting treatment and topography as fixed effects, and the site as a random effect. Posthoc Tukey pairwise multiple comparisons were performed for significant treatments.

2.6.2. Seedling Growth and Nutrient Concentrations

To analyse how seedling growth and nutrient concentrations were affected by harvesting treatment and topography, we carried out linear mixed models with harvesting treatment and topography as fixed effects, and the site as a random effect. Biomass values and fine root length were log-transformed to reach linear modelling assumptions. The date of emergence of the seedlings, which may have varied by a few weeks, could lead to differences in terms of seedling size and the allocation of carbon and nutrients are both known to vary with plant size and ontogeny [51]. Thus many seedling characteristics (root:shoot, height:diameter, specific root length, and plant nutrient concentrations) were expected to vary with seedling size [52–54]. Consequently, to take this effect due to seedling size into account when performing our analyses, we first carried out linear regressions between aboveground biomass and these variables. In a second step, when regressions were significant ($p < 0.001$ for R:S, H:D, N, P, K and Ca, $p = 0.97$ for Mg, $p = 0.72$ for SRL), the effects of harvesting treatment and topography were prospected on regression residuals and residual variance. Posthoc Tukey pairwise multiple comparisons were performed for significant factors.

2.6.3. Relationship between ECM and Seedling Response after Filtering out Effects of Harvesting Treatments and Topographical Positions

As ECM can improve access to soil resources for seedlings, analyses were only performed for the following variables related to soil resource availability and plant nutrition: root:shoot ratio (R:S), specific root length (SRL), and nutrient concentrations. The investigation of relationships between ECM and seedling characteristics cannot be carried out directly because seedling response can be influenced either as the result of mycorrhizal colonisation, or the impact of environmental variations on both seedlings and mycorrhizae. Our previous statistical models performed in 2.6.2 had taken the effects related to seedling size and local environment (harvest treatments and topography) into account. Therefore, using the residual variance of these models enabled us to focus on the potential remaining relationship between ECM and seedling properties alone, having filtered out the other aforementioned effects.

3. Results

3.1. Sites

The four exploration types were observed at all sites, and throughout most of the individual root systems examined; 73% and 25% of the seedlings had respectively four and three exploration types on their root system. ECM root colonization (%), number of root tips of each exploration type (nb m^{-1}), and total root tips differed in the five study sites (Figure 2; all p-values < 0.01). Root colonization was higher in F2. Contact exploration type was higher in S1 and S3, whereas short-distance exploration types were more prevalent in F1 and F2. The largest numbers of medium- and long-distance exploration types were found respectively in S2 and F1. The greatest number of root tips occurred in F1, followed by F2/S2 and S1/S3 (decreasing order; Figure 2f). Contact exploration types and total number of root tips were respectively positively and negatively linked to summer 2015 precipitations (Figure 3), but no effects of annual or decennial rainfall were observed. Medium-distance exploration types were positively correlated with soil pH ($r = 0.978$, $p = 0.004$).

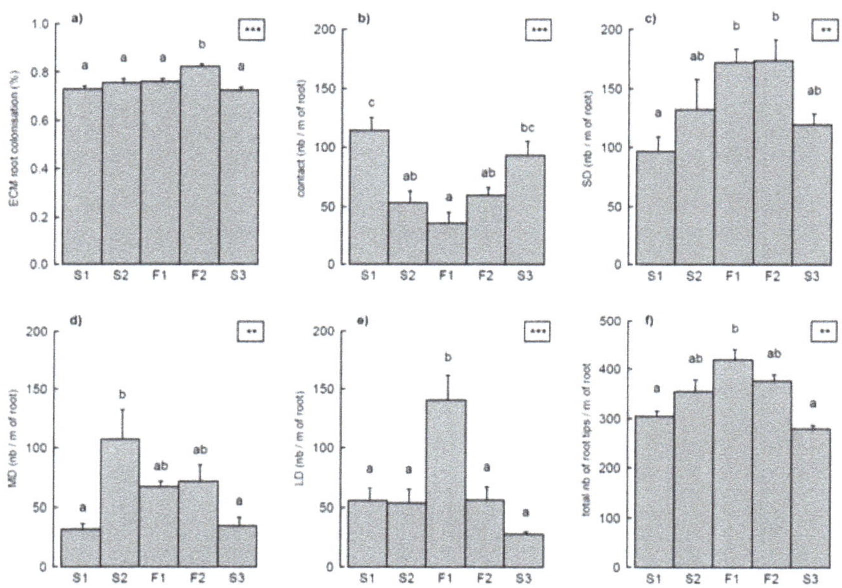

Figure 2. Differences in ectomycorrhizal (ECM) root colonization (**a**), the four ECM exploration types (**b**–**e**; expressed as the number of root tips per meter of root length), and total root tips (**f**) in the five study sites. Each bar represents mean ± SE (standard error) of nine values per site (harvesting treatments and topographical positions combined). Statistical significance is shown in the top right corner of each plot (***, $p < 0.001$; **, $p < 0.01$). Different lower-case letters indicate significant differences between sites ($p < 0.05$, Tukey test). SD, MD, and LD stand for short-distance, medium-distance, and long-distance exploration types, respectively.

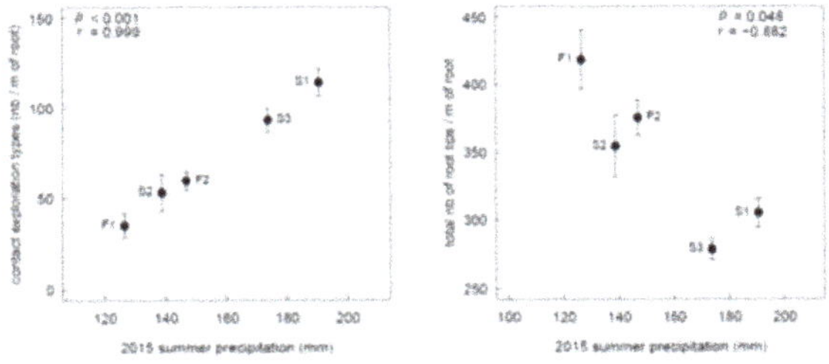

Figure 3. Significant ($p < 0.05$) Pearson correlations between ECM variables and summer precipitation (precipitation from July to September).

3.2. ECM Status Related to Harvesting Treatments and Topographical Positions

ECM root colonization ranged from 64.8% to 87.3% (mean 75.8 ± 0.7%) and was not impacted by harvesting treatment ($p = 0.288$) or by topography ($p = 0.619$). No significant relationship was observed between total number of root tips and harvesting treatments ($p = 0.131$) or topography ($p = 0.514$) either. Harvesting treatment (Figure 4a) and topography (Figure 4b) had a significant effect on ECM status. Contact ($p = 0.008$), short-distance ($p = 0.052$), and medium-distance ($p = 0.011$) exploration types were significantly affected by harvesting treatment (Figure 4a), but not long-distance exploration types ($p = 0.506$). Numbers of short- and medium-distance root tips were higher on seedlings in forests and smaller on those in clear-cuts. Seedlings in seed-trees had a greater number of contact exploration than those in forests. Topography significantly affected the number of contact ($p = 0.021$) and medium-distance ($p = 0.063$) exploration types (Figure 4b). Contact exploration decreased from bottom positions to top positions, whereas medium-distance exploration was higher in the bottom position than in the mid position.

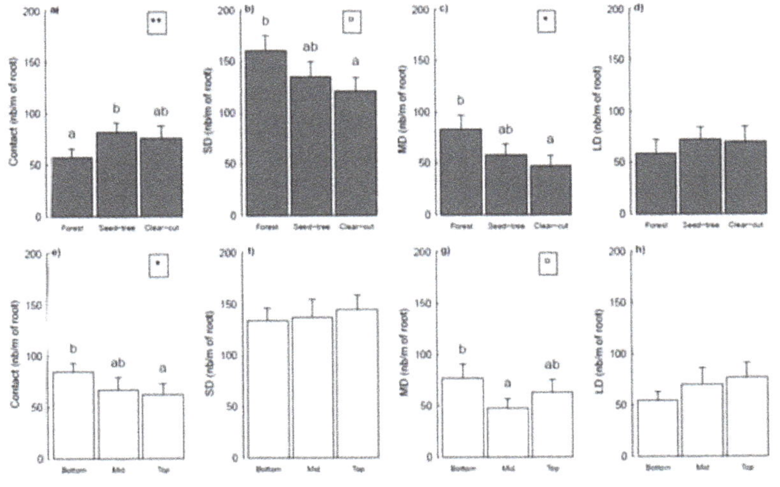

Figure 4. Number of root tips per meter of root length for the four ectomycorrhizal exploration types, under (**a**) three harvesting treatments and (**b**) three topographical positions. Each bar represents an average of 15 ± SE values. Significance of harvesting treatments or topographical positions is shown in the top right corner of each plot (**, $p < 0.01$; *, $p < 0.05$; ¤, $p < 0.1$). Different lower-case letters indicate significant differences for the same ECM exploration type between harvesting treatments (**a–d**) or topographical positions (**e–h**) at $p < 0.05$, Tukey test. SD, MD, and LD stand for short-distance, medium-distance, and long-distance exploration types, respectively.

3.3. Seedling Growth and Needle Nutrient Concentrations

All parameters except specific root length (SRL) and Mg-needle concentration were significantly affected by harvesting treatment (p = 0.430 for SRL, p = 0.712 for Mg, p < 0.001 for all others; Table 2).

Total seedling biomass and shoot biomass, stem diameter, and tap root length increased with a decreasing intensity of harvesting, from the forest to the clear-cut. Root biomass and fine root length were the lowest in forests, but no differences were found between seed-trees and clear-cuts. Seedling height was greater for seedlings growing in forests. Regarding needle nutrient concentrations, N and P were higher in seedlings in clear-cuts than in seed-trees, which were in turn higher than those in forests. K concentration was lower in forests than in seed-trees and clear-cuts. Ca concentration was higher in needles from seed-tree seedlings than in clear-cuts and forests. Topography only affected SRL (p = 0.017), seedling height (p = 0.038), and height:diameter ratio (p = 0.007), which were all greater at the mid position than the top position (Table 2). Regarding thresholds of sand culture studies, N and P concentrations in forest seedlings were under the threshold, while none of the other nutrients were considered as deficient in seed-trees or clear-cuts regardless of their topographical positions (Table 2). Using threshold values derived from pot trial studies, K and Ca concentrations were also deficient in the forest, whereas Ca was below the threshold for both seed-cuts and clear-cuts for all topographical positions.

Table 2. Effects of harvesting treatments and topographical positions on growth variables and needle nutrient concentrations.

	Harvesting Treatments						Topographical Positions					
	Forest		Seed-Tree		Clear-Cut		Bottom		Mid		Top	
Stem diameter (mm)	1.55 (0.05)	a	1.69 (0.06)	b	1.92 (0.06)	c	1.77 (0.07)		1.71 (0.08)		1.68 (0.06)	
Height (mm)	125.2 (4.7)	b	109.4 (3.2)	a	110.9 (2.7)	a	115.6 (3.4)	ab	120.2 (4.2)	b	109.8 (4.2)	a
Shoot biomass (g)	0.29 (0.02)	a	0.43 (0.03)	b	0.56 (0.05)	c	0.44 (0.05)		0.44 (0.05)		0.41 (0.05)	
Root biomass (g)	0.12 (0.01)	a	0.20 (0.02)	b	0.20 (0.02)	b	0.17 (0.02)		0.17 (0.02)		0.18 (0.02)	
Seedling biomass (g)	0.41 (0.03)	a	0.63 (0.05)	b	0.76 (0.06)	c	0.60 (0.06)		0.61 (0.07)		0.59 (0.06)	
Fine root length (cm)	99.4 (13.5)	a	187.7 (30.0)	b	163.5 (23.7)	b	148.1 (25.2)		168.6 (26.8)		133.9 (23.4)	
Tap root length (cm)	18.5 (0.71)	a	21.7 (0.82)	b	23.1 (0.56)	b	21.1 (0.84)		21.4 (0.80)		20.8 (0.95)	
Height:diameter (mm mm^{-1}) ·	82.2 (2.2)	c	67.0 (2.3)	b	59.8 (1.8)	a	67.6 (2.6)	a	74.0 (3.8)	b	67.3 (3.0)	a
Root:shoot (g g^{-1}) ·	0.42 (0.03)	a	0.50 (0.03)	b	0.39 (0.02)	a	0.42 (0.03)		0.42 (0.03)		0.47 (0.04)	
Specific root length (m g^{-1})	9.04 (0.86)		9.76 (1.04)		8.94 (0.92)		9.68 (1.03)	b	9.98 (0.73)	b	8.08 (0.97)	a
N-needle (mg g^{-1}) ·	11.5 (0.3) ‡	a	18.8 (0.6)	b	24.6 (0.6)	c	18.3 (1.5)		18.7 (1.4)		17.9 (1.6)	
P-needle (mg g^{-1}) ·	1.26 (0.05) ‡	a	2.36 (0.09)	b	3.12 (0.18)	c	2.20 (0.23)		2.38 (0.24)		2.16 (0.23)	
K-needle (mg g^{-1}) ·	5.82 (0.31) †	a	9.47 (0.34)	b	10.12 (0.31)	b	8.37 (0.60)		8.90 (0.60)		8.14 (0.57)	
Ca-needle (mg g^{-1}) ·	1.93 (0.05) †	a	2.65 (0.11) †	b	2.43 (0.08) †	a	2.32 (0.10) †		2.41 (0.15) †		2.27 (0.09) †	
Mg-needle (mg g^{-1})	1.58 (0.07)		1.59 (0.09)		1.65 (0.04)		1.60 (0.07)		1.57 (0.07)		1.65 (0.07)	

· indicates that analyses took effect of seedling size into consideration and focused on residuals of the regression between the variable and the aboveground biomass. Each value is a mean (±SE) of 15 values. Different lower-case letters indicate significant differences between harvesting treatments or topographical positions (p < 0.05, Tukey test). ‡ indicates values below the deficiency threshold in both sand culture and pot trials; † indicates values below the deficiency threshold only in pot trials ([48]; details in Table S1).

3.4. Effects of ECM on Seedling Root Properties and Nutrient Concentrations

SRL was highly affected by ECM composition but not root:shoot ratio (Table 3). The contact exploration type showed a positive effect on SRL, whereas other exploration types (short-, medium-, and long-distances) and total number of root tips were related to a decrease in SRL values. With regards to plant nutrient concentrations, several relationships were significant (Table 3), most of them showing an improvement in nutritive status of seedlings. Ca, K, and N needle concentrations significantly increased respectively with number of contacts, and short- and long-distance exploration types, whereas a higher number of the medium-distance exploration type corresponded to a decrease in Ca concentration. Total number of root tips was positively linked to higher N, K, and Mg concentrations.

Table 3. Results of linear regression between ECM colonization, root properties, and needle nutrient concentrations after removing effects due to harvesting treatments, topographical positions, and seedling size.

	ECM Root Colonization (%)	Contact (nb/m of Root)	SD (nb/m of Root)	MD (nb/m of Root)	LD (nb/m of Root)	Total nb of Root Tips/m of Root
Root:shoot (g g^{-1})		0.05 ¤				
Specific root length (m g^{-1})		0.32 ***	−0.06 ¤	−0.31 ***	−0.21 ***	−0.24 ***
N-needle (mg g^{-1})					0.17 **	0.07 *
P-needle (mg g^{-1})						
K-needle (mg g^{-1})			0.17 **			0.21 **
Ca-needle (mg g^{-1})		0.12 *		−0.11 *		
Mg-needle (mg g^{-1})						0.07 *

Values represent r^2 of regressions (n = 45). Symbols next to these values indicate the significance of regressions (***, $p < 0.001$; **, $p < 0.01$; *, $p < 0.05$; ¤, $p < 0.1$). Only significant regressions are shown. SD, MD, and LD stand for short-distance, medium-distance, and long-distance exploration types, respectively.

4. Discussion

4.1. Role of Abiotic Factors in ECM Composition

Two results confirmed the influence of soil moisture and hence water availability on ECM composition: (i) the influence of summer rainfall at the site and (ii) the influence of topography.

Significant differences in all ECM characteristics were observed among sites, suggesting that the local environment plays an important role in shaping ECM communities. Correlation between site environmental variables and ECM exploration types showed the positive influence of the summer rainfall during the first growing year on the number of contact exploration types and a negative effect on the total number of root tips (Figure 3). These results suggest that meteorological conditions during the first months of establishment are more important than the historical climatic conditions at the sites. Contact exploration types, which are less carbon costly for the seedling, take up water more efficiently during rainfall events thanks to their hydrophilic mantle [20]. Conversely, seedlings develop more root tips in drier areas [55], which may increase mycorrhizal associations. Rainfall and soil water availability have been shown to be an important factor governing fungal communities [25,56–58], but mainly by studying ECM species composition instead of exploration types. Jarvis et al. [29], who looked at both species and exploration types in *Pinus sylvestris* across Scotland, found that soil moisture and precipitation were the main drivers of ECM species composition, while temperature had an influence on exploration types. In our study, we could not test a link with temperature because of the small variation between our sites (less than 0.7 °C differences for all temperature variables). Conversely, we found that summer precipitation had a strong influence on ECM composition with higher contact exploration types at the wettest sites.

Regarding topography, a greater number of contact exploration types were shown at the bottom position than the top position. These results could be explained by higher water [43] and nutrient availabilities in the lower topographical positions [59], where the contact exploration types are more suitable because of their hydrophilic mantle, which allows them to be in closer contact with resources [20]. We expected the opposite pattern for long-distance exploration types which develop over a larger area and have greater mycelial expansion [20]. They are more likely to develop in stressful areas, but in our case, even though they tend to be more abundant at the top positions, the relationship was not significant. These results can be related to those observed by Bakker et al. [22] from two mature stands of maritime pines varying in distance from the water table and nutrient availability. They found a greater proportion of contact exploration types at the wet nutrient rich site and a higher proportion of long-distance types at the dry nutrient poor site.

Another positive significant correlation was found between pH and medium-distance explorations. A shift in ECM composition, from species developing high extramatrical mycelium instead of smooth types when pH increased, has already been reported in field survey studies (e.g., [60]) or liming experiments (e.g., [61]).

4.2. Impact of Forest Management on ECM Composition

In contradiction with our second hypothesis, we did not find any decrease of fungal root colonization with harvesting treatment, suggesting that the amount of fungal inoculum is still sufficient for the establishment of new seedlings the year after logging. It has been shown that forest practices similar to those used in our study may have a positive influence on ECM colonisation, and this may explain our results. More precisely: (i) mechanical soil disturbance leaving organic matter in the topsoil layers may have less impact than those that remove or bury the organic layer [33]; (ii) letting stumps in the forest stand during harvesting can have a positive impact on the maintenance of higher levels of fungal inoculum [34]; and (iii) tree harvesting occurring in late autumn or winter also allows mycorrhizae to remain active longer in the soil [32].

We expected a potential gradient of soil moisture which decreases from forests to clear-cuts due to the microenvironment created by the canopies of the trees, and that this soil moisture could impact ECM composition (leading to a greater number of contact exploration types in forests and long-distance exploration types in clear-cuts). Instead, differences found between forests and seed-trees/clear-cuts in ECM composition indicated that logging and site preparation influenced ECM composition in a different way. During our binocular observations, we often found patches of contact exploration types encrusted in small pieces of decaying wood. Contact exploration types are known to develop more in soil with high organic matter content, being able to degrade lignin directly from dead wood or rotting leaves to increase access to nutrients [20]. The mechanical tillage used in the study sites probably incorporated a supply of dead wood into the topsoil layers, which could explain the greater number of contact exploration types in seed-trees and clear-cuts. Another explanation for this increase in contact exploration is the higher resilience of these exploration types to disturbance, because they can easily regenerate their reduced system of extramatrical hyphae [62]. Conversely, short- and medium-distance exploration types with a lot of hyphae and some rhizomorphs decreased, due to the increasing harvesting intensity. In addition to the lower resilience of these ECM, the growth of nearby mature trees could influence ECM colonization for new seedlings. Trees may maintain greater mycelial networks from which fungi could vegetatively colonize new hosts due to the close vicinity of their roots [14,41,63]. This could be potentially more effective for exploration types with emanating hyphae. It has also been shown that ECM fungal propagules decreased sharply when isolated from a potential source, leading to a decrease in fungal colonization and diversity [64].

4.3. Impact of Forest Management on Seedling Characteristics

Morphological traits of seedlings were strongly impacted by harvesting treatment (15 out of the 17 variables studied showed significant differences; Table 2). As we had assumed, our results confirmed the light demanding characteristics of *Pinus pinaster* seedlings, which had higher above and below ground tissues, a greater stem diameter, and higher needle nutrient concentrations in clear-cuts than in forests. As shown by Robakowski et al. [65], seedlings growing under a higher level of light show greater net CO_2 assimilation rates and higher daily maximal photosynthetic rates, leading to a higher shoot biomass when light increases [66–70]. The observed increase of stem height and stem height:diameter ratio can also be interpreted as a shade avoidance response and are in agreement with previous results regarding these seedling traits [66,68,71,72].

Root biomass is also known to be improved in high light environments, but results regarding root:shoot ratio variations were heterogeneous, especially in the early stages. In *Pinus pinaster* seedlings, Rodríguez-García and Bravo [68] showed a higher allocation to roots when light increased in a garden experiment, whereas Ruano et al. [69] found no variations in root:shoot ratios in a field study with four harvesting intensities. These discrepancies may be due to differences in seedling size which affect root:shoot ratios [51]. The increase of tap root length from forests to clear-cuts is probably due to soil moisture differences between harvesting treatments; seedlings in dry areas improve their water foraging capacity by having long and deep roots [73,74].

Seedlings growing in forests had lower needle nutrient concentrations. Higher N and P concentrations in clear-cuts than in seed-trees could also be related to better nitrogen and phosphorus efficiency use in full light than in low light, as reported by Elliott and White [67]. In addition, the mechanical effect of logging and soil preparation carried out in clear-cuts and seed-trees could explain our results, as these can lead to an increase of mineralisation, thus causing a substantial release of mineral elements into the soil.

Overall, the sufficient level of nutrients observed in needles of our seedlings suggests that mycorrhizae may enable seedlings to overcome the nutritional stress of the local environment (Table S1).

4.4. Effects of ECM on Seedling Root Development and Nutritive Status

ECM colonisation is associated with a modification in fine root morphology according to the different exploration types rather than a greater investment to the roots (Table 3). Mycorrhizal colonisation usually increases both shoot and root biomass, but the root:shoot ratios could be lower or higher in conifer seedlings depending on the amount of fungal tissue present [13,75]. Indeed, the ECM fungal identity is the main factor determining fine root morphology [76] and aboveground biomass variations [77]. ECM colonisation increased fine root diameter and decreased SRL, especially due to the mycelial mantle surrounding fine roots [12,13]. High SRL values suggest fast growth and intensive soil exploration. Thus, seedlings with lower SRL values will need more root tips and exploration types with a lot of hyphae or rhizomorphs to compensate for the lower soil exploration area.

Plant nutrient concentration is greater with more root tips, and each exploration type is significantly associated with a single nutrient (Table 3). These results are consistent with our third hypothesis, and suggest that different functional types appear to be complementary for access to different nutrient sources. Several studies looking at functional diversity showed relationships with soil properties and nutrient availability, especially N [21,23,29,30]. Long-distance exploration types are able to prevent resources from leaching during transport with their hydrophobic rhizomorphs and would have a strong ability to acquire organic N [23]. However, this strategy is expensive in energy for the plant and thus such a strategy is competitively dominant only in lower nutrient environments where resources are rare and patchy, as in our sites. Our findings can be related to those of de Witte et al. [30], who investigated ECM exploration types in several beech forests and found many correlations between exploration types and foliar concentrations or soil properties. Similar to our results, they showed a positive relationship between contact exploration types and soil Ca (soil and foliar Ca were positively correlated in the study), together with a negative relationship between medium-distance exploration types and soil Ca. Furthermore, they found that a higher abundance of medium- and long-distance types was negatively associated with foliar N, and suggested that this might be due to the non N-limited environment in their study.

4.5. Can ECM Explain Failures of Forest Regeneration?

In the case where mycorrhization could be a factor involved in regeneration failures, we would expect that sites in failure areas (F1 and F2) would have insufficient mycorrhization (in terms of root colonization or number of root tips), or would have a clearly different composition of ECM communities. However, failure sites have higher numbers of root tips and higher or equivalent root colonization, which is contrary to our first hypothesis. The four exploration types were found on seedlings at all sites and only the short-distance ones discriminated failure areas from successful areas in terms of exploration type composition (Figure 2c). However, this exploration type was not specific of failure sites because it corresponded to the more prevalent exploration type at four of the five sites (and the second highest in S1). The prevalence of the short-distance exploration type could be explained by the fungal identity which is the most representative of this exploration type, *Cenococcum geophilum*, and by the summer drought conditions occurring in our region. This species is known to be drought-tolerant and extremely resilient after drought [78], allowing an early uptake of water and nutrients in the post-drought period [79]. In addition, fast colonization of new seedlings could also be the result of frequent disturbances within these forests (fires, storms), which has been shown to lead to an early-seral ECM community (of which *Cenococcum* is a part) necessary to promote seedling establishment [35].

A high diversity of exploration types was observed at all sites and throughout most of the individual root systems. This diversity may promote a higher resilience of ECM communities to environmental changes, and suggests that fungal association is essential for the survival of young seedlings in the region even though it does not explain the failure of regeneration in some specific areas. However, we should consider that by harvesting the seedlings during their first winter, we only have seedlings that have survived the dry summer conditions, which is probably the main cause of mortality in the early establishment of maritime pine seedlings [38]. Seedlings with lower mycorrhizal associations in both quantity and/or exploration type diversity (and *Cenococum geophilum* in particular) may have died during the summer season and are thus missing from our sample, and this may be the main limit of our study.

5. Conclusions

Our results showed that ECM composition was affected by both harvesting and topography. In turn, mycorrhization appeared to be essential for early seedling establishment by improving the nutritive status of seedlings. However, in our context, ECM colonisation one year after logging does not seem to be linked to regeneration failures that occur in some specific areas. Other kinds of biotic interactions such as facilitation/competition with other plants or herbivory should also be explored to understand such regeneration failures.

Supplementary Materials: The following are available online at http://www.mdpi.com/1999-4907/9//245/s1. Table S1: nutrient threshold for pine seedlings (A) in sand culture or (B) in pot trial, from van der Burg [48].

Author Contributions: A.G. and M.R.B. designed and performed the experiments; A.G. and F.D. analyzed the data; A.G., F.D., M.G., L.A., and M.R.B. wrote the paper.

Acknowledgments: We thank Nathalie Gallegos and Cathy Lambrot for soil and nutrient concentration analyses, Météo-France and Benoît Persyn for meteorological data acquisition, Aldyth Nys for revising the English, and three anonymous reviewers for constructive comments on the manuscript. We are very grateful to the "Office National des Forêts" (ONF), in particular Didier Canteloup, Francis Maugard, and all staff working on the management of the study sites for the valuable help provided. The ONF staff were in charge of the implementation of seed-tree and clear-cut harvesting treatments. A. Guignabert was funded by a PhD grant by the Région Nouvelle-Aquitaine and Bordeaux Sciences Agro.

Conflicts of Interest: The authors declare no conflict of interest.

References

1. IFN Inventaire Forestier National. Available online: https://inventaire-forestier.ign.fr/ (accessed on 18 December 2017).
2. Sardin, T. *Guide des Sylvicultures Forêts Littorales Atlantiques Dunaires*, ONF ed.; Office National des Forêts: Paris, France, 2009; p. 175, ISBN 9782842073374.
3. Ouallet, P. Quels Peuvent-être les Facteurs Écologiques Responsables des Échecs de Régénération Naturelle du pin Maritime sur les Dunes Littorales des Forêts Domaniales de Biscarrosse et de Sainte-Eulalie? Master's Thesis, Bordeaux Sciences Agro et ONF, Gradignan, France, 2012.
4. Clark, J.S.; Macklin, E.; Wood, L. Stages and Spatial Scales of Recruitment Limitation in Southern Apalachians forests. *Ecol. Monogr.* **1998**, *68*, 213–235. [CrossRef]
5. Kitajima, K.; Fenner, M. Ecology of seedling regeneration. In *Seeds: The Ecology of Regeneration in Plant Communities*; CABI: Wallingford, UK, 2000; pp. 331–359, ISBN 0851994326.
6. Harvey, A.E.; Jurgensen, M.F.; Larsen, M.J.; Graham, R.T. Relationships among soil microsite, ectomycorrhizae, and natural conifer regeneration of old-growth forests in western Montana. *Can. J. For. Res.* **1987**, *17*, 58–62. [CrossRef]
7. Harmer, R. The effect of plant competition and simulated summer browsing by deer on tree regeneration. *J. Appl. Ecol.* **2001**, *38*, 1094–1103. [CrossRef]

8. Rodríguez-García, E.; Gratzer, G.; Bravo, F. Climatic variability and other site factor influences on natural regeneration of *Pinus pinaster* Ait. in Mediterranean forests. *Ann. For. Sci.* **2011**, *68*, 811–823. [CrossRef]
9. Perry, D.A.; Molina, R.; Amaranthus, M.P. Mycorrhizae, mycorrhizospheres, and reforestation: Current knowledge and research needs. *Can. J. For. Res.* **1987**, *17*, 929–940. [CrossRef]
10. Miller, S.L.; McClean, T.M.; Stanton, N.L.; Williams, S.E. Mycorrhization, physiognomy, and first-year survivability of conifer seedlings following natural fire in Grand Teton National Park. *Can. J. For. Res.* **1998**, *28*, 115–122. [CrossRef]
11. Barker, J.S.; Simard, S.W.; Jones, M.D.; Durall, D.M. Ectomycorrhizal fungal community assembly on regenerating Douglas-fir after wildfire and clearcut harvesting. *Oecologia* **2013**, *172*, 1179–1189. [CrossRef] [PubMed]
12. Rousseau, J.V.D.; Sylvia, D.M.; Fox, A.J. Contribution of ectomycorrhiza to the potential nutrient-absorbing surface of pine. *New Phytol.* **1994**, *128*, 639–644. [CrossRef]
13. Smith, S.E.; Read, D. *Mycorrhizal Symbiosis*, 3rd ed.; Academic Press: Cambridge, MA, USA, 2008; ISBN 9780123705266.
14. Teste, F.P.; Simard, S.W.; Durall, D.M.; Guy, R.D.; Jones, M.D.; Schoonmaker, A.L. Access to mycorrhizal networks and roots of trees: Importance for seedling survival and resource transfer. *Ecology* **2009**, *90*, 2808–2822. [CrossRef] [PubMed]
15. Perrin, R.; Garbaye, J. Influence of ectomycorrhizae on infectivity of Pythium-infested soils and substrates. *Plant Soil* **1983**, *71*, 345–351. [CrossRef]
16. Van Tichelen, K.K.; Colpaert, J.V.; Vangronsveld, J. Ectomycorrhizal protection of *Pinus sylvestris* against copper toxicity *New Phytol.* **2001**, *150*, 203–213. [CrossRef]
17. Lehto, T.; Zwiazek, J.J. Ectomycorrhizas and water relations of trees: A review. *Mycorrhiza* **2011**, *21*, 71–90. [CrossRef] [PubMed]
18. Mousain, D. Effets de deux facteurs edaphiques (teneur en phosphore et qualite de la matiere organique des sols) sur l'etablissement de la symbiose ectomycorhizienne du pin maritime (*Pinus pinaster* Soland. *in* Ait.). *Rev. Ecol. Biol. Sol* **1975**, *12*, 239–259.
19. Read, D.J.; Leake, J.R.; Perez-Moreno, J. Mycorrhizal fungi as drivers of ecosystem processes in heathland and boreal forest biomes. *Can. J. Bot.* **2004**, *82*, 1243–1263. [CrossRef]
20. Agerer, R. Exploration types of ectomycorrhizae: A proposal to classify ectomycorrhizal mycelial systems according to their patterns of differentiation and putative ecological importance. *Mycorrhiza* **2001**. [CrossRef]
21. Lilleskov, E.A.; Hobbie, E.A.; Horton, T.R. Conservation of ectomycorrhizal fungi: Exploring the linkages between functional and taxonomic responses to anthropogenic N deposition. *Fungal Ecol.* **2011**, *4*, 174–183. [CrossRef]
22. Bakker, M.R.; Augusto, L.; Achat, D.L. Fine root distribution of trees and understory in mature stands of maritime pine (*Pinus pinaster*) on dry and humid sites. *Plant Soil* **2006**, *286*, 37–51. [CrossRef]
23. Hobbie, E.A.; Agerer, R. Nitrogen isotopes in ectomycorrhizal sporocarps correspond to belowground exploration types. *Plant Soil* **2010**, *327*, 71–83. [CrossRef]
24. Slankis, V. Soil Factors Influencing Formation of Mycorrhizae. *Annu. Rev. Phytopathol.* **1974**, *12*, 437–457. [CrossRef]
25. Miyamoto, Y.; Sakai, A.; Hattori, M.; Nara, K. Strong effect of climate on ectomycorrhizal fungal composition: Evidence from range overlap between two mountains. *ISME J.* **2015**, *9*, 1870–1879. [CrossRef] [PubMed]
26. Cox, F.; Barsoum, N.; Lilleskov, E.A.; Bidartondo, M.I. Nitrogen availability is a primary determinant of conifer mycorrhizas across complex environmental gradients. *Ecol. Lett.* **2010**, *13*, 1103–1113. [CrossRef] [PubMed]
27. Kilpeläinen, J.; Barbero-López, A.; Vestberg, M.; Heiskanen, J.; Lehto, T. Does severe soil drought have after-effects on arbuscular and ectomycorrhizal root colonisation and plant nutrition? *Plant Soil* **2017**. [CrossRef]
28. Fernandez, C.W.; Nguyen, N.H.; Stefanski, A.; Han, Y.; Hobbie, S.E.; Montgomery, R.A.; Reich, P.B.; Kennedy, P.G. Ectomycorrhizal fungal response to warming is linked to poor host performance at the boreal-temperate ecotone. *Glob. Chang. Biol.* **2017**, *23*, 1598–1609. [CrossRef] [PubMed]
29. Jarvis, S.; Woodward, S.; Alexander, I.J.; Taylor, A.F.S. Regional scale gradients of climate and nitrogen deposition drive variation in ectomycorrhizal fungal communities associated with native *Scots pine*. *Glob. Chang. Biol.* **2013**, *19*, 1688–1696. [CrossRef] [PubMed]

30. De Witte, L.C.; Rosenstock, N.P.; van der Linde, S.; Braun, S. Nitrogen deposition changes ectomycorrhizal communities in Swiss beech forests. *Sci. Total Environ.* **2017**, *605–606*, 1083–1096. [CrossRef] [PubMed]
31. Harvey, A.E.; Jurgensen, M.F.; Larsen, M.J.; Schlieter, J.A. *Distribution of Active Ectomycorrhizal Short Roots of the Inland Northwest: Effects of Site and Disturbance*; INT-374; Intermountain Research Station: Ogden, UT, USA, 1986; p. 8.
32. Jones, M.D.; Durall, D.M.; Cairney, J.W.G. Ectomycorrhizal fungal communities in young forest stands regenerating after clearcut logging. *New Phytol.* **2003**, *157*, 399–422. [CrossRef]
33. Lazaruk, L.W.; Macdonald, S.E.; Kernaghan, G. The effect of mechanical site preparation on ectomycorrhizae of planted white spruce seedlings in conifer-dominated boreal mixedwood forest. *Can. J. For. Res.* **2008**, *38*, 2072–2079. [CrossRef]
34. Page-Dumroese, D.S.; Harvey, A.E.; Jurgensen, M.F.; Amaranthus, M.P. Impacts of soil compaction and tree stump removal on soil properties and outplanted seedlings in northern Idaho, USA. *Can. J. Soil Sci.* **1998**, *78*, 29–34. [CrossRef]
35. Kranabetter, J.M.; Haeussler, S.; Wood, C. Vulnerability of boreal indicators (ground-dwelling beetles, understory plants and ectomycorrhizal fungi) to severe forest soil disturbance. *For. Ecol. Manag.* **2017**, *402*, 213–222. [CrossRef]
36. Lazaruk, L.W.; Kernaghan, G.; Macdonald, S.E.; Khasa, D. Effects of partial cutting on the ectomycorrhizae of *Picea glauca* forests in northwestern Alberta. *Can. J. For. Res.* **2005**, *35*, 1442–1454. [CrossRef]
37. Jones, M.D.; Twieg, B.D.; Durall, D.M.; Berch, S.M. Location relative to a retention patch affects the ECM fungal community more than patch size in the first season after timber harvesting on Vancouver Island, British Columbia. *For. Ecol. Manag.* **2008**, *255*, 1342–1352. [CrossRef]
38. Rodríguez-García, E.; Bravo, F.; Spies, T.A. Effects of overstorey canopy, plant–plant interactions and soil properties on Mediterranean maritime pine seedling dynamics. *For. Ecol. Manag.* **2011**, *262*, 244–251. [CrossRef]
39. Bingham, M.A.; Simard, S. Ectomycorrhizal Networks of *Pseudotsuga menziesii* var. *glauca* Trees Facilitate Establishment of Conspecific Seedlings under Drought. *Ecosystems* **2012**, *15*, 188–199. [CrossRef]
40. Warren, J.M.; Brooks, J.R.; Meinzer, F.C.; Eberhart, J.L. Hydraulic redistribution of water from *Pinus ponderosa* trees to seedlings: Evidence for an ectomycorrhizal pathway. *New Phytol.* **2008**, *178*, 382–394. [CrossRef] [PubMed]
41. Jonsson, L.; Dahlberg, A.; Nilsson, M.C.; Kårén, O.; Zackrisson, O. Continuity of ectomycorrhizal fungi in self-regenerating boreal *Pinus sylvestris* forests studied by comparing mycobiont diversity on seedlings and mature trees. *New Phytol.* **1999**, *142*, 151–162. [CrossRef]
42. Dahlberg, A.; Schimmel, J.; Taylor, A.F.S.; Johannesson, H. Post-fire legacy of ectomycorrhizal fungal communities in the Swedish boreal forest in relation to fire severity and logging intensity. *Biol. Conserv.* **2001**, *100*, 151–161. [CrossRef]
43. Pitaud, G. L'hydrogéologie et la Végétation dans les Dunes du Littoral Aquitain. Ph.D. Thesis, University Bordeaux, Nouvelle-Aquitaine, France, 1967.
44. Augusto, L.; Bakker, M.R.; Morel, C.; Meredieu, C.; Trichet, P.; Badeau, V.; Arrouays, D.; Plassard, C.; Achat, D.L.; Gallet-Budynek, A.; et al. Is "grey literature" a reliable source of data to characterize soils at the scale of a region? A case study in a maritime pine forest in southwestern France. *Eur. J. Soil Sci.* **2010**, *61*, 807–822. [CrossRef]
45. Porté, A.; Bosc, A.; Champion, I.; Loustau, D. Estimating the foliage area of Maritime pine (*Pinus pinaster* Aït.) branches and crowns with application to modelling the foliage area distribution in the crown. *Ann. For. Sci.* **2000**, *57*, 73–86. [CrossRef]
46. Berbigier, P.; Bonnefond, J. Measurement and modelling of radiation transmission within a stand of maritime pine (*Pinus pinaster* Ait). *Ann. Sci. For.* **1995**, *52*, 23–42. [CrossRef]
47. Tennant, D. A test of a modified line intersect method of estimating root length. *J. Ecol.* **1975**, *63*, 995–1001. [CrossRef]
48. Van den Burg, J. *Foliar Analysis for Determination of Tree Nutrient Status: A Compilation of Literature Data*; Rijksinstituut voor Onderzoek in de Bos-en Landschapsbouw de Dorschkamp: Wageningen, The Netherlands, 1985.
49. R Core Team. *R: A Language and Environment for Statistical Computing*; R Foundation for Statistical Computing: Vienna, Austria, 2017; Available online: https://www.R-project.org/ (accessed on 3 July 2017).

50. Warton, D.I.; Hui, F.K.C. The arcsine is asinine: The analysis of proportions in ecology. *Ecology* **2011**, *92*, 3–10. [CrossRef] [PubMed]
51. McConnaughay, K.D.M.; Coleman, J.S. Biomass Allocation in Plants: Ontogeny or Optimality? A Test along Three Resource Gradients. *Ecology* **1999**, *80*, 2581–2593. [CrossRef]
52. Achat, D.L.; Deleuze, C.; Landmann, G.; Pousse, N.; Ranger, J.; Augusto, L. Quantifying consequences of removing harvesting residues on forest soils and tree growth—A meta-analysis. *For. Ecol. Manag.* **2015**, *348*, 124–141. [CrossRef]
53. Augusto, L.; Ranger, J.; Ponette, Q.; Rapp, M. Relationships between forest tree species, stand production and stand nutrient amount. *Ann. For. Sci.* **2000**, *57*, 313–324. [CrossRef]
54. Ritson, P.; Sochacki, S. Measurement and prediction of biomass and carbon content of *Pinus pinaster* trees in farm forestry plantations, south-western Australia. *For. Ecol. Manag.* **2003**, *175*, 103–117. [CrossRef]
55. Pena, R.; Simon, J.; Rennenberg, H.; Polle, A. Ectomycorrhiza affect architecture and nitrogen partitioning of beech (*Fagus sylvatica* L.) seedlings under shade and drought. *Environ. Exp. Bot.* **2013**, *87*, 207–217. [CrossRef]
56. Tedersoo, L.; Bahram, M.; Toots, M.; Diédhiou, A.G.; Henkel, T.W.; Kjøller, R.; Morris, M.H.; Nara, K.; Nouhra, E.; Peay, K.G.; et al. Towards global patterns in the diversity and community structure of ectomycorrhizal fungi. *Mol. Ecol.* **2012**, *21*, 4160–4170. [CrossRef] [PubMed]
57. Bahram, M.; Põlme, S.; Kõljalg, U.; Zarre, S.; Tedersoo, L. Regional and local patterns of ectomycorrhizal fungal diversity and community structure along an altitudinal gradient in the Hyrcanian forests of northern Iran. *New Phytol.* **2012**, *193*, 465–473. [CrossRef] [PubMed]
58. Shi, L.; Guttenberger, M.; Kottke, I.; Hampp, R. The effect of drought on mycorrhizas of beech (*Fagus sylvatica* L.): Changes in community structure, and the content of carbohydrates and nitrogen storage bodies of the fungi. *Mycorrhiza* **2002**, *12*, 303–311. [CrossRef] [PubMed]
59. Sewerniak, P.; Jankowski, M. Topographically-controlled site conditions drive vegetation pattern on inland dunes in Poland. *Acta Oecol.* **2017**, *82*, 52–60. [CrossRef]
60. Bahr, A.; Ellström, M.; Akselsson, C.; Ekblad, A.; Mikusinska, A.; Wallander, H. Growth of ectomycorrhizal fungal mycelium along a Norway spruce forest nitrogen deposition gradient and its effect on nitrogen leakage. *Soil Biol. Biochem.* **2013**, *59*, 38–48. [CrossRef]
61. Bakker, M.; Garbaye, J.; Nys, C. Effect of liming on the ectomycorrhizal status of oak. *For. Ecol. Manag.* **2000**, *126*, 121–131. [CrossRef]
62. Tedersoo, L.; Smith, M.E. Lineages of ectomycorrhizal fungi revisited: Foraging strategies and novel lineages revealed by sequences from belowground. *Fungal Biol. Rev.* **2013**, *27*, 83–99. [CrossRef]
63. Cline, E.T.; Ammirati, J.F.; Edmonds, R.L. Does proximity to mature trees influence ectomycorrhizal fungus communities of Douglas-fir seedlings? *New Phytol.* **2005**, *166*, 993–1009. [CrossRef] [PubMed]
64. Peay, K.G.; Schubert, M.G.; Nguyen, N.H.; Bruns, T.D. Measuring ectomycorrhizal fungal dispersal: Macroecological patterns driven by microscopic propagules. *Mol. Ecol.* **2012**, *21*, 4122–4136. [CrossRef] [PubMed]
65. Robakowski, P.; Wyka, T.; Samardakiewicz, S.; Kierzkowski, D. Growth, photosynthesis, and needle structure of silver fir (*Abies alba* Mill.) seedlings under different canopies. *For. Ecol. Manag.* **2004**, *201*, 211–227. [CrossRef]
66. Chen, H.Y.; Klinka, K. Survival, growth, and allometry of planted *Larix occidentalis* seedlings in relation to light availability. *For. Ecol. Manag.* **1998**, *106*, 169–179. [CrossRef]
67. Elliott, K.J.; White, A.S. Effects of light, nitrogen, and phosphorus on red pine seedling growth and nutrient use efficiency. *For. Sci.* **1994**, *40*, 47–58.
68. Rodríguez-García, E.; Bravo, F. Plasticity in *Pinus pinaster* populations of diverse origins: Comparative seedling responses to light and Nitrogen availability. *For. Ecol. Manag.* **2013**, *307*, 196–205. [CrossRef]
69. Ruano, I.; Pando, V.; Bravo, F. How do light and water influence *Pinus pinaster* Ait. germination and early seedling development? *For. Ecol. Manag.* **2009**, *258*, 2647–2653. [CrossRef]
70. Ingleby, K.; Munro, R.C.; Noor, M.; Mason, P.A.; Clearwater, M.J. Ectomycorrhizal populations and growth of *Shorea parvifolia* (Dipterocarpaceae) seedlings regenerating under three different forest canopies following logging. *For. Ecol. Manag.* **1998**, *111*, 171–179. [CrossRef]
71. Zavala, M.A.; Espelta, J.M.; Caspersen, J.; Retana, J. Interspecific differences in sapling performance with respect to light and aridity gradients in Mediterranean pine–oak forests: Implications for species coexistence. *Can. J. For. Res.* **2011**, *41*, 1432–1444. [CrossRef]

72. Wang, G.G.; Qian, H.; Klinka, K. Growth of *Thuja plicata* seedlings along a light gradient. *Can. J. Bot.* **1994**, *72*, 1749–1757. [CrossRef]
73. Markesteijn, L.; Poorter, L. Seedling root morphology and biomass allocation of 62 tropical tree species in relation to drought- and shade-tolerance. *J. Ecol.* **2009**, *97*, 311–325. [CrossRef]
74. Brunner, I.; Herzog, C.; Dawes, M.A.; Arend, M.; Sperisen, C. How tree roots respond to drought. *Front. Plant Sci.* **2015**, *6*, 1–16. [CrossRef] [PubMed]
75. Alexander, I.J. The *Picea sitchensis* + *Lactarius rufus* mycorrhizal association and its effects on seedling growth and development. *Trans. Br. Mycol. Soc.* **1981**, *76*, 417–423. [CrossRef]
76. Ostonen, I.; Tedersoo, L.; Suvi, T.; Lõhmus, K. Does a fungal species drive ectomycorrhizal root traits in *Alnus* spp.? *Can. J. For. Res.* **2009**, *39*, 1787–1796. [CrossRef]
77. Nara, K. Ectomycorrhizal networks and seedling establishment during early primary succession. *New Phytol.* **2006**, *169*, 169–178. [CrossRef] [PubMed]
78. Pigott, C.D. Survival of mycorrhiza formed by *Cenococcum geophilum* Fr. in dry soils. *New Phytol.* **1982**, *92*, 513–517. [CrossRef]
79. Kreuzwieser, J.; Gessler, A. Global climate change and tree nutrition: Influence of water availability. *Tree Physiol.* **2010**, *30*, 1221–1234. [CrossRef] [PubMed]

forests

MDPI

Article

Toward Sustainable Cultivation of *Pinus occidentalis* Swartz in Haiti: Effects of Alternative Growing Media and Containers on Seedling Growth and Foliar Chemistry

Kyrstan L. Hubbel [1], Amy L. Ross-Davis [2], Jeremiah R. Pinto [3], Owen T. Burney [4] and Anthony S. Davis [2,*]

1 Center for Forest Nursery and Seedling Research, University of Idaho, 1025 Plant Science Road, Moscow, ID 83843, USA; 123khubbel@gmail.com
2 College of Forestry, Oregon State University, 109 Richardson Hall, Corvallis, OR 97331, USA; rossdava@oregonstate.edu
3 USDA Forest Service, Rocky Mountain Research Station, 1221 South Main Street, Moscow, ID 83843, USA; jpinto@fs.fed.us
4 College of Agricultural, Consumer, and Environmental Sciences, New Mexico State University, John T. Harrington Forestry Research Center, 3021 Highway 518, Mora, NM 87732, USA; oburney@nmsu.edu
* Correspondence: anthony.davis@oregonstate.edu

Received: 8 June 2018; Accepted: 10 July 2018; Published: 13 July 2018

Abstract: Haiti has suffered great losses from deforestation, with little forest cover remaining today. Current reforestation efforts focus on seedling quantity rather than quality. This study examined limitations to the production of high-quality seedlings of the endemic Hispaniolan pine (*Pinus occidentalis* Swartz). Recognizing the importance of applying sustainable development principles to pine forest restoration, the effects of growing media and container types on seedling growth were evaluated with the goal of developing a propagation protocol to produce high-quality seedlings using economically feasible nursery practices. With regard to growing media, seedlings grew best in compost-based media amended with sand. Topsoil, widely used in nurseries throughout Haiti, produced the smallest seedlings overall. Despite a low water holding capacity and limited manganese, compost-based media provided adequate levels of essential mineral nutrients (particularly nitrogen), which allowed for sufficient seedling nutrition. Seedling shoot and root growth, as well as the ratio of shoot biomass to root biomass, were greater in polybags relative to D40s. Results indicate that economically feasible improvements to existing nursery practices in Haiti can improve the early growth rates of *P. occidentalis* seedlings.

Keywords: compost; foliar nutrients; Hispaniolan pine; pine forests; seedling quality; sustainable development principles; tropical forest nursery

1. Introduction

Haiti, a tropical country with a landmass that was at one time 60% forested, has suffered great losses from deforestation [1], with little forest cover remaining today [2]. While generating new forests may seem a daunting task [3], the production of high-quality seedlings is an imperative first step. These seedlings must be grown specifically for reforestation, be economically accessible to local people, and be produced using locally available materials [4]. Nursery practices in Haiti, however, focus primarily on seedling quantity rather than quality. Multiple studies have shown that outplanting performance on reforestation sites correlates highly with seedling quality [5–7]. In a nursery system, seedling quality is often quantified by several morphological and physiological measurements. No single measurement can reliably predict performance; however, many studies suggest that seedling root-collar-diameter, shoot height, root volume, and the ratio of shoot-to-root biomass correlate highly with outplanting success [8–10].

Proper nursery culture has the greatest impact on seedling quality. In modern forest seedling nurseries, growers use high-quality growing media in concert with cutting-edge container technology and a wide array of fertilizers in a controlled environment. These ideal growing conditions and associated resources allow modern nurseries to easily implement the Target Plant Concept, which is defined as the specific physiological and morphological plant characteristics cultivated in the nursery that lead to the growth and survival of the outplanted seedling at a particular site [4]. The majority of nurseries located in areas of tropical deforestation, including Haiti, lack access to many of these resources; these nurseries must produce healthy seedlings using economically practical methods and available resources [11]. Effective use of the Target Plant Concept [4] in these resource-limited nurseries to better connect nursery cultural practices, and the resultant seedlings, with the anticipated field conditions, empowers reforestation managers with a framework that can lead to measurable success in post-planting seedling establishment.

Access to moisture and nutrients are critical to seedling quality [12,13] and are often managed through growing media. In Haiti, like many developing countries, topsoil is the primary component of available potting media [14,15], despite evidence that using topsoil in container nurseries often results in low outplanting success [16]. In addition to issues associated with poor drainage and compaction, topsoil is an unsustainable resource and is particularly valuable in heavily deforested regions already suffering from erosional soil losses [14]. Various alternative sources of growing media may be used [17], such as rice hulls, sand, compost, sawdust, or pine bark [5]. Incorporating compost as a component of growing media can prove highly valuable but requires batch testing to account for variability among feedstock sources [15,18–20]. Given the emerging compost industry in Haiti [21], local sources may be available. Amending potting media with vermicompost improved germination of container-grown maritime pine (*Pinus pinaster* Aiton) [18], and both germination and seedling growth of container-grown alleppo pine (*Pinus halepensis* Miller) were greater when activated sewage sludge was incorporated into peat-based media [19].

A second factor to consider when producing nursery-grown seedlings is container type [22,23]. Forest seedling nursery containers must perform a combination of functions, and the right container choice will vary by species [24]. Size (volume and depth) and design features work to mitigate root spiraling and influence overall root architecture [25]. In many developing countries, including Haiti, nursery growers typically use small plastic bags (i.e., polybags) as containers for growing tree seedlings. This container type is widely available and is lightweight and collapsible, which greatly reduces shipping costs. However, polybags have been associated with several plant growth and development issues, including root malformation, which can reduce outplanting success [26,27].

Hispaniolan pine (*Pinus occidentalis* Swartz), an endangered tree species native to the island of Hispaniola (comprising Haiti and the Dominican Republic), has been recognized as a species for over 200 years but has received limited scientific attention [28]. Given the location of the remaining pine forests at high elevation [29], the species also represents a critical component of restoration programs aimed at conserving soil and reducing damage during heavy rains. Since container type and growing media are two important considerations for the production of high-quality seedlings essential for reforestation success, the objective of this study was to examine the influence of growing media and container type on the development of Hispaniolan pine seedlings. We hypothesized that seedlings grown in nutrient rich, compost-based growing media would exhibit sufficient foliar nutrient levels and greater early seedling growth rates relative to unamended peat-based media or topsoil. We also hypothesized that seedling growth would be greater in rigid-walled D40 containers relative to polybags.

2. Materials and Methods

2.1. Experimental Design

The experiment followed a randomized complete block design (RCBD) with a factorial structure (five media treatments × two container types) containing five replicates (i.e., blocks) per treatment. Each tray representing a growing medium and container treatment combination was considered as a block, and containers within each block were randomized weekly to minimize the effects of the nursery environment.

Five growing media mixtures were compared in this study: (1) 100% peat-based [Pe] (an unamended media consisting of 45% Canadian sphagnum peat moss, 45% coarse-grade vermiculite, and 10% fine-aged bark by volume; SunGro® Metro Mix, Agawam, MA, USA); (2) 100% "topsoil" [T] (60% topsoil, 20% animal manure, and 20% bark mulch; NuLife Topsoil, Waupaca, WI, USA); (3) 80:20 compost:topsoil [CT] (municipal biosolids and yard waste feedstock; Eko Compost, Lewiston, ID, USA); (4) 80:20 compost:grit [CG] (Grit size medium, 1.2–4.8 mm, Target® Forestry Nursery Grit, Burnaby, BC, USA); (5) 70:20:10 compost:topsoil:grit [CTG]. Pe is widely used in commercial forest tree seedling production nurseries in the US as well as other developed countries [30]. Topsoil is widely used in container nursery systems in developing countries, including Haiti [11,14,15]; however, our mix was a commercially available product. The three compost-based media types serve as potential alternatives.

Hispaniolan pine seeds from a government-funded seed bank in the Dominican Republic (Nigua Seed Bank, Santo Domingo, Dominican Republic, 18°22′34.644″ N, 70°4′9.7674″ W, provenance unknown) were soaked in distilled water for 12 h prior to sowing, as recommended by the Nigua Seed Bank. Since the number of seeds obtained was lower than anticipated and to ensure that all available seeds were used in the experiment and that each treatment received an equal number of seeds, 1–2 seeds were directly sown into each container on 12 June 2014 at the University of Idaho Pitkin Forest Nursery in Moscow, Idaho (46°43′32.0″ N, 116°57′20.4″ W). Plants were propagated in a greenhouse with daytime temperatures ranging from 10–27 °C and nighttime temperatures ranging from 5–16 °C; relative humidity ranged from 15%–100% over the growing season. No supplemental lighting was provided and daylength ranged from 9 h 31 min to 15 h 51 min over the course of the experiment. Seeds were sown into D40 (656 mL, 6.4 cm diam, 25.4 cm height; Stuewe and Sons Inc., Tangent, OR, USA) and polybag (946 mL, 7.6 cm diam, 19.1 cm height; Peaceful Valley Farm Supply, Grass Valley, CA, USA) containers. D40 containers are designed for growing tree seedlings and are rigid containers made of recycled polypropylene resin with internal longitudinal ridges and five bottom drainage holes. Polybag containers were modified to the same volume of D40 containers using a heat sealer (Uline, Pleasant Prairie, WI, USA). The heat sealer was used to close off excess container space vertically while avoiding loss of container depth. Each of the five growing media types was premixed and used to fill 100 containers of each container type, for a total of 1000 containers. After direct sowing, containers were covered with Deluxe Seed Guard germination cloth (Dewitt Company Inc., Sikeston, MO, USA) and irrigated using an overhead boom system three times daily with 3 passes per application until germination ceased. Seedlings in each treatment combination were then irrigated when block weights (one tray consisting of 5 to 20 seedlings) reached 80% of the weight at field capacity via the nursery manager method [31]. Using this method, D40-grown seedlings received 18, 9, 10, 6, and 7 irrigation events and polybag-grown seedlings received 15, 16, 12, 11, and 10 irrigation events over the 22-week growing period across Pe, T, CT, CG, and CTG treatments, respectively. No fertilizer was added at any point throughout the growing regime, representative of many situations where fertilizer is difficult to obtain in developing countries.

2.2. Sampling

Destructive sampling occurred during the week of 15 December 2014 for all seedlings. Measurements included morphological plant growth metrics of height (HT) and root-collar diameter (RCD), root volume (RV), root dry mass (RDM), and shoot dry mass (SDM). First, root systems were carefully washed clean of all growing media. Second, RV was determined by water displacement [32]. Next, seedlings were severed at the root-collar, and roots and shoots were dried separately in paper bags at 70 °C for 72 h. Following drying, SDM and RDM were used to determine seedling shoot-to-root ratios (S:R). Tissue samples were collected at the time of destructive sampling from the entire shoot of each seedling and analyzed for nutrient concentrations (A & L Great Lakes Laboratories, Fort Wayne, IN, USA).

Media samples (n = 5) from each treatment were also analyzed for nutrient concentrations (C, N, NO_3, P, K, Ca, Mg, and Na), pH, and electrical conductivity (EC) at the beginning and end of the growing season (A & L Great Lakes Laboratories, Fort Wayne, IN, USA). Nutrient concentrations were determined via the saturated media extract (SME) method, whereby growing media samples were saturated with distilled water and allowed to equilibrate for one hour. After equilibration, pH measurements were taken directly from the media slurry. All other analyses were performed on the extracted leachate from the slurry obtained via a Buchner funnel lined with filter paper [33].

Media bulk density was approximated for each treatment by filling five of each container type for each media type and then oven-drying the media from each container in paper bags at 100 °C for 48 h [34] prior to recording weights. Bulk density (g cm^{-3}) was calculated by dividing the dry weight of the media (g) by the volume of the media (cm^3).

The media water holding capacity (WHC) was calculated separately at the University of Idaho's Soil laboratory. A high-range pressure system with ceramic plates was used to determine the water holding capacity for all five media types at two water potentials: −0.033 MPa (field capacity) and −1.5 MPa (wilting point) [35]. Five samples of each media type were analyzed for field capacity and wilting point. Bulk density and the gravimetric water content of each sample were obtained. Gravimetric soil water content (SWC) was calculated as

$$\text{SWC} = (\text{Db} \times \theta\text{m})/\text{Dw} \tag{1}$$

where Db = media bulk density, θm = gravimetric water content, and Dw = water density (Dw = 1 g cm^{-3}). Once SWC was determined for each media type, both at field capacity and wilting point, WHC was calculated:

$$\text{WHC} = \text{field capacity SWC} - \text{wilting point SWC} \tag{2}$$

2.3. Statistical Analysis

Data were analyzed using SAS software (version 9.4; SAS Institute, Cary, NC, USA) via PROC GLIMMIX. Models included the main effects of growing media and container type, as well as their interaction, with replicate included as a random effect. Where main effects did not interact ($p > 0.05$), the interaction term was omitted from the model. Treatment comparisons were evaluated at $\alpha = 0.05$.

3. Results

3.1. Media Characterization

Bulk density (BD) differed significantly across media types ($p < 0.0001$) but not between container types ($p = 0.8870$) and the main effects did not interact ($p = 0.0919$). BD ranked as follows: Pe < T = CT < CG = CTG (0.131, 0.375, 0.398, 0.545, 0.546 g cm^{-3}, respectively). Water holding capacity (WHC) varied across growing media types with Pe having the highest water-holding capacity at 62%, followed by T at 32%, CT at 30%, CTG at 16%, and finally CG at 7%.

Analyses revealed that Pe was low in NO_3, K, Ca, and Mg but initially provided acceptable levels of P, although this declined to inadequate levels in the absence of fertilizer by the end of the growing season (Table 1). While T initially provided optimal levels of NO_3 and acceptable levels of P, K, Ca, and Mg, all nutrient levels were inadequate by the end of the growing season. Compost-based media amended with topsoil (CT) or grit (CG) initially provided acceptable levels of NO_3, very high levels of P, high levels of K, and low levels of Ca and Mg. Compost-based media amended with both topsoil and grit (CTG) initially provided optimal levels of NO_3, very high levels of P, high levels of K, and acceptable levels of Ca and Mg. By the end of the growing season, NO_3 levels had become inadequate for all compost-based media, while P levels remained high, levels of K were adequate, and Ca and Mg were inadequate.

Table 1. Chemical properties of media used to grow Hispaniolan pine seedlings measured before (top row) and after (bottom row) seedling production. Note: C, N, and C:N were assessed before seedling production but not after. Means (SE) are presented (n = 5); different letters within a row indicate a significant difference across growing media types at α = 0.05. EC = electrical conductivity; Pe = peat; T = topsoil; CT = compost–topsoil; CG = compost–grit; CTG = compost–topsoil–grit.

	F Ratio (p Value)	Pe	T	CT	CG	CTG
pH	69.92 (<0.0001)	5.16 (0.14) b	5.50 (0.06) b	6.98 (0.10) a	7.16 (0.07) a	7.00 (0.16) a
	196.52 (<0.0001)	6.04 (0.06) c	6.70 (0.03) b	7.40 (0.03) a	7.38 (0.06) a	7.44 (0.02) a
EC (mmho cm^{-1})	5.17 (0.0050)	0.10 (<0.01) b	1.22 (0.29) ab	1.05 (0.28) ab	1.17 (0.14) ab	2.11 (0.56) a
	80.38 (<0.0001)	0.12 (<0.01) b	0.16 (<0.01) b	0.59 (0.04) a	0.66 (0.04) a	0.58 (0.02) a
C (%)	65.77 (<0.0001)	29.45 (1.51) a	12.52 (0.55) c	19.25 (0.41) b	18.36 (0.27) b	16.35 (0.42) b
N (%)	523.52 (<0.0001)	0.48 (0.03) d	0.42 (0.01) d	1.75 (0.03) b	1.97 (0.04) a	1.54 (0.05) c
NO_3^- (ppm)	3.58 (0.0233)	1.2 (0.20) b	184.0 (14.60) ab	51.6 (28.61) ab	56.0 (14.11) ab	198.0 (97.36) a
	9.97 (0.0001)	1.0 (<0.01) b	1.0 (<0.01) b	9.2 (3.48) a	15.2 (1.62) a	7.2 (1.80) ab
P—$H_2PO_4^-$ (ppm)	13.86 (<0.0001)	3.52 (0.33) b	3.38 (0.30) b	31.86 (6.20) a	37.94 (5.27) a	27.16 (5.39) a
	49.08 (<0.0001)	1.18 (0.02) b	1.64 (0.07) b	16.58 (0.69) a	18.08 (2.00) a	15.54 (1.67) a
K^+ (ppm)	12.47 (<0.0001)	13.0 (1.14) c	82.8 (4.49) bc	300.0 (71.64) ab	351.6 (45.28) a	551.8 (107.68) a
	36.45 (<0.0001)	11.8 (0.20) b	14.6 (0.40) b	141.8 (14.74) a	163.80 (18.11) a	132.0 (14.04) a
Ca^{2+} (ppm)	7.48 (0.0007)	3.4 (0.51) c	110.8 (12.50) a	45.6 (9.47) bc	49.4 (2.46) abc	86.8 (29.78) ab
	96.27 (<0.0001)	7.2 (0.80) c	22.4 (0.81) b	42.2 (2.54) a	38.4 (1.36) a	38.4 (1.29) a
Mg^{2+} (ppm)	3.80 (0.0187)	2.4 (0.24) b	46.6 (4.77) a	19.4 (5.44) ab	18.2 (1.59) ab	44.6 (20.40) a
	68.23 (<0.0001)	3.8 (0.20) c	7.0 (0.32) b	16.4 (0.98) a	17.2 (0.97) a	17.6 (1.03) a
Na^+ (ppm)	20.60 (<0.0001)	22.4 (0.93) c	295.0 (18.28) a	112.6 (25.49) bc	105.2 (9.00) bc	191.6 (38.76) b
	20.79 (<0.0001)	57.6 (0.93) b	58.8 (0.80) b	85.2 (5.95) a	92.0 (4.49) a	81.6 (1.83) a
C:N	3583.72 (<0.0001)	61.60 (0.23) a	29.74 (0.77) b	11.02 (0.15) c	9.34 (0.05) d	10.60 (0.15) cd

Significant differences were found in chemical properties across growing media types (Table 1). Prior to seedling production, Pe was characterized by low pH, very low EC, high C, low N, NO_3, P, K, Ca, Mg, and Na, and a very high C:N relative to other media types. T was characterized by low pH, intermediate EC, low C, N, P, and K, and high NO_3, Ca, Mg, and Na, with an intermediate C:N relative to other media types. CT was characterized by neutral pH, intermediate EC, high P, and intermediate C, N, NO_3, K, Ca, Mg, and Na, and a low C:N relative to other media types. Similarly, CG was characterized by neutral pH, intermediate EC, high N, P, and K, intermediate C, NO_3, Ca, Mg, and Na, and a low C:N relative to other media types. Finally, CTG was also characterized by neutral pH, high EC, high NO_3, P, K, and Mg, intermediate C, N, Ca, and Na, and a low C:N relative to other media types. Trends across media types persisted following seedling production (Table 1). Relative to initial values, pH of all media types increased. For Pe, EC and concentrations of Ca, Mg, and Na increased but for all other growing media types, EC and concentrations of NO_3, P, K, Ca, Mg, and Na declined following seedling production.

3.2. Seedling Morphology

Seedling morphology differed significantly between container types (with the exception of RV) and across growing media types, but the main effects did not interact ($p \geq 0.2969$; Table 2). Seedlings grown in polybags were 10% taller and 11% thicker in RCD compared to seedlings grown in D40s. SDM was 38% greater and RDM was 8% greater for seedlings grown in polybags relative to those grown in D40s. S:R was 47% greater for seedlings grown in polybags compared to those grown in D40 containers.

With regard to growing media, seedling height and SDM were greatest when grown in CG, followed by CTG, CT, and Pe, with the shortest and lightest seedlings grown in T. Similarly, RCD and RDM of seedlings grown in CG were significantly greater than seedlings grown in all other media types. RV was greatest for seedlings grown in CG, followed by those grown in Pe, with seedlings grown in T, CT, and CTG having the smallest RV. S:R was 115% higher for seedlings grown in CTG compared to seedlings grown in T.

Table 2. Nursery growth of Hispaniolan pine seedlings across growing media and container treatments. Means (SE) are presented; different letters within a main effect indicate a significant difference at $\alpha = 0.05$. HT = height; RCD = root-collar diameter; RV = root volume; RDM = root dry mass; SDM = shoot dry mass; S:R = SDM/RDM; Pe = peat; T = topsoil; CT = compost–topsoil; CG = compost–grit; CTG = compost–topsoil–grit.

		HT (cm)	RCD (mm)	RV (cm^3)	RDM (g)	SDM (g)	S:R
Container	D40s	3.94 (0.05) b	0.98 (0.01) b	0.91 (0.03)	0.13 (0.01) b	0.13 (0.01) b	1.13 (0.05) b
	Polybags	4.32 (0.07) a	1.09 (0.02) a	0.96 (0.04)	0.14 (0.01) a	0.18 (0.01) a	1.58 (0.07) a
Media	Pe	3.80 (0.07) bc	0.93 (0.02) b	1.03 (0.04) b	0.09 (0.01) b	0.09 (0.01) bc	1.32 (0.10) ab
	T	3.68 (0.07) c	0.91 (0.02) b	0.72 (0.03) c	0.11 (0.01) b	0.08 (0.01) c	0.83 (0.04) b
	CT	4.04 (0.10) b	0.94 (0.03) b	0.72 (0.04) c	0.11 (0.01) b	0.14 (0.01) b	1.57 (0.08) ab
	CG	4.92 (0.08) a	1.40 (0.03) a	1.56 (0.05) a	0.27 (0.01) a	0.33 (0.01) a	1.35 (0.04) ab
	CTG	4.24 (0.11) b	1.02 (0.03) b	0.61 (0.04) c	0.12 (0.01) b	0.15 (0.01) b	1.72 (0.15) a
Type III tests of fixed effects		F ratio (*p* value)	F ratio (*p* value)	F ratio (*p* value)	F ratio (*p* value)	F ratio (*p* value)	F ratio (*p* value)
Container	1/40 [1]	24.30 (<0.0001)	26.19 (<0.0001)	3.13 (0.0846)	4.71 (0.0360)	21.43 (<0.0001)	8.87 (0.0049)
Media	4/40	28.29 (<0.0001)	48.99 (<0.0001)	30.84 (<0.0001)	48.26 (<0.0001)	46.91 (<0.0001)	3.24 (0.0215)

[1] degrees of freedom.

3.3. Foliar Chemistry

Plant tissue analyses revealed significant differences in the levels of all nutrients examined among growing media types, and in levels of most nutrients examined between container types, with significant interactions between the two main effects for Na, B, Fe, and Cu (Table 3 and Figures 1–3). Seedlings grown in polybags yielded significantly greater foliar concentrations of N, K, S and Zn and lower foliar concentrations of Mg and Ca relative to those grown in D40s, with no significant difference in foliar P, Mn, or Al concentrations between container types (Figure 1).

Table 3. Effects of growing media and container type on Hispaniolan pine seedling foliar nutrient concentrations.

	Media F Ratio (p Value)	Container F Ratio (p Value)	Media I Container F Ratio (p Value)
N (%)	5.06 (0.0019)	14.57 (0.0004)	ns *
P (%)	16.12 (<0.0001)	2.70 (0.1076)	ns
K (%)	9.90 (<0.0001)	9.53 (0.0035)	ns
S (%)	14.42 (<0.0001)	14.64 (0.0004)	ns
Mg (%)	26.40 (<0.0001)	14.78 (0.0004)	ns
Ca (%)	45.12 (<0.0001)	5.61 (0.0223)	ns
Na (%)	35.69 (<0.0001)	0.70 (0.4081)	6.37 (0.0005)
B (ppm)	24.02 (<0.0001)	10.46 (0.0025)	3.79 (0.0106)
Zn (ppm)	6.08 (0.0006)	28.17 (<0.0001)	ns
Mn (ppm)	210.98 (<0.0001)	2.19 (0.1457)	ns
Fe (ppm)	37.42 (<0.0001)	7.41 (0.0096)	5.36 (0.0015)
Cu (ppm)	20.55 (<0.0001)	19.04 (<0.0001)	2.82 (0.0377)
Al (ppm)	19.85 (<0.0001)	0.20 (0.6607)	ns

* ns = interaction term was not significant at $\alpha = 0.05$ and therefore dropped from the model.

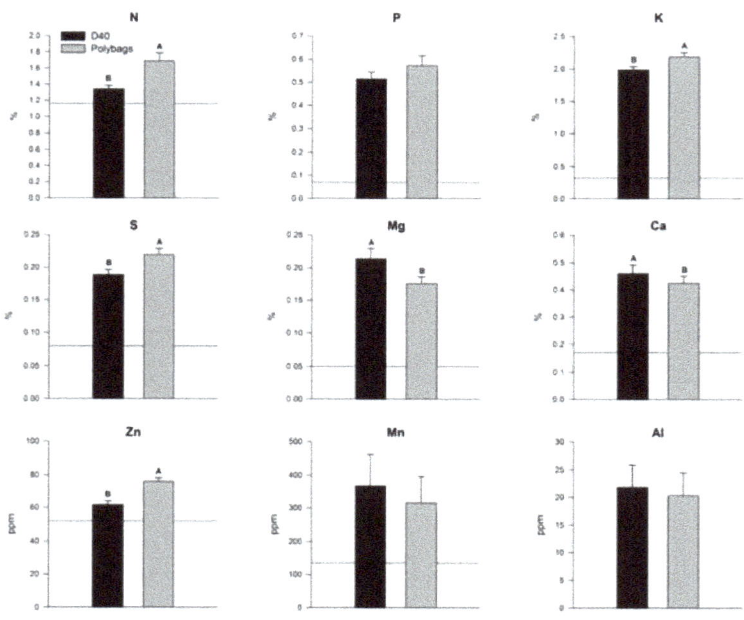

Figure 1. Foliar nutrient concentrations of Hispaniolan pine seedlings between container types. Horizontal lines indicate normal nutrient levels reported for *Pinus strobus* L. from http://agsci.psu.edu/aasl/plant-analysis/plant-tissue-total-analysis/interpretive-nutrient-levels-for-plant-analysis/pine-white (accessed 2 February 2018).

With regard to growing media (Figure 2), foliar N was greater in seedlings grown in CT and CG relative to T. Foliar P was greater in seedlings grown in CT and CTG relative to Pe, T, and CG. Foliar K was greater in seedlings grown in T, CT, CG, and CTG relative to Pe. Foliar S was greatest in seedlings grown in CT and CTG, followed by CG, then Pe, and finally T. Foliar Mg was greatest in seedlings grown in T and Pe, followed by those grown in CT and CTG, and finally those grown in CG. Foliar Ca was greater in seedlings grown in T, followed by those grown in Pe, CT, and CTG, and finally those grown in CG. Foliar Mn was greater in seedlings grown in Pe, followed by those grown in T, with those grown in CT, CG, and CTG yielding the lowest levels of Mn. Foliar Al was greater in seedlings grown in Pe and T relative to those grown in CT, CG, and CTG.

The main effects of growing media and container type interacted to impact foliar concentrations of Na, B, Fe, and Cu (Figure 3). Foliar Na and foliar Fe were greatest among seedlings grown in Pe in D40s or in T in either D40s or polybags. Foliar B was greatest among seedlings grown in polybags containing compost-amended growing media (CT, CG, and CTG). Foliar Cu was greatest among seedlings grown in Pe (in either D40s or polybags) and in polybags containing T or CT.

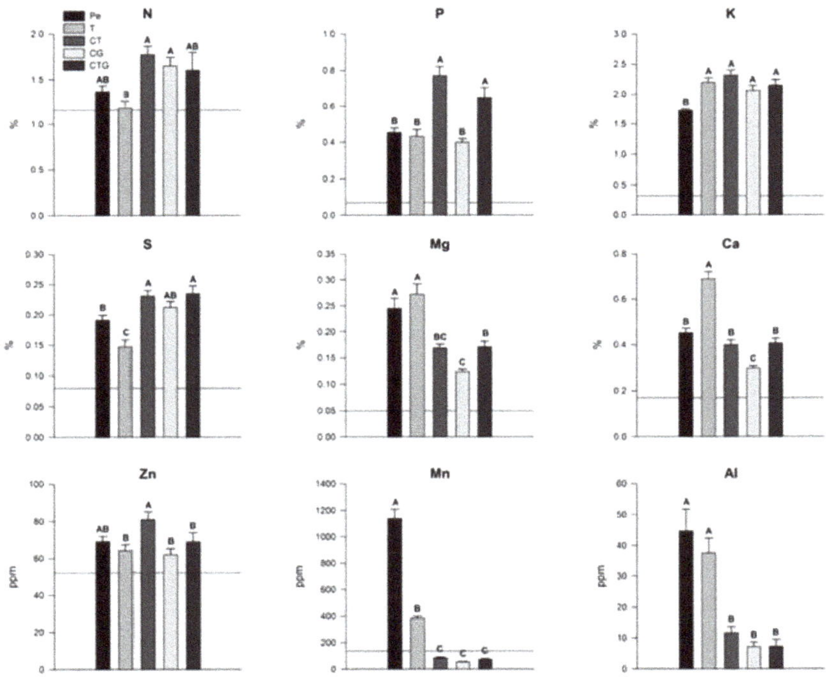

Figure 2. Foliar nutrient concentrations of Hispaniolan pine seedlings among growing media treatments. Pe = peat; T = topsoil; CT = compost–topsoil; CG = compost–grit; CTG = compost–topsoil–grit. Horizontal lines indicate normal nutrient levels reported for *Pinus strobus* from http://agsci.psu.edu/aasl/plant-analysis/plant-tissue-total-analysis/interpretive-nutrient-levels-for-plant-analysis/pine-white (accessed 2 February 2018).

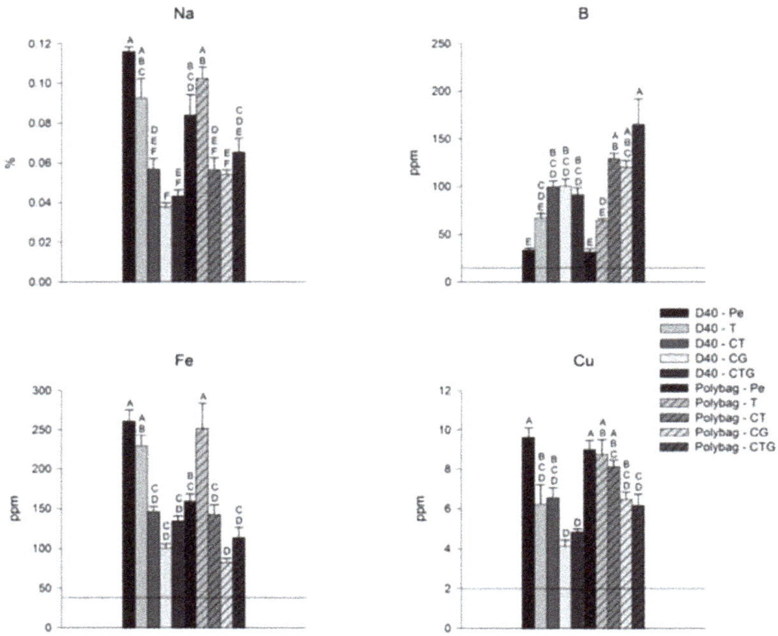

Figure 3. Foliar nutrient concentrations of Hispaniolan pine seedlings across interacting container and growing media treatment levels where significant ($\alpha = 0.05$). Horizontal lines indicate normal nutrient levels reported for *Pinus strobus* from http://agsci.psu.edu/aasl/plant-analysis/plant-tissue-total-analysis/interpretive-nutrient-levels-for-plant-analysis/pine-white (accessed 2 February 2018).

4. Discussion

The growing media and container type did not interact to affect seedling morphology, but each independently influenced seedling growth, providing evidence that these nursery cultural practices can be used to cultivate *P. occidentalis* seedlings suited to particular outplanting conditions. It is well documented that seedling morphology and early outplanting performance can differ among container types [36,37], but it remains unknown exactly why Hispaniolan pine seedling growth differed between polybags and D40s. Although the diameter of both container types was similar, seedlings grown in polybags had greater shoot and root growth, but also higher S:R. Despite the known growth and development issues associated with the use of polybags [16], Hispaniolan pine seedling morphology and foliar chemistry indicate that this container is a suitable choice for the species. This is promising given that polybags are currently widely used, and a conversion to rigid plastic containers would likely be impractical as it would require a concomitant shift in the entire nursery system, from the use of uniform, artificial growing media and the consequent need for fertilization and frequent irrigation to the need for raised benches to promote root pruning [16].

One of the main issues with the use of polybags for the production of tree seedlings relates to poor root growth, particularly when seedlings are held in their containers for too long [16]. The higher S:R suggests that the rate of root growth relative to shoot growth was lower for seedlings grown in polybags compared to those grown in D40s, perhaps indicating that root growth was constrained in polybags by the end of the 22-week growing season. Moreover, research suggests that seedlings with low S:R tend to have increased survival rates when outplanted to harsh sites, such as those in Haiti, because of the increased uptake of water and nutrients afforded by the larger root system relative to lower demands made by the smaller shoot [38]. Thus, even though seedlings grown in D40s were smaller overall, the S:R was potentially more favorable for seedlings that are likely to encounter periodic dry conditions following outplanting. Further study into how specific characteristics of different container types influence growth of seedlings of this species, and how this relates to outplanting performance, is warranted.

With regard to growing media, differences in chemical and physical properties among types were likely the primary drivers for the noted differences in seedling response variables. Soil bulk density and water holding capacity influence the amount of water available for plant uptake [39] as well as fine root proliferation. Soils that are highly compressed tend to lack the pore spaces necessary for holding water and air. Thus, limited pore space may lead to stunted growth [40]. Field-sourced soils can be highly variable in their physical characteristics and tropical soils in particular can contain high amounts of clay and silt, which have lower macropore space that may limit gas exchange, water drainage, and potentially plant-available water, since clays hold more water at high tension [17]. In some instances, packing topsoil in containers and the settling of the soils in the container may contribute to higher bulk densities, thereby making rooting a problem [17]. None of the growing media types had bulk densities beyond that of root penetration, which is approximately 1.5 g cm^{-3} [41], so bulk density likely did not inhibit root penetration. Studies have shown that as bulk density increases, available water holding capacity decreases [42]. Media water holding capacity also varied across treatments, with an 8-fold difference between CG and Pe (7% and 62%, respectively). Media types can be selected and designed to have higher water holding capacities and provide increased moisture availability [43]; however, depending on soil texture and other properties, additional factors such as oxygen availability may be affected [30].

Nonetheless, early seedling growth was not better in Pe but rather in CG, which may be explained by media chemistry. Although seedlings likely had adequate foliar nutrient levels across all growing media, based on recommended nutrient levels for *Pinus strobus* L. (http://agsci.psu.edu/aasl/plant-analysis/plant-tissue-total-analysis/interpretive-nutrient-levels-for-plant-analysis/pine-white), both Pe and T were low in %N, and seedlings grown using these media types likely bordered on nitrogen deficient. While T may have had a relatively high amount of NO_3 initially (most likely due to the manure and bark mulch; Table 1), we suspect a large fraction of it leached out before the roots could exploit the full capacity of the container. Nitrogen is a macronutrient which is essential to all plant physiological processes, and a lack of access to this nutrient may have contributed to the stunted growth of these seedlings. While amending T with compost improved growing conditions, it was when seedlings were grown in a combination of compost and grit (CG) that better growth was observed. CG was characterized by neutral pH, higher levels of N, P, and K, and lower C:N compared to Pe which had a slightly acidic pH, lower levels of N, P, and K, and very high C:N. Yet, seedlings grown in compost-based growing media (CT, CG, and CTG) showed low levels of foliar manganese, which is essential for the synthesis of chlorophyll and also serves as an enzyme activator [44]. Manganese absorption may have been inhibited by pH [45] and high concentrations of iron, calcium and aluminum in compost-based media types [46] (Figure 2). Similar results have been found for *Pinus sylvestris* seedling tissue nutrient concentrations grown in compost [47].

5. Conclusions

Improvements can be made to existing nursery practices in Haiti to enhance the production of *P. occidentalis* seedlings, particularly if viewed through the framework of the Target Plant Concept [4]. Compost-based growing media, particularly when amended with forestry grit or similar coarse-grained sand, show promise for early seedling growth. Despite low water holding capacity and limited manganese, this growing medium provided adequate levels of essential mineral nutrients (particularly nitrogen). Seedling morphology and foliar chemistry indicate that polybags remain a suitable choice for the species, although higher S:R among seedlings grown in polybags relative to D40s suggest that further work is needed to determine the specific container characteristics that are optimal (i.e., meet the target specifications for a given objective) for the production of *P. occidentalis* seedlings for reforestation and restoration in Haiti. Given that the seedlings in this study were relatively small, an examination of the nutritional needs of *P. occidentalis* as well as a field component to determine what are the most important seedling attributes that influence post-planting establishment is needed to advance reforestation success in Haiti.

Author Contributions: K.L.H., J.R.P., O.T.B. and A.S.D. conceived and designed the research; K.L.H. performed the experiments and collected the data; K.L.H. and A.L.R.-D. analyzed the data; K.L.H., J.R.P., O.T.B., A.S.D. and A.L.R.-D. wrote and edited the manuscript.

Funding: This work was funded in part by private donors and through the Center for Forest Nursery and Seedling Research.

Acknowledgments: The authors gratefully acknowledge contributions from Sabine Deristen and Pierre-Paul Jules of SOHADERK in Haiti, Wilman Placido of the Nigua Seed Bank, Santo Domingo, Dominican Republic, Marla Schwartz, the University of Idaho's Center for Forest Seedling and Nursery Research, Layla Dunlap, Rebecca Sheridan, Matt Fisk, Becca Lieberg, and Guy Knudsen at the University of Idaho.

Conflicts of Interest: The authors declare no conflict of interest.

References

1. Foxx, R.M. Te Terre a Fatige "The Earth is Tired": Reversing Deforestation in Haiti. *Behav. Interv.* **2012**, *108*, 105–108. [CrossRef]
2. Churches, C.E.; Wampler, P.J.; Sun, W.; Smith, A.J. Evaluation of forest cover estimates for Haiti using supervised classification of Landsat data. *Int. J. Appl. Earth Obs. Geoinform.* **2014**, *30*, 203–216. [CrossRef]
3. Williams, V.J. A case study of the desertification of Haiti. *J. Sustain. Dev.* **2011**, *4*, 20–31. [CrossRef]
4. Haase, D.L.; Davis, A.S. Developing and supporting quality nursery facilities and staff are necessary to meet global forest and landscape restoration needs. *Reforesta* **2017**, *4*, 69–93.
5. Jacobs, D.F.; Landis, T.D.; Luna, T. Chapter 5 Growing Media. In *Nursery Manual for Native Plants: A Guide for Tribal Nurseries, Volume 1: Nursery Management, Agriculture Handbook 730*; Dumroese, R.K., Luna, T., Landis, T.D., Eds.; U.S. Department of Agriculture, Forest Service: Washington, DC, USA, 2009; pp. 77–94.
6. Grossnickle, S.C. Importance of root growth in overcoming planting stress. *New For.* **2005**, *30*, 273–294. [CrossRef]
7. Haase, D.L.; Landis, T.D.; Dumroese, R.K. Chapter 17 Outplanting. In *Tropical Nursery Manual: A Guide to Starting and Operating a Nursery for Native and Traditional Plants, Agriculture Handbook 732*; Wilkinson, K.M., Landis, T.D., Haase, D.L., Daley, B.F., Dumroese, R.K., Eds.; U.S. Department of Agriculture, Forest Service: Washington, DC, USA, 2014; pp. 312–337.
8. Mattsson, A. Predicting field performance using seedling quality assessment. *New For.* **1997**, *13*, 223–248. [CrossRef]
9. Davis, A.S.; Jacobs, D.F. Quantifying root system quality of nursery seedlings and relationship to outplanting performance. *New For.* **2005**, *30*, 295–311. [CrossRef]
10. Mexal, J.; Rangel, R.C.; Landis, T. Reforestation success in Central Mexico: Factors determining survival and early growth. *Tree Plant. Notes* **2008**, *53*, 16–22.
11. Liegel, L.H.; Venator, C.R. *A Technical Guide for Forest Nursery Management in the Caribbean and Latin America*; General Technical Report. SO-67; U.S. Department of Agriculture, Forest Service, Southern Forest Experiment Station: New Orleans, LA, USA, 1987; 156p.
12. Landis, T.D.; Morgan, N. Growing Media Alternatives for Forest and Native Plant Nurseries. In *National Proceedings: Forest and Conservation Nursery Associations—2008, RMRS-P-58*; Dumroese, R.K., Riley, L.E., Eds.; U.S. Department of Agriculture, Forest Service, Rocky Mountain Research Station: Fort Collins, CO, USA, 2009; pp. 26–31.
13. Wolken, J.M.; Landhäusser, S.M.; Lieffers, V.J.; Dyck, M.F. Differences in initial root development and soil conditions affect establishment of trembling aspen and balsam poplar seedlings. *Botany* **2010**, *88*, 275–285. [CrossRef]
14. Mexal, J. Forest Nursery Activities in Mexico. In *National Proceedings: Forest and Conservation Nursery Associations—1996, PNW-GTR-389*; Landis, T.D., South, D.B., Eds.; U.S. Department of Agriculture, Forest Service, Pacific Northwest Research Station: Portland, OR, USA, 1997; pp. 228–232.

15. Akpo, E.; Stomph, T.J.; Kossou, D.K.; Omore, A.O.; Struik, P.C. Effects of nursery management practices on morphological quality attributes of tree seedlings at planting: The case of oil palm (*Elaeis guineensis* Jacq.). *For. Ecol. Manag.* **2014**, *324*, 28–36. [CrossRef]
16. Landis, T.D. Improving Polybag Culture for Sustainable Nurseries. *For. Nurs. Notes* **1995**, 6–8.
17. Landis, T.D.; Jacobs, D.F.; Wilkinson, K.M.; Luna, T. Chapter 6 Growing Media. In *Tropical Nursery Manual: A Guide to Starting and Operating A Nursery for Native and Traditional Plants, Agriculture Handbook 732*; Wilkinson, K.M., Landis, T.D., Haase, D.L., Daley, B.F., Dumroese, R.K., Eds.; U.S. Department of Agriculture, Forest Service: Washington, DC, USA, 2014; pp. 100–121.
18. Lazcano, C.; Sampedro, L.; Zas, R.; Domínguez, J. Vermicompost enhances germination of the maritime pine (*Pinus pinaster* ait.). *New For.* **2010**, *39*, 387–400. [CrossRef]
19. Mañas, P.; Castro, E.; Vila, P.; Heras, J. Use of waste materials as nursery growing media for Pinus halepensis production. *Eur. J. For. Res.* **2010**, *129*, 521–530. [CrossRef]
20. Avramidou, P.; Evangelou, A.; Komilis, D. Use of municipal solid waste compost as a growth media for an energy plant (rapeseed). *J. Environ. Manag.* **2013**, *121*, 152–159. [CrossRef] [PubMed]
21. Preneta, N.; Kramer, S.; Magloire, B.; Noel, J.M. Thermophilic co-composting of human wastes in Haiti. *J. Water Sanit. Hyg. Dev.* **2013**, *3*, 649–654. [CrossRef]
22. Budy, J.D.; Miller, E.L. Survival, Growth, and Root Form of Containerized Jeffrey Pines Ten Years after Outplanting. In *The Challenge of Producing Native Plants for the Intermountain Area, General Technical Report INT-168, Proceedings of the Intermountain Nurseryman's Association, Las Vegas, NV, USA, 8–11 August 1983*; Murphy, P.M., Ed.; U.S. Department of Agriculture, Forest Service, Intermountain Forest and Range Experiment Station: Ogden, UT, USA, 1984; pp. 82–88.
23. Amoroso, G.; Frangi, P.; Piatti, R.; Ferrini, F.; Fini, A.; Faoro, M. Effect of container design on plant growth and root deformation of littleleaf linden and field elm. *HortScience* **2010**, *45*, 1824–1829.
24. Aphalo, P.; Rikala, R. Field performance of silver-birch planting-stock grown at different spacing and in containers of different volume. *New For.* **2003**, *25*, 93–108. [CrossRef]
25. Aldrete, A.; Mexal, J.G. Chemical Root Pruning of Conifer Seedlings in Mexico. In *National Proceedings: Forest and Conservation Nursery Associations—1999, 2000, 2001, RMRS-P-24*; Dumroese, R.K., Riley, L.E., Landis, T.D., Eds.; U.S. Department of Agriculture, Forest Service, Rocky Mountain Research Station: Ogden, UT, USA, 2002; pp. 160–164.
26. Landis, T.D. Chapter 1 Containers: Types and Functions. In *The Container Tree Nursery Manual—Volume 2—Containers and Growing Media, Agricultural Handbook 674*; Landis, T.D., Tinus, R.W., McDonald, S.E., Barnett, J.P., Eds.; U.S. Department of Agriculture, Forest Service: Washington, DC, USA, 1990; pp. 1–39.
27. Khurram, S.; Burney, O.T.; Morrissey, R.C.; Jacobs, D.F. Bottles to trees: Plastic beverage bottles as an alternative nursery growing container for reforestation in developing countries. *PLoS ONE* **2017**, *12*, e0177904. [CrossRef] [PubMed]
28. Darrow, K.; Zanoni, T. Hispaniolan pine (*Pinus occidentalis* Swartz) a little known subtropical pine of economic potential. *Commonw. For. Rev.* **1990**, *69*, 133–146.
29. Kennedy, L.; Horn, S. Postfire Vegetation Recovery in Highland Pine Forests of the Dominican Republic. *Biotropica* **2008**, *40*, 412–421. [CrossRef]
30. Landis, T.D. Chapter 2 Growing Media. In *The Container Tree Nursery Manual—Volume 2—Containers and Growing Media, Agricultural Handbook 674*; Landis, T.D., Tinus, R.W., McDonald, S.E., Barnett, J.P., Eds.; U.S. Department of Agriculture, Forest Service: Washington, DC, USA, 1990; pp. 41–86.
31. Dumroese, R.K.; Montville, M.E.; Pinto, J.R. Using container weights to determine irrigation needs: A simple method. *Native Plants J.* **2015**, *16*, 67–71. [CrossRef]
32. Burdett, A.N. A nondestructive method for measuring the volume of intact plant parts. *Can. J. For.* **1979**, *9*, 120–122. [CrossRef]
33. Warnke, D. Chapter 13 Recommended Test Procedures for Greenhouse Growth Media. In *Recommended Soil Testing Procedures for the Northeastern United States, Northeastern Regional Publication No. 493*, 3rd ed.; Agricultural Experiment Station, University of Delaware: Newark, DE, USA, 2009; pp. 103–110.
34. Doran, J.W.; Parkin, T.B. Quantitative Indicators of Soil Quality: A Minimum Data Set. In *Methods for Assessing Soil Quality*; Soil Science Society of America: Madison, WI, USA, 1996; pp. 25–37.

35. Klute, A. Chapter 26 Water Retention: Laboratory Methods. In *Methods of Soil Analysis: Part 1—Physical and Mineralogical Methods*; Soil Science Society of America, American Society of Agronomy: Madiosn, WI, USA, 1986; pp. 635–662.
36. Dominguez-Lerena, S.; Sierra, N.H.; Manzano, I.C.; Bueno, L.O.; Rubira, J.P.; Mexal, J.G. Container characteristics influence *Pinus* pinea seedling development in the nursery and field. *For. Ecol. Manag.* **2006**, *221*, 63–71. [CrossRef]
37. Pinto, J.R.; Marshall, J.D.; Dumroese, R.K.; Davis, A.S.; Cobos, D.R. Establishment and growth of container seedlings for reforestation: A function of stocktype and edaphic conditions. *For. Ecol. Manag.* **2011**, *261*, 1876–1884. [CrossRef]
38. Mokany, K.; Raison, R.J.; Prokushkin, A.S. Critical analysis of root: Shoot ratios in terrestrial biomes. *Glob. Chang. Biol.* **2006**, *12*, 84–96. [CrossRef]
39. Pires, A.; Kay, B.D.; Perfect, E. Management versus inherent soil properties effects on bulk density and relative compaction. *Soil Tillage Res.* **1997**, *44*, 81–93.
40. Tracy, S.; Black, C.; Roberts, J.; Mooney, S. Exploring the interacting effect of soil texture and bulk density on root system development in tomato (*Solanum lycopersicum* L.). *Exp. Bot.* **2013**, *91*, 38–47. [CrossRef]
41. Brady, N.C.; Weil, R.R. *Elements of the Nature and Properties of Soils*, 2nd ed.; Pearson Prentice-Hall: Upper Saddle River, NJ, USA, 2004; 606p, ISBN 13-978-0130480385.
42. Davey, C.B. Nursery Soil Management-Organic Amendments. In *National Proceedings: Forest and Conservation Nursery Associations—1996, PNW-GTR-389*; Landis, T.D., South, D.B., Eds.; U.S. Department of Agriculture, Forest Service, Pacific Northwest Research Station: Portland, OR, USA, 1997; pp. 6–18.
43. Chirino, E.; Vilagrosa, A.; Vallejo, V.R. Using hydrogel and clay to improve the water status of seedlings for dryland restoration. *Plant Soil* **2011**, *344*, 99–110. [CrossRef]
44. Landis, T.D. Mineral Nutrition as an Index of Seeedling Quality. In *Proceedings: Evaluating Seedling Quality: Principles, Procedures, and Predictive Abilities of Major Tests, Corvallis, OR, USA, 16–18 October 1984*; Duryea, M.L., Ed.; Forest Research Laboratory, Oregon State University: Corvallis, OR, USA, 1985; pp. 29–48.
45. Peterson, J.C. Effects of pH upon nutrient availability in a commercial soilless root medium utilized for floral crop production. *Ohio Agric. Res. Dev. Center Res. Cirucular* **1982**, *268*, 16–19.
46. Lombard, K.; O'Neill, M.; Heyduck, R.; Onken, B.L.; Ulery, A.; Mexal, J.; Unc, A. Composted biosolids as a source of iron for hybrid poplars (*Populus* sp.) grown in northwest New Mexico. *Agrofor. Syst.* **2010**, *81*, 45–56. [CrossRef]
47. Selivanovskya, S.Y.; Latypova, V.Z. Effects of composted sewage sludge on microbial biomass, activity and pine seedlings in nursery forest. *Waste Manag.* **2006**, *26*, 1253–1258. [CrossRef] [PubMed]

forests

MDPI

Article

Comparing Thinning System Effects on Ecosystem Services Provision in Artificial Black Pine (*Pinus nigra* J. F. Arnold) Forests

Maurizio Marchi [1,*], Alessandro Paletto [1], Paolo Cantiani [1], Elisa Bianchetto [2] and Isabella De Meo [2]

1 CREA—Research Centre for Forestry and Wood, I-52100 Arezzo, Italy; alessandro.paletto@crea.gov.it (A.P.); paolo.cantiani@crea.gov.it (P.C.)

2 CREA—Research Centre for Agriculture and Environment, I-50121 Florence, Italy; elisa.bianchetto@crea.gov.it (E.B.); isabella.demeo@crea.gov.it (I.D.M.)

* Correspondence: maurizio.marchi@crea.gov.it; Tel.: +39-349-838-7082

Received: 26 February 2018; Accepted: 3 April 2018; Published: 5 April 2018

Abstract: Provision of forest ecosystem services is influenced by site and stand characteristics as well as forest management practices. In order to evaluate the influence of forest management on ecosystem services provision, two artificial black pine forests located in Central Italy were studied where two different thinning approaches (traditional and selective) were applied under the SelPiBio LIFE project. Four main ecosystem services were selected and assessed: timber and bioenergy production, carbon sequestration, forest stand stability-protection, and biodiversity conservation. Even if not supported by statistical evidence, results highlighted an interesting trend just 2 years after treatment. The selective thinning was able to enhance the majority of ecosystem services compared to the traditional one. A higher growth rate of selected crop trees was measured (i.e., carbon sequestration). The slenderness ratio was sensibly reduced (i.e., mechanical stability) with a positive implication on soil retention and the prevention of landslides. Moreover, valuable and interesting commercial assortments have been proven to be retrieved from the stands with the selective approach. Larger and also better formed trees were harvested, given the impact of selective thinning on the co-dominant class. The Shannon index increased only with the selective thinning intervention. In conclusion, the provided results and methods are encouraging and might represent the basis for novel and longer monitoring efforts.

Keywords: selective thinning; thinning from below; forest management practices; planted forests; forest multifunctionality; Mediterranean area

1. Introduction

According to the Millennium Ecosystem Assessment (MEA) and the definition given in 2005, an "ecosystem service" (ES) can be defined as the benefits provided by ecosystems that contribute to making human life both possible and worth living [1,2]. ES can be classified into four categories, namely, (i) provisioning services; (ii) regulating services; (iii) supporting services; and (iv) cultural services. With special regard to forests, provisioning services include wood and non-wood forest products; regulating services include carbon sequestration, water regulation, natural hazard protection; supporting services include plant production, biodiversity and nutrient cycling; and cultural services include recreational opportunities, historical and spiritual values [3–5].

Since the ES concept has been developed, both the scientific community and forest managers started to investigate the relationship between forest management practices and provision of ESs [6–8]. Currently, one of the most important research challenges is how to manage forests for timber and bioenergy production while maintaining and/or improving other ESs such as habitat and biodiversity conservation, natural hazard protection, carbon sequestration, water regulation, and recreation [9]. Forest management trajectories and strategies can generate ES trade-offs, particularly if interactions among ESs are not well known [10,11]. Trade-offs can occur when—due to a forest management choice—the provision of one ES is reduced as consequence of the increased use of another one [12]. On the contrary, in some cases the relationship between various forest ESs might be synergistic and complementary [13].

Forest management choices that can affect the provision of ESs include the silvicultural system (i.e., forest structure: high forest or coppice, even or uneven-aged), rotation period and thinning regime. In particular, silvicultural treatments affect the level of biodiversity, the water cycle components, but also recreational services, influencing forest species composition, horizontal and vertical stand structure, stand density and age [14]. Furthermore, silvicultural treatments can modify the natural cycle of elements. For instance, the mineralization of carbon and nutrients (cycles of the elements) can be highly influenced by the amount and the spatial distribution of solar radiation that hits the ground [15]. The result is higher micro-climatic variability, reflected by a higher level of soil biodiversity (e.g., fungi and bacteria) especially at the understory level. Moreover, recent studies have evaluated the utility of forest management as a tool to mitigate the effects of climate change on ESs as well as to maintain high growth rates (i.e., carbon sequestration) in living trees [8,16–18]. In this view, the relationship between forest management and ES provision represents a focal point for future development of many forested zones and a new challenge to cope with a changing environment [19,20]. Maintaining and balancing the ESs supplied by forests require thorough assessment and evaluation at different spatial and temporal scales [21]. In the international literature, two main approaches are generally used to assess ESs from the biophysical point of view: the first one is based on a qualitative assessment of ESs, using experts' opinion and stakeholders' evaluation [22]. The second one focuses on the quantitative assessment of ESs through the measurement of field-based biophysical outcomes [23].

During the last decades and especially after the First and Second World War, many afforestation programmes were planned in most of the European countries. Such activities were seen as an important strategy for people's safety and wellbeing [24–26]. Artificial stands were established with pioneer conifers (e.g., *Pinus nigra* spp., *Pinus pinaster* Aiton, *Pinus halepensis* Mill.) not only in degraded or abandoned lands by farmers but also in mountain zones at high elevation and coastal areas. The purposes ranged from soil protection to recreation and from dune protection to scenic beauty. According to the second Italian National Forest Inventory (2005), black pine (*Pinus nigra* J.F. Arnold spp.) stands cover an area of 236,467 ha, corresponding to 2.5% of the total national forest area. Generally, such artificial stands are characterized by a low biodiversity level, due to their reduced species composition and spatial structure (i.e., structural homogeneity). Actually, these stands represent the most simplified forest systems in Italy, mostly occurring in pure stands of even-aged forests (50 years old on average), established during the 19th and early 20th century, on bare or overexploited soils for protective purposes, following a rigid planting scheme [26]. The squared design was the most applied for planting, with 2500 4-year-old seedlings per hectare (2 × 2 m) and canopy closure occurring at an early age, forecasting a first pre-commercial thinning at age 15, followed by additional thinning every 15 years [27,28]. This treatment was rarely applied, as well as other scheduled interventions. Despite the low value of timber production, a new interest in artificial ecosystems is currently rising in the whole Europe, mainly thanks to their valuable ecological importance, and provision of ESs such as carbon sequestration and hydrological safeguarding [26,29,30]. Managing such stands may represent a challenging opportunity for forest managers, aiming to guide natural evolution to more complex and stable systems (e.g., climax specific composition) and testing innovative management practices [20]. On the basis of the positive results obtained by thinning from below of medium-heavy intensity [15,31–33], selective thinning has been discovered as an interesting opportunity even in artificial forest stands. In fact, thinning from below of heavy intensity is able to influence the structure of the forest even in the dominant layer, usually the focus of selective thinning. When applied to artificial black pine stands, the species reacted positively, even to late-thinning, and both timber harvesting and ES provisions were improved [34,35]. Selective thinning, widely used in many forest systems (i.e., beech forests, mixed forests) [36] has been rarely applied in artificial pine forests. Conversely, thinning from below (i.e., removing the smaller, weaker and poorer quality trees to concentrate growth on the better trees remaining and according to a specific dbh or basal area threshold) still remains the most applied treatment. Its impact on the stands' structures has been demonstrated to be very low, especially concerning the carbon cycle, soil biodiversity and improvement of ecological dynamics (i.e., natural mortality). In addition, economic sustainability is rarely achieved [6,8,35]. In selective thinning, the choice of the trees to be cut is based on a positive selection of candidate trees (i.e., crop tree). In other words, candidate trees are first selected and then valorized by removing their direct competitors at crown level. The average number of candidate trees per hectare must be balanced according to species and rotation age (i.e., maximum potential crown width), selected among the most vigorous and stable ones. With selective thinning, 30–40% of basal area is removed and all crown competitors trees are harvested, including standing dead trees and lying deadwood slightly decomposed [37,38].

Starting from these considerations, the aim of this study is to evaluate the effects of two different forest management practices (selective thinning and traditional thinning) on three categories of ESs (provisioning services, regulating services and supporting services) analysing trade-offs and synergies that are generated. The effects of forest management practices on ESs are analysed through a combined approach of quantitative and qualitative biophysical assessment of ESs [39]. The study is conducted in two case studies in Central Italy (Amiata and Pratomagno) characterized by different site and stand features.

2. Materials and Methods

2.1. Study Area

2.1.1. Pratomagno Study Area

The first study area is named "Pratomagno" (Figure 1) and is located in the north-west of the Arezzo province in Tuscany region (43°39′ N 11°39′ E). The Pratomagno study area covers around 3000 ha and the main tree species are European beech (*Fagus sylvatica* L.) and Turkey oak (*Quercus cerris* L.), while black pine stands cover a surface of about 800 ha. The latter is the result of a reforestation programme, which began in 1954 and ended in the late 1980's. Among the numerous black pine subspecies, the main used in the early years of reforestation in Italy was laricio pine (*Pinus nigra* ssp. *laricio*). However, Austrian pine (*Pinus nigra* ssp. *nigra*) was commonly used but preferred in areas with low fertility (especially pasture ridge). The study area of Pratomagno has an average elevation of 1150 m a.s.l., a prevailing south-west aspect and average slope of 40%. From the lithological point of view, Pratomagno is characterized by quartz-feldspar sandstones alternated by siltstones and argillites. The argillites and siltstones provide a very thin layer ranging from a few up to 15 cm, while the thickness of the sandstone layers is more considerable, exceeding half a meter; this implies the emersion of large banks of thick sandstone whose heads are well visible.

The average annual temperature is 10.5 °C (maximum of 19 °C in July and minimum of 1.5 °C in January), while the average rainfall is 997 mm with a maximum peak in autumn and minimum precipitation in June.

2.1.2. Amiata Study Area

The second study area is Amiata (Figure 1), located in the Castiglione d'Orcia municipality (42°53′ N 11°37′ E, Siena province in Tuscany region). Similar to the previous area, artificial pine forests are the result of 50 years of reforestation programs and currently cover an area of 115 ha. Concerning the forest area (approximately 1930 ha), the main tree species are Austrian pine and Turkey oak, followed by Downey oak (*Quercus pubescens* L.), hedge maple (*Acer campestre* L.), and silver fir (*Abies alba* Mill.).

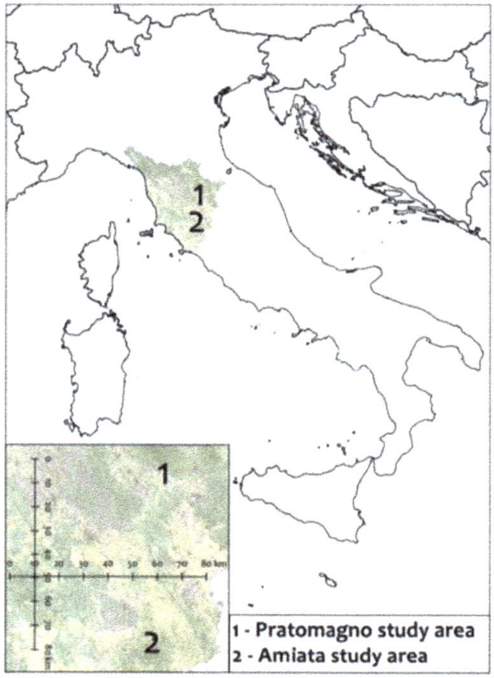

Figure 1. Geographic position of the two study areas.

From the lithological point of view, the Amiata study area is characterized by fissile clays, silty clays, marly clays with sporadic inclusions of limestone, basic limestone. This lithotype forms morphologies consisting of long, wavy sides with slope mainly moderate to strong, subject to erosion by channelled water and mass movement.

The average annual temperature is 12.5 °C and the average rainfall is 687 mm. July is the driest month with 28 mm, while January receives the max rain-snowfall (average of 88 mm). July is the hottest month with an average temperature of 21.7 °C, while January has an average temperature of 4.5 °C, the lowest in the year.

2.2. Field Measurements

The data were collected in 36 circular fixed area sampling plots (15-m radius), half of which were located in each study area. In particular, in each study area, 18 sampling plots randomly located in 6 forest monitoring sectors of 1 ha in size (3 plots in each forest monitoring sector) were identified. A monitoring sector is a thinning replicate whose treatment was randomly assigned. With this experimental design, each plot corresponded to a sub-replicate.

In each sampling plot, the main dendrometric data—i.e., tree height and diameter at breast height (dbh) for all standing living trees, number of stems, canopy cover (percentage of ground covered by crown projection dissolving intersections between polygons), height and diameter at breast height (dbh) for all standing dead trees—were collected before and two years after the silvicultural treatments. All data were stored in an open access and freely available dataset in ESRI shapefile format before the application of the silvicultural treatment [37]. In addition, in-field operational stages were supervised: all the activities of forest enterprises assigned for timber harvesting were controlled to prevent discrepancies between stored information and applied silvicultural treatment.

Subsequently, in each study area, 3 forest monitoring sectors were managed by selective thinning (3 ha in total, 9 plots per study area) and 3 forest monitoring sectors were managed by traditional thinning (3 ha in total, 9 plots per study area). The selective and traditional thinnings were conducted between spring and summer 2015 in both study areas (Figure 2). In the selective thinning, 100 trees per hectare were selected from among the better formed and mechanically-stable trees. During cutting, all crown competitors of target trees were harvested to increase their growth (positive selection). All standing dead trees and lying deadwood slightly decomposed were also removed (1st and 2nd decay classes). In the traditional thinning, only dominated, small or standing dead trees were harvested (negative selection) during in-field operations. This was done up to 40% of total trees per hectare. In this thinning, lying deadwood is not removed from the forest. The traditional thinning represents the most common silvicultural treatment applied in Central Italy in both artificial and natural stands, developed according with regional forest laws.

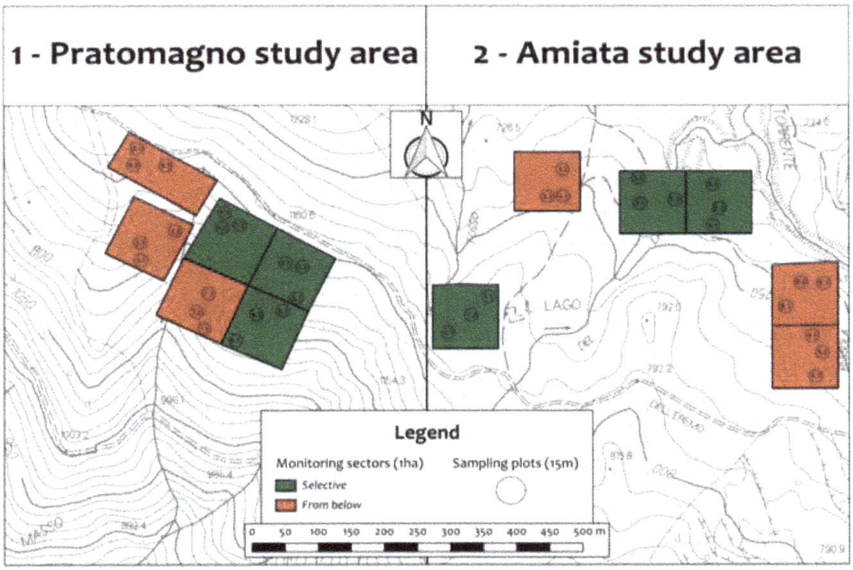

Figure 2. Spatial distribution of monitoring sectors and sampling plots in both of the study areas. The different thinning systems applied are marked with different colours.

2.3. Assessment of Ecosystem Services

The ecosystem services were evaluated using the data collected in the field and the information provided by local forest enterprises. The biophysical assessment of ESs after the silvicultural treatments was done using a combined approach of quantitative and qualitative information.

In order to evaluate the effects of thinning on ES provision, three categories of ESs were assessed: provisioning services (wood assortments), regulating services (mechanical stability of the forest system, carbon sequestration), and supporting services (tree species diversity, floristic diversity).

At the end of the ES assessment, a matrix of the effects of silvicultural treatments on ESs in the two study areas was produced in order to compare traditional thinning and selective thinning in black pine forests.

2.3.1. Provisioning Services

The volume of trees harvested was quantified using the most updated volume tables for black pine, considering the harvesting rate applied with the traditional and selective thinning. Total harvested timber has been estimated using data measured in the field and volume tables provided by the second Italian National Forest inventory [40], using Equation (1):

$$V = b_1 + b_2 \times d^2 \times h + b_3 \times d \tag{1}$$

where the total Volume (V) of the stem and large branches is provided in cubic decimeters, the diameter at breast height (d) and total tree height (h) are expressed in centimeters and meters respectively, and b_1, b_2 and b_3 are species-specific coefficients.

Moreover, the proportions of different wood assortments (e.g., roundwood, poles and woodchips) were estimated by means of a local assortment table [41]. Finally, a check between the potential wood assortments and those effectively sold by the forest enterprise was done.

2.3.2. Regulating Services

Two regulating services were assessed in the present study: mechanical stability of the forest system and carbon sequestration.

2.4. Mechanical Stability of the Forest System

The slenderness ratio (height/diameter) is widely acknowledged as the main indicator of single-tree mechanical stability, especially in artificial stands and for conifers, where the dense planting scheme often influences trees growth [26,42]. Even if this indicator has been rarely used at stand-level, being the "average" stability a concept quite far away from a simple arithmetic mean of single-tree values, it can be considered a fair proxy of the protective capacity (i.e., hydrogeological and natural hazard protection) of forest stand [18,26,43]. However, variations in H/D ratios are largely a result of spacing and, consequently, of stand density resulting from silvicultural treatments. Actually, recent studies [44] show that a high H/D can indicate that a tree has grown in a dense stand under the influence of close mutual support; besides, trees with a high value can be more vulnerable because their stems have not been able to develop to conditions of high mechanical perturbation. In addition, the H/D is directly related to the ability of the forest stand to protect from landslides and snow damage [45].

In the present study, an average H/D value for the dominant trees only (i.e., the trees we would select as candidates to be the final goal of our treatment) and around 100 trees ha^{-1} has been used as a proxy to evaluate the ability of the system to maintain the protective function and the general (mechanical) stability. This was done to avoid biases in H/D values due to the treatment. Indeed, the two thinning systems are well-known to change the average H/D ratio differently for the whole stand and due to the different diameter classes, which are going to be cut. By doing so, just the H/D values of released trees whose growth we would like to maximise and on which we are going to concentrate the carbon stocking were considered. The H/D ratios were calculated before and after the traditional thinning and selective thinning application in each plot and then averaged. For each plot, 7 well shaped and dominant trees (i.e., $7 \times 3 \times 3 = 63$ trees per treatment in each study area) were marked and measured before and two years after treatment.

2.5. Carbon Sequestration

The method used to estimate the annual carbon sequestration is based on the IPCC "Good Practice Guidance for Land use, land-use Change and Forestry" [46]. The annual forest capacity to transform atmospheric carbon into biomass was estimated considering two carbon pools (above-ground biomass and below-ground biomass); while the other three carbon pools (litter, soil, and deadwood) were not considered as the changes in the annual increment of carbon stock are negligible. In accordance with the approach proposed by [38,47], the biophysical assessment of carbon sequestration (C) in above-ground and below-ground biomass of the two study areas was estimated using the annual increment of tree volume (m^3 $ha^{-1}\cdot yr^{-1}$) measured after the two silvicultural treatments.

The carbon sequestration was estimated using as variables the annual increment of volume before thinning (I_b), the annual increment of volume after the traditional thinning (I_t) and the annual increment of volume after the selective thinning (I_s). Annual increments of volume were estimated by taking random wood samples with a Pressler borer from trees distributed across all diameter classes in the forest monitoring sectors and estimated at plot level. The other parameters considered in the estimation of C sequestration in above-ground and below-ground biomass were wood basal density, biomass expansion factor, and root/shoot ratio of black pine [38,47].

The changes in carbon sequestration before and after the two types of thinning were calculated using the following equations (Equations (2) and (3)):

$$\Delta_{ct} = C_t - C_b \quad (2)$$

$$\Delta_{ct} = C_S - C_b \quad (3)$$

where C_b is the annual carbon sequestration before thinning (tC $ha^{-1}\cdot yr^{-1}$), C_t is the annual carbon sequestration after the traditional thinning (tC $ha^{-1}\cdot yr^{-1}$), C_s is the annual carbon sequestration after the selective thinning (tC $ha^{-1}\cdot yr^{-1}$), Δ_{ct} is the change in carbon sequestration in the traditional thinning scenario, Δ_{cs} is the change in carbon sequestration in the selective thinning scenario.

2.6. Supporting Services

Two supporting services were assessed in the present study: tree species diversity and floristic diversity.

2.6.1. Tree Species Diversity

The field measurements were also used to evaluate the species diversity. The overall diversity level of tree species in the two studied areas was evaluated by means of the Shannon diversity index (H') (Equation (4)):

$$H' = -\Sigma pi \times log_z(pi) \quad (4)$$

where pi represents the relative frequency of the species (i.e., the relative abundance in each sampling plot). This index was calculated to assess the alpha diversity (α-diversity) of each sampling plot and treatment, intended as the mean species diversity in sites or habitats at a local scale (Whittaker 1972). Firstly, a before-treatment value was calculated for the 18 sampling plots. Then, this parameter was measured again after the thinning interventions.

2.6.2. Floristic Diversity

The floristic diversity was evaluated in accordance with the Braun-Blanquet phytosociological method based on the estimation of plant cover and number of individual plants [48,49]. The floristic sampling was carried out every year during spring and summer, when vegetation is flowering and species identification is easier. Within each sampling plot, vegetation species were identified and their abundance-dominance was assessed in a sub-sampling plot of 10-m radius using the Braun-Blanquet scale amended by from 1 to 5 (class 1 = cover 1–5%, class 2 = cover 6–25%, class 3 = cover 26–50%, class 4 = cover 51–75%, class 5 = cover 76–100%). The symbol "+" was given to species with cover <1%, and "r" was used for rare species. The floristic list of the species was then compiled and a value of ground coverage was assigned to each species through a visual estimation. Data analysis was conducted after transformation of the abundance-dominance list obtained with the Braun Blanquet method in specific frequencies to calculate percentage contribution, according to the Van der Maarel scale [48]. Then, similar to tree species diversity, the floristic diversity was evaluated using the Shannon index and the α-diversity of each sampling plot was calculated before and after the thinning interventions.

2.7. Statistical Analysis

All the studied biological processes were tested for statistical significance. Given the different nature of collected data, two different statistical analyses were run. Firstly, the total volume of timber harvested was analyzed by means of a classic parametric ANOVA between treatments. Then, differences in radial increment (RI), H/D ratio, tree species diversity (tH') and floristic diversity (fH') were evaluated by means of Linear Mixed-models, given the intrinsic autocorrelation between measurements (longitudinal study). While site and treatment sector were used as replicates, the two thinning systems (treatment) and sampling year were the fixed effects, with the sampling plots as sub-samples. Only the zone (i.e., Amiata and Pratomagno) was set as a random effect. Actually, we were not interested in evaluating the effect of the treatment at each site but to test whether the treatment was somehow significant for artificial black pine stands. Linear mixed models were run using the *lme4* package [50] available in the R statistical environment [51].

3. Results

3.1. Provisioning Services

The results show that the harvested rate in the Amiata study area was 18.5% of total standing volume after the traditional thinning and 30.1% after the selective thinning, while the harvested rate in the Pratomagno study area was 19.4% of total standing volume after the traditional thinning and 29.6% after the selective thinning. Consequently, in both study areas, the harvested volume was higher with the selective thinning than with the traditional thinning. Such differences were also statistically significant (p-value < 0.05) both globally (selective versus traditional) and even within the same study area (i.e., analyzing the two study areas separately).

In the Amiata study area, 100% of wood products was provided as woodchips, while in Pratomagno, roundwood and pole production contributed for 70% of the total value of provisioning services in traditional thinning and 78% in selective thinning (Table 1). Results show that only in the Pratomagno study area were real and potential production of assortments similar, while in Amiata the opportunity for timber production has not been taken into account by local forest enterprises.

Table 1. Provisioning services before and after thinning in the two study areas. The potential percentage of wood assortment is reported in parentheses.

	Volume before Thinning (m^3 ha^{-1})	Harvested Volume (m^3 ha^{-1})	Wood Assortments: Timber (Roundwood & Poles)	Wood Assortments: Woodchips
		Amiata study area		
Traditional thinning	362.9	67.3	0% (68%)	100% (32%)
Selective thinning	456.6	137.4	0% (74%)	100% (26%)
		Pratomagno study area		
Traditional thinning	721.1	139.6	70% (75%)	30% (25%)
Selective thinning	586.6	173.9	78% (79%)	22% (21%)

3.2. Regulating Services

3.2.1. Mechanical Stability of the Forest System

The mixed model showed that differences in H/D between the two thinning treatments were not statistically different after two years. The H/D ratio slightly decreased for both types of interventions.

In this sense the detected annual variation was higher after the selective thinning, suggesting this treatment has a higher capacity to increase the mechanical stability of the stand compared to the traditional thinning. After the selective thinning, the annual variation was −1.3% in the Amiata study area and −1.0% in the Pratomagno study area, while after the traditional thinning, the annual variation was −1.0% in Amiata and −0.89% in Pratomagno (Table 2).

Table 2. H/D ratios before and after thinning in the two study areas.

	H/D Ratio (before Thinning)	H/D Ratio (Two Years after Thinning)	Annual Variation (%)
	Amiata study area		
Traditional thinning	63.15	61.93	−0.979
Selective thinning	66.32	64.62	−1.284
	Pratomagno study area		
Traditional thinning	61.05	59.96	−0.889
Selective thinning	52.81	51.74	−1.012

3.2.2. Carbon Sequestration

The mixed model did not show statistically significant differences in the annual increment of volume and carbon sequestration between the two treatments after the thinning interventions. Results showed that in the Pratomagno study area, after the thinning, the annual increment of volume was 1.52 m^3 $ha^{-1}{\cdot}yr^{-1}$ and 0.71 m^3 $ha^{-1}{\cdot}yr^{-1}$ for the selectively and traditionally thinned plots, respectively. In the Amiata study area, the values were 1.11 m^3 $ha^{-1}{\cdot}yr^{-1}$ and 1.55 m^3 $ha^{-1}{\cdot}yr^{-1}$ (Table 3). The change in carbon sequestration after the traditional thinning was 0.27 tCO_{2eq} $ha^{-1}{\cdot}yr^{-1}$ in Amiata and 0.12 tCO_{2eq} $ha^{-1}{\cdot}yr^{-1}$ in Pratomagno, while the change in carbon sequestration after the selective thinning was 0.47 tCO_{2eq} $ha^{-1}{\cdot}yr^{-1}$ in Amiata and 0.37 tCO_{2eq} $ha^{-1}{\cdot}yr^{-1}$ in Pratomagno. Even if not supported by statistical evidence, in both study areas, the radial growth of crop trees that were selectively thinned and their annual carbon sequestration were stimulated more than under traditional thinning.

Table 3. Change in carbon sequestration (Δ) after thinning in the two study areas.

Silvicultural Treatments	Annual Increment (m^3 $ha^{-1}{\cdot}yr^{-1}$)		C (tCO_{2eq} $ha^{-1}{\cdot}yr^{-1}$)		Δ (tCO_{2eq} $ha^{-1}{\cdot}yr^{-1}$)
	Before	After	Before	After	
		Amiata study area			
Traditional thinning	1.37	1.55	1.14	1.61	0.2682
Selective thinning	0.78	1.11	1.99	2.26	0.4693
		Pratomagno study area			
Traditional thinning	0.63	0.71	0.92	1.04	0.1195
Selective thinning	1.27	1.52	1.85	2.22	0.3746

3.3. *Supporting Services*

3.3.1. Tree Species Diversity

The results of the Shannon index for tree species are reported in Table 4. The mixed model was not statistically significant (only a significant intercept term was found). *Pinus nigra* ssp. *laricio* was largely detected as the most abundant species in the Amiata study area, with a homogeneous distribution across both the forest monitoring sectors (91% each) with just a few hardwood species (*Quercus cerris* L. and *Quercus pubescens* Willd., 1805) covering less than 4% each. Concerning the Pratomagno study area, *Pinus nigra* ssp. *laricio* is still the most important species, but with different percentages between forest monitoring sectors. Actually, while a value similar to Pratomagno was recorded for the selective treatment (92%), a lower percentage (73%) was found in the zones under traditional thinning, with a relevant presence of *Abies alba* Mill. (25%). This is the main reason for a quite different starting H' value detected in the Pratomagno study area (Table 4). The main variation is observed in the Amiata study area where both thinning operations increased the species diversity. Indeed, the H' variation recorded in both cases is +0.11. On the contrary, a different situation is observed in the Pratomagno study area. While the selective thinning seems to increase the overall species diversity (+0.07), a lower H' value is observed in the zones where the traditional thinning is applied (−0.02).

Table 4. Tree species' Shannon index (tH') before and after thinning in the two study areas.

Silvicultural Treatments	tH' before Thinning	tH' after Thinning	Variation
	Amiata study area		
Traditional thinning	0.53	0.64	+0.11
Selective thinning	0.50	0.61	+0.11
	Pratomagno study area		
Traditional thinning	0.91	0.89	−0.02
Selective thinning	0.40	0.47	+0.07

3.3.2. Floristic Diversity

As far as the Shannon index is concerned, the mixed model showed that differences between selective and traditional treatments two years after the thinning were not statistically significant. The measurements highlight an increase in the Shannon index after the thinning in both study areas (Table 5). The increase is higher in the sampling plots under selective thinning (variation +0.2 in Amiata study area and +0.3 in Pratomagno). The floristic composition is characterized by the dominance of the endemic species *Brachypodium rupestre* (Host) Roem. & Schult.; its presence has not been influenced by the thinning intervention. Conversely, the occurrence of heliophilous species belonging to the *Leguminosae* family is increased by the thinning due to the higher penetration of light to the soil surface. In the Pratomagno study area, the species are those typical of a pine stand herbaceous layer, while in the Amiata study area, there is a relevant presence of species typical of open grass-land-habitats. The increase in the mean number of species is higher for sampling plots under selective thinning and more evident in Amiata study area, where the number of species show an increase of 38, while in the Pratomagno study area, the increase is 7 species after the selective thinning.

Table 5. Change in the floristic Shannon index (fH') and species richness after thinning in the two study areas.

Silvicultural Treatments	N° Species before Thinning	N° Species after Thinning	Variation	fH' before Thinning	fH' after Thinning	Variation
		Amiata study area				
Traditional thinning	78	92	+14	3.1	3.2	+0.1
Selective thinning	54	92	+38	2.9	3.1	+0.2
		Pratomagno study area				
Traditional thinning	38	42	+4	2.1	2.2	+0.1
Selective thinning	38	45	+7	2.2	2.5	+0.3

At the end of the assessment of three categories of ESs, a matrix was prepared in order to compare the results for the study area and thinning intervention (Table 6). The matrix shows that in both study areas, the selective thinning increased the ES provisions: wood provision, mechanical stability of the forest system, carbon sequestration, tree species biodiversity and floristic biodiversity.

Table 6. Comparison among ecosystem services in the two study areas.

	Provisioning	Regulating		Supporting	
	Harvested Volume (m^3 ha^{-1})	Mechanical Stability of the Forest System (Annual Variation H/D Ratio)	Δ Carbon Sequestration (tCO_{2eq} $ha^{-1} \cdot yr^{-1}$)	Floristic Biodiversity Variation fH'	Tree Species Biodiversity Variation tH'
		Amiata study area			
Traditional thinning	67.3	−0.969	0.2682	+0.1	+0.11
Selective thinning	137.4	−1.284	0.4693	+0.2	+0.11
		Pratomagno study area			
Traditional thinning	139.6	−0.889	0.1195	+0.1	−0.02
Selective thinning	173.9	−1.012	0.3746	+0.3	+0.07

4. Discussion

Thinning is a strategy for forest managers to modify forest structure and composition according to forest management goals, regulating inter- and intra-specific competition to concentrate biomass allocation on a few target trees [16,17]. In Italy, thinning from below is generally recognized as the most important early-stage operation carried out between canopy closure and the final harvest and is also suggested as the unique approach for artificial stand management [26,35]. Despite that, even if our results were not able to demonstrate the lower performance of thinning from below in ES delivery in artificial black pine stands (none of the analyzed ESs were supported by statistical evidence), the detected trends possibly suggest an interesting starting point just 2 years after treatment. Therefore, even if no conclusive evidence of increased ES provisioning generated by selective thinning is found, early trends suggest that selective thinning treatment may show differences in ES provisioning over longer monitoring periods [9,13,39]. Moreover, even if energy wood harvesting has often been evaluated as one of the main products of artificial softwood stands [30], and especially those located in the Italian central Apennines [26], roundwood and many other ESs might be successfully obtained from such stands if managed properly. It is well known that the diameter growth of forest species is directly connected to early thinning, especially in conifer stands [18,28]. Moreover, the use of an early implementation of selective thinning has also been advised as one of the main tools for stand stability and high quality wood production [27].

Concerning wood products and despite the main protective goal of such stands, our results confirmed that different thinning interventions could influence the provisioning services in terms of wood assortment production. Concerning wood production, the selective thinning yielded a higher harvesting rate and a higher percentage of roundwood compared with woodchips. The results are comparable to those in a black pine peri-urban forest in Central Italy where the harvesting rate is higher with selective thinning [38]. Even if the Italian market is mainly directed toward bioenergy production, more valuable wood assortments (i.e., roundwood or poles) can be obtained and sold on the market. The results are in accordance with the review by Cameron [27], demonstrating that selective thinning is an investment from an economic point of view because it improves future timber quality and economic value. Moreover, as highlighted by Macdonald and Hubert [52] in Sitka spruce (*Picea sitchensis* (Bong.) Carrière) stands in the United Kingdom, selective thinning improves straightness and branching characteristics and is the recommended thinning intervention for sawlog production. Actually, once the potential assortments of a harvesting activity are evaluated, knowledge of the market and skills of local enterprises represent the main gap to be filled by public research. Indeed, with almost the same stand type and structure, the two forest enterprises (one per study area) worked with a completely different harvesting scheme and profit. In particular, the main variables that influenced the allocation of wood assortments on the market are the prices of different wood products on the local market, the quality of wood, and the local communities' customs and dynamics. In such a framework, the ability of forest enterprises to place their products on the market is the main driver [53].

When regulating services are concerned, it is recognized that the mechanical stability of trees is a fundamental issue in public safety and erosion control. In this view, even if not supported by statistical evidence, our results suggested that black pine is a very plastic and reactive species, even when a late selective thinning is applied (>40 years old). Selective thinning is also very suitable for improving stand structure and stability, more than traditional thinning. A positive reaction was observed on all candidate trees (i.e., a decreasing H/D ratio and higher current radial increment). The H/D ratio of taller trees (i.e., candidate trees) was reduced (i.e., more stable trees) just 2 years after the treatment and in line with previous studies in similar zones [35]. Consequently, positive implications on soil retention and the prevention of landslides might be attributed to the selective thinning, thanks to higher ground coverage and survival of healthy and well-formed trees [54] The positive effect of selective thinning on stands' stability due to the reduction of the H/D ratio of higher trees is also reported by Cameron [27] for Sitka spruce stands in Scotland. In this sense, del Río et al. [55] highlighted the importance of the kind and intensity of thinning for stand stability. In particular, for Scots pine (*Pinus sylvestris* L.) stands, the long-term positive effect of thinning from above on stand stability was found compared to thinning from below.

These positive results found for selective thinning seem to be interesting, even concerning the higher growth rate trends measured on selected trees and mainly concentrated on diameter increment, which also includes higher carbon sequestration. Even if not significant, our carbon sequestration results are not different from those derived for other forest systems. For instance, in a long-term Mediterranean thinning experiment, Ruiz-Peinado et al. [56] reported mean values that ranged from 128 Mg C ha^{-1} to 193 Mg C ha^{-1} at the age of 52 years depending on the thinning treatment, with the greatest value in unthinned plots. Del Río et al. [55] highlight how in Mediterranean pine forests carbon sequestration changes over time, and with different forest management strategies: carbon sequestration rates are influenced by the rotation length, thinning intensity, stand composition, with different results amongst species. However, for Mediterranean maritime pine, heavy thinning increased carbon sequestration when carbon fixed in removed wood was also considered. In general, our results are in accordance with literature showing that the total carbon sequestration potential usually increases with thinning [56,57].

When considering supporting services, the Shannon index expressing tree species diversity tended to increase in both study areas with both tested treatments and mainly with the selective thinning intervention. This aspect is in line with expected results, with selective thinning being much more plastic than traditional thinning and able to preserve all small broadleaves species which must be removed with thinning from below. However, *tH'* increment values are also highly correlated to before-treatment vertical and horizontal stand structures, and while a homogeneous starting point (0.5) was found in the Amiata study area, quite a different situation was found in the Pratomagno area with 0.9 and 0.4 *tH'* starting values. Such differences are also connected to the genesis of the stands: rural mid-elevation areas in Amiata with broadleaves trees often occur in plots; high elevation and steep zones in Pratomagno with species admixture occur only in some areas. Actually, starting differences were mainly due to silver fir or European beech trees, planted in mixture with black pine in more fertile zones only. However, in the end, the results demonstrated that the two systems are almost interchangeable in Amiata, where trees biodiversity is balanced; instead in Pratomagno, in an environment with low species richness, thinning from below is not able to improve tree species diversity. This fact is probably due to the rigid thinning scheme to be applied, which cut trees according to their dbh, removing all dominant trees despite their taxonomy.

Concerning the understory level, floristic diversity—expressed by the Shannon index—increased with both thinning systems and more so with the selective than with the traditional one. Without statistical significance, such results must be considered as a trend to be confirmed with additional data in the next few years. The results showed that the number of species (particularly light-demanding ones) increased with thinning intervention. The effect was higher in plots under the selective regime with "new" species coming from the surroundings of the plots we studied. Consequently, our results can be interpreted as an activation of available seeds in the soil. In other words, the activation of existing potential. The results of the present study are comparable with other studies investigating the effects of silvicultural systems on understory plant diversity, including species composition, structural attributes and functional organization. Studies demonstrated that, in general, the species pool was higher for selectively cut areas, and a high proportion of light reaching the forest floor induced the spread of light-demanding species and the detriment of true forest species [15,58].

The matrix synthesizing the effects of silvicultural treatments on ESs shows that in both study areas the selective thinning suggested a promising effect on all ESs investigated (apart from tree species biodiversity in Amiata). It is a fact that the thinning regime influences competition relationships among plants, changing the overall ecosystem complexity. Thus, it is important to support experiments on different thinning regimes where the objective is to improve the forest multi-functionality (biodiversity, wood production, soil protection) [17]. Even if our results report data only for two years after treatments, the trends described might justify further monitoring activities.

5. Conclusions

Artificial pine forests in Italy are an extremely simplified system, poorly managed and characterized by a low biodiversity level. Their original function was to improve soil nutrients and to catalyze ecological succession on bare and overexploited soils and to prepare soil for more demanding species. This shift from an artificial conifer stand to a mixed forest type through natural regeneration, attributed to autochthonous species, was and currently represents the final goal. However, while in a former concept the substitution was planned to be realized by means of clear-cutting of pines and artificial plantation of autochthonous trees, current evidence demonstrates that a gradual substitution can be encouraged by thinning [20], also improving ESs delivered by such stands. In such a framework, the selective thinning can be successfully implemented even in older structures for a gradual enhancement of species composition and natural regeneration where crop trees can represent those to be favored in order to improve their seed dispersal capacity.

From the methodological point of view, the method used to assess and compare the effects of thinning on ES is simple and easy to apply. The small time period between the treatment and post-harvesting surveys probably represents the major issue in our experimental design. However, funding time is always rigid in European Union (EU) projects and this may be the cause for a lack of statistical significance. The results are encouraging and can represent the basis for novel survey campaigns or longer projects. In future, the possibility to derive single-tree statistics, e.g., with dendrometers and terrestrial laser scanning techniques, could represent an interesting approach.

The matrix on the effects of silvicultural treatments on ESs provides useful information to decision makers to choose the most appropriate forest management strategy based on the forest stand. Conversely, the applied method provides a partial analysis of the effects of silvicultural treatments on ESs because only one sub-set of all ES forests was assessed. Therefore, future findings of this study will consist of assessing the effects of different thinning regimes on a larger number of ESs, including cultural services (e.g., recreational activities).

Acknowledgments: This study was carried out and funded by the SelPiBio LIFE project (Innovative silvicultural treatments to enhance soil biodiversity in artificial black pine stands, i.e., LIFE13 BIO/IT/000282) for demonstration of innovative silvicultural treatments in artificial black pine stands. We really appreciated the useful suggestions provided by the reviewers and we are very grateful to all of them for their contribution during the editorial process of this paper.

Author Contributions: P.C. and I.D.M. conceptualized the SelPiBio LIFE project and designed the experiment. M.M., I.D.M. and P.C. collected all the mensurational data and performed the analysis. A.P. was responsible for ecosystem services assessment. E.B. was responsible for the floristic part. M.M. ran the statistical analysis. M.M. and A.P. wrote the paper. All the Authors revised the paper and approved the final version.

Conflicts of Interest: The authors declare no conflict of interest.

References

1. Nahlik, A.M.; Kentula, M.E.; Fennessy, M.S.; Landers, D.H. Where is the consensus? A proposed foundation for moving ecosystem service concepts into practice. *Ecol. Econ.* **2012**, *77*, 27–35. [CrossRef]
2. Seppelt, R.; Dormann, C.F.; Eppink, F.V.; Lautenbach, S.; Schmidt, S. A quantitative review of ecosystem service studies: Approaches, shortcomings and the road ahead. *J. Appl. Ecol.* **2011**, *48*, 630–636. [CrossRef]
3. Hein, L.; van Koppen, K.; de Groot, R.S.; van Ierland, E.C. Spatial scales, stakeholders and the valuation of ecosystem services. *Ecol. Econ.* **2006**, *57*, 209–228. [CrossRef]
4. Melaku Canu, D.; Ghermandi, A.; Nunes, P.A.L.D.; Lazzari, P.; Cossarini, G.; Solidoro, C. Estimating the value of carbon sequestration ecosystem services in the Mediterranean Sea: An ecological economics approach. *Glob. Environ. Chang.* **2015**, *32*, 87–95. [CrossRef]
5. Paletto, A.; Geitner, C.; Grilli, G.; Hastik, R.; Pastorella, F.; Garcìa, L.R. Mapping the value of ecosystem services: A case study from the Austrian Alps. *Ann. For. Res.* **2015**, *58*, 157–175. [CrossRef]
6. Boscolo, M.; Vincent, J.R. Nonconvexities in the production of timber, biodiversity, and carbon sequestration. *J. Environ. Econ. Manag.* **2003**, *46*, 251–268. [CrossRef]
7. Ruddell, S.; Sampson, R.; Smith, M.; Giffen, R.; Cathcart, J.; Hagan, J.; Sosland, D.; Godbee, J.; Heissenbuttel, J.; Lovett, S.; et al. The role for sustainably managed forests in climate change mitigation. *J. For.* **2007**, *105*, 314–319.
8. Ray, D.; Petr, M.; Mullett, M.; Bathgate, S.; Marchi, M.; Beauchamp, K. A simulation-based approach to assess forest policy options under biotic and abiotic climate change impacts: A case study on Scotland's National Forest Estate. *For. Policy Econ.* **2017**. [CrossRef]
9. Duncker, P.S.; Raulund-Rasmussen, K.; Gundersen, P.; Katzensteiner, K.; De Jong, J.; Ravn, H.P.; Smith, M.; Eckmüllner, O.; Spiecker, H. How forest management affects ecosystem services, including timber production and economic return: Synergies and trade-offs. *Ecol. Soc.* **2012**, *17*. [CrossRef]
10. Ricketts, T.H.; Lonsdorf, E. Mapping the margin: Comparing marginal values of tropical forest remnants for pollination services. *Ecol. Appl.* **2013**, *23*, 1113–1123. [CrossRef] [PubMed]
11. Walker, B.; Carpenter, S.; Anderies, J.; Abel, N.; Cumming, G.; Janssen, M.; Lebel, L.; Norberg, J.; Peterson, G.D.; Pritchard, R. Resilience management in social-ecological systems: A workinghypothesis for a participatory approach. *Conserv. Ecol.* **2002**, *6*, 14. [CrossRef]
12. Rodríguez García, L.; Curetti, G.; Garegnani, G.; Grilli, G.; Pastorella, F.; Paletto, A. La valoración de los servicios ecosistémicos en los ecosistemas forestales: Un caso de estudio en Los Alpes Italianos. *Bosque* **2016**, *37*, 41–52. [CrossRef]
13. Häyhä, T.; Franzese, P.P.; Paletto, A.; Fath, B.D. Assessing, valuing, and mapping ecosystem services in Alpine forests. *Ecosyst. Serv.* **2015**, *14*, 12–23. [CrossRef]
14. Fernandes, T.J.G.; Del Campo, A.D.; Herrera, R.; Molina, A.J. Simultaneous assessment, through sap flow and stable isotopes, of water use efficiency (WUE) in thinned pines shows improvement in growth, tree-climate sensitivity and WUE, but not in WUEi. *For. Ecol. Manag.* **2016**, *361*, 298–308. [CrossRef]
15. Tomao, A.; Antonio, J.; Martínez, J.; Aragón, D. Is silviculture able to enhance wild forest mushroom resources? Current knowledge and future perspectives. *For. Ecol. Manag.* **2017**, *402*, 102–114. [CrossRef]
16. Cabon, A.; Mouillot, F.; Lempereur, M.; Ourcival, J.-M.; Simioni, G.; Limousin, J.-M. Thinning increases tree growth by delaying drought-induced growth cessation in a Mediterranean evergreen oak coppice. *For. Ecol. Manag.* **2018**, *409*, 333–342. [CrossRef]
17. Jonard, M.; Misson, L.; Ponette, Q. Long-term thinning effects on the forest floor and the foliar nutrient status of Norway spruce stands in the Belgian Ardennes. *Can. J. For. Res.* **2006**, *36*, 2684–2695. [CrossRef]
18. Peltola, H.; Miina, J.; Rouvinen, I.; Kellomäki, S. Effect of early thinning on the diameter growth distribution along the stem of Scots pine. *Silva Fenn.* **2002**, *36*, 813–825. [CrossRef]
19. Lukina, N.V.; Orlova, M.A.; Gornov, A.V.; Kryshen', A.M.; Kuznetsov, P.V.; Knyazeva, S.V.; Smirnov, V.E.; Bakhmet, O.N.; Eydlina, S.P.; Ershov, V.V.; et al. Assessment of sustainable forest management criteria using indicators of the international programme ICP forests. *Contemp. Probl. Ecol.* **2013**, *6*, 734–745. [CrossRef]
20. Brang, P.; Spathelf, P.; Larsen, J.B.; Bauhus, J.; Bonc Ina, A.; Chauvin, C.; Drossler, L.; Garcia-Guemes, C.; Heiri, C.; Kerr, G.; et al. Suitability of close-to-nature silviculture for adapting temperate European forests to climate change. *Forestry* **2014**, *87*, 492–503. [CrossRef]

21. Baral, H.; Guariguata, M.R.; Keenan, R.J. A proposed framework for assessing ecosystem goods and services from planted forests. *Ecosyst. Serv.* **2016**, *22*, 260–268. [CrossRef]
22. Baral, H.; Keenan, R.J.; Stork, N.E.; Kasel, S. Measuring and managing ecosystem goods and services in changing landscapes: A south-east Australian perspective. *J. Environ. Plan. Manag.* **2014**, *57*, 961–983. [CrossRef]
23. Burkhard, B.; Kroll, F.; Nedkov, S.; Müller, F. Mapping ecosystem service supply, demand and budgets. *Ecol. Indic.* **2012**, *21*, 17–29. [CrossRef]
24. Enescu, C.; de Rigo, D.; Caudullo, G.; Mauri, A.; Houston Durrant, T. *Pinus nigra* in Europe: Distribution, habitat, usage and threats. In *European Atlas of Forest Tree Species*; San-Miguel-Ayanz, J., de Rigo, D., Caudullo, G., Houston Durrant, T., Mauri, A., Eds.; European Commission: Luxembourg, 2016; Volume 6, pp. 126–127.
25. Marchi, M.; Nocentini, S.; Ducci, F. Future scenarios and conservation strategies for a rear-edge marginal population of *Pinus nigra* Arnold in Italian central Apennines. *For. Syst.* **2016**, *25*, e072. [CrossRef]
26. Cantiani, P.; Chiavetta, U. Estimating the mechanical stability of *Pinus nigra* Arn. using an alternative approach across several plantations in central Italy. *iForest Biogeosci. For.* **2015**, *8*, 846–852. [CrossRef]
27. Cameron, A.D. Importance of early selective thinning in the development of long-term stand stability and improved log quality: A review. *Forestry* **2002**, *75*, 25–35. [CrossRef]
28. Novák, J.; Dušek, D.; Slodičák, M.; Kacálek, D. Importance of the first thinning in young mixed Norway spruce and European beech stands. *J. For. Sci.* **2017**, *63*, 254–262.
29. Emer, B.; Grigolato, S.; Lubello, D.; Cavalli, R. Comparison of biomass feedstock supply and demand in Northeast Italy. *Biomass Bioenergy* **2011**, *35*, 3309–3317. [CrossRef]
30. Lehtimäki, J.; Nurmi, J. Energy wood harvesting productivity of three harvesting methods in first thinning of scots pine (*Pinus sylvestris* L.). *Biomass Bioenergy* **2011**, *35*, 3383–3388. [CrossRef]
31. Bonet, J.A.; De-Miguel, S.; Martínez de Aragón, J.; Pukkala, T.; Palahí, M. Immediate effect of thinning on the yield of Lactarius group deliciosus in *Pinus pinaster* forests in Northeastern Spain. *For. Ecol. Manag.* **2012**, *265*, 211–217. [CrossRef]
32. Lindgren, P.M.: Ransome, D.B.; Sullivan, D.S.; Sullivan, T.P. Plant community attributes 12 to 14 years following precommercial thinning in a young lodgepole pine forest. *Can. J. For. Res.* **2006**, *36*, 48–61. [CrossRef]
33. Taki, H.; Inoue, T.; Tanaka, H.; Makihara, H.; Sueyoshi, M.; Isono, M.; Okabe, K. Responses of community structure, diversity, and abundance of understory plants and insect assemblages to thinning in plantations. *For. Ecol. Manag.* **2010**, *259*, 607–613. [CrossRef]
34. Cantiani, P.; De Meo, I.; Becagli, C.; Bianchetto, E.; Cazau, C.; Mocali, S.; Salerni, E. Effects of thinnings on plants and fungi biodiversity in a *Pinus nigra* plantation: A case study in central Italy. *For. Ideas* **2015**, *21*, 149–162.
35. Kellogg, L.D.; Bettinger, P. Thinning productivity and cost for a mechanized cut-to-length system in the Northwest Pacific Coast Region of the USA. *J. For. Eng.* **1994**, *5*, 43–54. [CrossRef]
36. Di Salvatore, U.; Ferretti, F.; Cantiani, P.; Paletto, A.; De Meo, I.; Chiavetta, U. Multifunctionality assessment in forest planning at landscape level. The study case of Matese Mountain Community (Italy). *Ann. Silvic. Res.* **2013**, *37*, 45–54.
37. Cantiani, P.; Marchi, M. A spatial dataset of forest mensuration collected in black pine plantations in central Italy. *Ann. For. Sci.* **2017**, *74*, 50. [CrossRef]
38. Paletto, A.; De Meo, I.; Grilli, G.; Nikodinoska, N. Selective and traditional forest management options for black pine forests in Central Italy: Effects on ecosystem services. *Ann. For. Res.* **2017**, *60*, 313–326. [CrossRef]
39. Cimon-Morin, J.Ô.; Darveau, M.; Poulin, M. Fostering synergies between ecosystem services and biodiversity in conservation planning: A review. *Biol. Conserv.* **2013**, *166*, 144–154. [CrossRef]
40. Tabacchi, G.; Di Cosmo, L.; Gasparini, P.; Morelli, S. *Stima del Volume e della Fitomassa delle Principali Specie Forestali Italiene, Equazioni di Previsione, Tavole del Volume e Tavole della Fitomassa Arborea Epigea*; Consiglio per la Ricerca e la Sperimentazionein Agricoltura: Roma, Italy, 2011; ISBN 9788897081111.
41. Rinaldini, G.; Marchi, M.; Cantiani, P. Primi passi verso una tavola assortimentale per il pino nero. *Sherwood* **2017**, *227*, 7–10.
42. Marchi, M.; Chiavetta, U.; Cantiani, P. Assessing the mechanical stability of trees in artificial plantations of *Pinus nigra* J. F. Arnold using the LWN tool under different site indexes. *Ann. Silvic. Res.* **2017**, *41*, 48–53.

43. Bošel'A, M.; Konôpka, B.; Šebeň, V.; Vladovič, J.; Tobin, B. Modelling height to diameter ratio—An opportunity to increase Norway spruce stand stability in the Western Carpathians. *For. J.* **2014**, *60*, 71–80.
44. Valinger, E.; Fridman, J. Modelling probability of snow and wind damage in Scots pine stands using tree characteristics. *For. Ecol. Manag.* **1997**, *97*, 215–222. [CrossRef]
45. Cremer, K.W.; Borough, C.J.; McKinnell, F.H.; Carter, P.R. Effects of stocking and thinning on wind damage in plantations. *N. Z. J. For. Sci.* **1982**, *12*, 244–268.
46. Intergovernmental Panel on Climate Change (IPCC). *IPCC Climate Change 2014: Impacts, Adaptation, and Vulnerability*; Field, C.B.V.R., Barros, D.J., Dokken, K.J., Eds.; Part A: Global and Sectoral Aspects; Contribution of Working Group II to the Fifth Assessment Report of the Intergovernmental Panel on Climate Change; Cambridge University Press: Cambridge, UK; New York, NY, USA, 2014; 1132p.
47. Grilli, G.; Nikodinoska, N.; Paletto, A.; De Meo, I. Stakeholders' preferences and economic value of forest ecosystem services: An example in the Italian alps. *Balt. For.* **2015**, *21*, 298–307.
48. Van Der Maarel, E. The Braun-Blanquet approach in perspective. *Vegetatio* **1975**, *30*, 213–219. [CrossRef]
49. .Enright, N.J.; Nuñez, M.A. The Braun-Blanquet reviews in Plant Ecology: In honour of our founding editor, Josias Braun-Blanquet. *Plant Ecol.* **2013**, *214*, 1417–1418. [CrossRef]
50. Bates, D.; Mächler, M.; Bolker, B.M.; Walker, S.C. Fitting linear mixed-effects models using lme4. *J. Stat. Softw.* **2015**, *67*, 1–48. [CrossRef]
51. R Core Team. *R: A Language and Environment for Statistical Computing*; R Foundation for Statistical Computing: Vienna, Austria, 2018.
52. Macdonald, E.; Hubert, J. A review of the effects of silviculture on timber quality of Sitka spruce. *Forestry* **2002**, *75*, 107–138. [CrossRef]
53. Pra, A.; Pettenella, D. Consumption of wood biomass for energy in Italy: A strategic role based on weak knowledge. *Ital. J. For. Mount. Environ.* **2016**, *71*, 49–62. [CrossRef]
54. Jactel, H.; Nicoll, B.C.; Branco, M.; Gonzalez-Olabarria, J.R.; Grodzki, W.; Långström, B.; Moreira, F.; Netherer, S.; Orazio, C.; Piou, D.; et al. The influences of forest stand management on biotic and abiotic risks of damage. *Ann. For. Sci.* **2009**, *66*, 701. [CrossRef]
55. Del Río, M.; Bravo-Oviedo, A.; Pretzsch, H.; Löf, M.; Ruiz-Peinado, R. A review of thinning effects on Scots pine stands: From growth and yield to new challenges under global change. *For. Syst.* **2017**, *26*, eR03S. [CrossRef]
56. Ruiz-Peinado, R.; Bravo-Oviedo, A.; López-Senespleda, E.; Montero, G.; Río, M. Do thinnings influence biomass and soil carbon stocks in Mediterranean maritime pinewoods? *Eur. J. For. Res.* **2013**, *132*, 253–262. [CrossRef]
57. Balboa-Murias, M.Á.; Rodríguez-Soalleiro, R.; Merino, A.; Álvarez-González, J.G. Temporal variations and distribution of carbon stocks in aboveground biomass of radiata pine and maritime pine pure stands under different silvicultural alternatives. *For. Ecol. Manag.* **2006**, *237*, 29–38. [CrossRef]
58. Decocq, G.; Aubert, M.; Dupont, F.; Alard, D.; Saguez, R.; Wattez-Franger, A.; De Foucault, B.; Delelis-Dusollier, A.; Bardat, J. Plant diversity in a managed temperate deciduous forest: Understorey response to two silvicultural systems. *J. Appl. Ecol.* **2004**, *41*, 1065–1079. [CrossRef]

forests

MDPI

Article

Designing Thinning Operations in 2nd Age Class Pine Stands—Economic and Environmental Implications

Piotr S. Mederski [1,*], Rachele Venanzi [2], Mariusz Bembenek [1], Zbigniew Karaszewski [3], Martyna Rosińska [1], Zenon Pilarek [4], Irene Luchenti [2] and Michał Surus [1]

1 Faculty of Forestry, Department of Forest Utilisation, Poznań University of Life Sciences (PULS), ul. Wojska Polskiego 71A, 60-625 Poznań, Poland; mariusz.bembenek@up.poznan.pl (M.B.); martyna.rosinska@up.poznan.pl (M.R.); surus.michal@gmail.com (M.S.)

2 Department of Agriculture and Forest Sciences (DAFNE), Tuscia University, Via S. Camillo de Lellis, 01100 Viterbo, Italy; venanzi@unitus.it (R.V.); ireneluchenti@hotmail.it (I.L.)

3 Department of Wood Investigation and Application, Wood Technology Institute, ul. Winiarska 1, 60-654 Poznań, Poland; z_karaszewski@itd.poznan.pl

4 Faculty of Forestry, Department of Forest Technology, Poznań University of Life Sciences (PULS), ul. Wojska Polskiego 71C, 60-625 Poznań, Poland; zpilarek@up.poznan.pl

* Correspondence: piotr.mederski@up.poznan.pl; Tel.: +48-61-848-7761

Received: 28 May 2018; Accepted: 4 June 2018; Published: 7 June 2018

Abstract: The application of a harvester for thinning operations in young stands can pose several dilemmas. Firstly, the value of the timber obtained should be higher than the cost of the thinning operation—this is especially difficult with young stands, where the timber is of a small size and productivity is low. In addition, small harvesters used for thinning usually have short cranes, less than 10 m in length, which is rather impractical in stands where the distance between strip roads is a standard 20 m. The aim of this research was to select the best mechanised thinning operation, taking into account economic suitability and some environmental aspects (damage to the remaining stand and density of strip roads). The research was carried out in a 31-year-old pine stand (11.27 ha) in which three different harvester thinning (T) operations were designed: TM1) with midfield and one harvester pass, TM2) with midfield and two harvester passes and TWM) without midfield and one harvester pass. In all the proposed operations, a Vimek 404 T5 harvester (with a 4.6 m-long crane) and a Vimek 606 TT forwarder were used. The most economic operation was TWM, with a total cost of €13.73 m^{-3}, while TM1 was 13% more expensive at €15.51 m^{-3}. The lowest level of damage, 1.5%, was recorded in TWM, while in TM1 the level was 2.1%. The TWM operation required a net of strip roads twice as dense as in TM1. Taking into account all the analysed aspects, a thinning operation with midfield and one harvester pass is recommended when using the Vimek 404 T5 harvester and the Vimek 606 TT forwarder in the first thinning of a pine stand.

Keywords: early thinning; strip roads; productivity; costs; damage; Vimek

1. Introduction

The effective use of a mechanised thinning operation in a young pine stand is challenging from an economic perspective. When using a harvester and forwarder, it is best to start with the first thinning when the value of the timber obtained is higher than the cost of the thinning operation. However, this is especially difficult in young stands, where the timber is of a small size, as this may lead to low productivity. This results in the rare application of mechanised early thinnings, but there is also a growing interest in suitable mechanised silvicultural treatments in artificial ecosystems in Europe [1,2].

To obtain the best compromise between economic and ecological aspects, particularly in the first early thinning, geometrically selective thinning may be applied when possible, with respect to selective treatments when ecological aspects are of paramount importance [3,4].

In Central Europe, the first commercial thinning of pine (*Pinus sylvestris* L.) can be carried out in ca. 30-y.o. stands and the timber obtained is usually used in the pulp and paper industry or for chips as renewable energy biomass. Young stands ready for the first thinning need to be accessible for forest operation machines. To a certain degree, harvester parameters determine the distribution of strip roads in the forest. The crane length impacts on the distance between the strip roads, while the width of the harvester (but also of a forwarder) determines the width of the strip roads. In the majority of cases, the distance between strip roads is 20 m (measuring between the axis of the strip roads) and the width is approximately 4 m. This is due to the fact that nowadays, harvester cranes are ca. 10 m in length and the machines themselves are no wider than ca 3 m. However, some manufacturers of cut-to-length (CTL) technology may vary these measurements. Smaller harvesters can be equipped with cranes 4 to 6 m long and their width may be no bigger than 2 m. Available early thinning machines of these sizes are manufactured, for example, by Vimek. Vimek harvesters are equipped with a 4.6 m long crane and the machine is 1.8 m wide. Forwarders have the same long crane and machine width. These parameters present obvious dilemmas when the first thinning is expected and new strip roads are planned. There are two solutions: (1) the establishment of strip roads with a distance of 9–10 m between them; or (2) a distance of 20 m between strip roads, but a 10-m-wide midfield in the middle, on which trees are cut with a chainsaw and felled towards the closest strip road [5]. Taking into account the two variants of the distance between strip roads, 10 or 20 m, it was hypothesised that a greater distance would, in the end, give a higher productivity from harvester/forwarder application and consequently lead to lower costs. It was also hypothesised that the forwarder, in particular, would be more efficient due to a higher timber concentration after thinning, if the larger 20 m distance between strip roads was applied. It was also considered that there may be differences in the level of damage to the remaining trees on plots where the distance between the strip roads was 10 and 20 m.

Therefore, the objective of this research was to analyse: (1) the productivity and costs of the Vimek harvester and forwarder in a very young pine stand undergoing its first commercial thinning with respect to different strip road design and machine application; (2) the level of tree damage due to the mechanised thinning operation on plots with a distance between strip roads of between 10 and 20 m; and (3) the total area per hectare of the stand which needs to be excluded from production in order to establish strip roads.

2. Materials and Methods

The research was carried out in a 31-year-old pine (*Pinus sylvestris* L.) stand of 11.27 ha, in compartment 108 h in the North of Poland (Bobolice Forest District, Table 1). According to the forest inventory data from 2015 (year of thinning operation), the average DBH of pine was 13 cm and the mean height 11 m. Silver birch (*Betula pendula* Roth) was also represented in the stand (10%, Table 1) with the same age and mean DBH, but with a slightly greater height of 12 m.

Table 1. Stand characteristics before thinning; general forest inventory data valid on 1 Jan 2015.

Com-Partment	Area, ha	Species	Share, %	Age, Years	DBH, Mean, cm	H, Mean, m	Stocking, m^3 ha^{-1}	Understorey Species
108h	11.27	pine	90	31	13	11	137	birch
		birch	10	31	13	12	10	spruce
		spruce	sporadically	-				juniper
		larch	sporadically	-				

The species composition was enriched with sporadic appearances of spruce and larch. In addition, birch, spruce and juniper occurred intermittently in the understorey. The stocking of the main stand layer before thinning was 147 m^3 ha^{-1}. This stocking was achieved after a late cleaning was carried out in 2003, as well as pre-commercial thinning in 2007, when strip roads were also established by removing one tree row every ca 20 m. During the pre-commercial thinning, 160.85 m^3 of energy wood was harvested (a maximum top diameter of 7 cm over bark) and 19.70 m^3 of pulp wood (a minimum top diameter of 7 cm over bark).

In 2015, three different harvester thinning (T) operations were designed: TM1) with midfield and one harvester pass, TM2) with midfield and two harvester passes, and TWM) without midfield and one harvester pass (Figure 1).

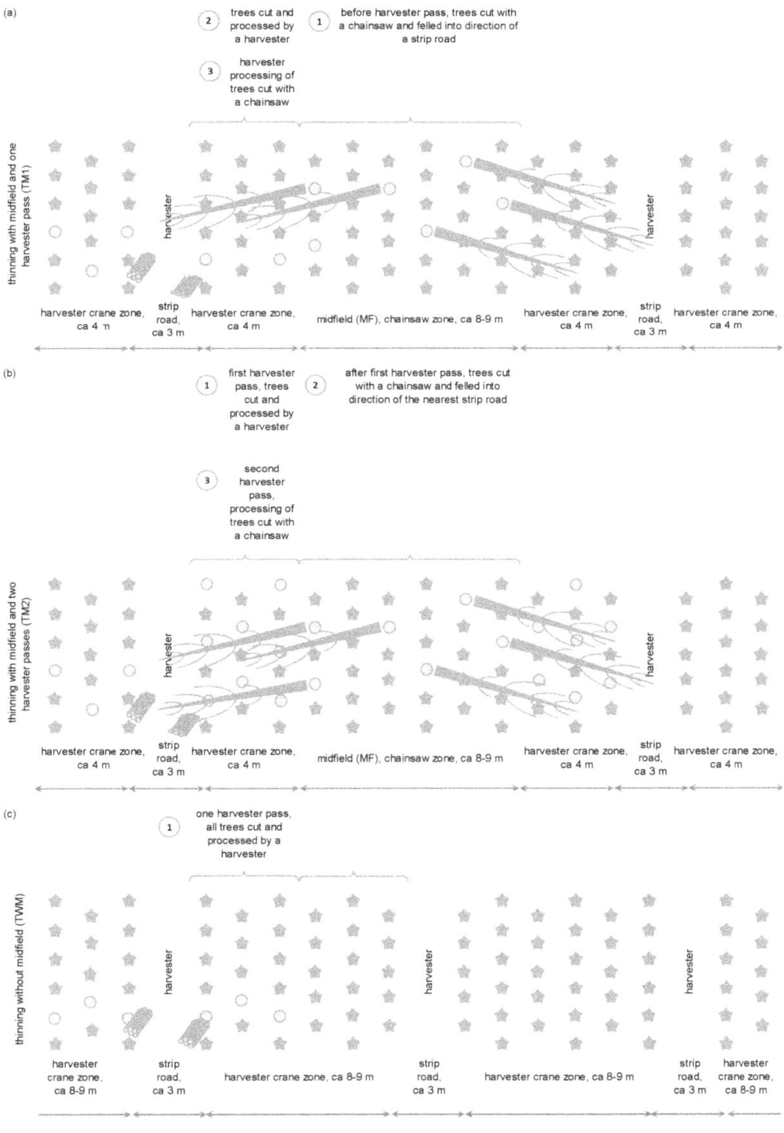

Figure 1. Design of analysed thinning operations: (**a**) Thinning with midfield and one harvester pass (TM1); (**b**) Thinning with midfield and two harvester passes (TM2); (**c**) Thinning without midfield (TWM). Distance between tree rows was from 1.4 to 1.6 m.

The thinning operations with midfield and without have been described in detail in previous research [5]. In TM1, the distance between the strip roads was ca 20 m (between the axis of the strip roads). In the first stage, on the midfield (a ca 8–9 m-wide zone between the strip roads), the trees were cut with a chainsaw and felled towards the nearest strip road. In stage two, the harvester cut the trees on a 4 m-wide zone next to the strip road and processed them. Finally, in stage three (within the same pass), the harvester processed the trees cut by the chainsaw. In TM2, the distance between the strip roads was also ca 20 m. In stage one, the harvester cut and processed the trees in a 4 m-wide zone next to the strip road (as in step two in TM1). In stage two, the chainsaw operator felled the trees towards the nearest strip road. In stage three, in the second pass, the harvester processed the trees cut by the chainsaw. Finally, in TWM, the distance between the strip roads was ca 10–12 m; the harvester cut and processed the trees from the 4 m-wide zones next to the strip roads. In all the thinning operations, the strip roads were ca 3 m wide, established by removal of one tree row during pre-commercial thinning in 2007. The distance between the tree rows ranged from 1.4 to 1.6 m.

Each thinning operation was carried out along two strip roads, which were ca 175 m long. In each type of operation, 20 sample plots were marked (10 m × 10 m), on which all trees for harvesting were measured. The DBH was measured twice perpendicularly using a Haglöf Mantax Blue manual caliper (Haglöf Sweden AB, Långsele, Sweden), with an accuracy of 1 mm, while the heights were measured using a Haglöf Vertex Laser height measurer (Haglöf Sweden AB, Långsele, Sweden), with an accuracy of 0.1 m. The trees selected for harvesting in all the operations had a similar mean DBH and height with no statistically significant differences (Table 2, K-W test, $p < 0.05$).

Table 2. Characteristics of trees selected for thinning operations.

Feature	TM1 DBH, cm	TM1 H, m	TM2 DBH, cm	TM2 H, m	TWM DBH, cm	TWM H, m
Mean	12.3 ^{a}	12.5 ^{a}	11.5 ^{a}	11.9 ^{a}	10.9 ^{a}	11.8 ^{a}
Minimum	7.0	7.8	7.0	6.3	7.0	8.2
Maximum	25.0	17.8	24.0	15.8	23.0	16.4
SD	3.9	1.8	3.9	2.0	3.4	1.7
N	100	100	112	112	110	105

Different letters next to mean values indicate statistically significant differences. DBH: diameter at breast height, H: height.

All the trees for harvesting were marked with individual numbers for further tree recognition during the time studies and further time analysis per one tree when possible: for harvester as well as for chainsaw felling. For forwarding, only general productivity was analysed in m^3 per hour. For the harvesting, forwarding, and chainsaw felling, a time study was carried out with an accuracy of 1 s. As the thinning operations were of a particular design, the typical categories of time were split up in order to determine the reasons for differences in productivities. For the harvester, three categories of time were distinguished: (A) preparation (tidying residues), machine moving, crane out, head positioning, felling; (B) delimbing and bucking; (C) delays. Time A was of particular interest, as it was hypothesised that it would be the longest in operation TM1, of medium duration in TM2, and the shortest in TWM. At the same time, it was presumed that time B would be the same in all three operations. Delay times were initially recorded, but eventually for the cost calculations, machine utilisation (MU) was considered more representative for machine use and chainsaw operator work. This decision was made for two reasons: (1) delay times were either accidental in this case study; or (2) compulsory in the harvesting and forwarding on account of the scheduled refuelling at the end of each strip road.

MU was obtained from engine working time (EWT), recorded by machine meter (total time from engine ignition until switch off). EWT was finally reduced by time of engine-idle speed time and time spent on machine relocation. Machine relocation on wheels (without a low-loading trailer) happened occasionally and was accepted up to ca 10 km. Otherwise, a low-loading trailer was used, the costs of which were considered in suitable calculations. This reduction used for calculation of MU was based on the expert decision of the machine owner. EWT was taken as total hours from beginning of machine use until the day of analysis, divided by total number of shift hours in the respective period of time.

For forwarding, four time categories were considered: (A) driving empty; (B) driving and loading; (C) driving loaded; (D) unloading. It was expected that loading in TM1 and TM2 would be more effective than in TWM due to a higher timber concentration next to the strip roads. When forwarding, delay times were linked to refuelling after finishing extraction on one strip road. These breaks would not take place in a normal day shift; therefore MU was eventually considered, and it was obtained using the same methodological approach as in case of the harvester.

For the chainsaw operation, three time categories were applied: (A) walking from one tree to another; (B) felling, and (C) delays. It was hypothesised that felling would take less time in TM2 than in TM1, due to the opening up of the space between the trees by the harvester in TM2 (fewer hanging trees). Delay times were also recorded. Again, as it was a case study, in the final analysis, the average daily delay times were used to calculate the MU.

For the productivity and cost measurements, all the harvested trees were considered. In all the proposed operations, a Vimek 404 T5 harvester and a Vimek 606 TT forwarder were used (Table 3).

Table 3. Harvester and forwarder characteristics.

Feature	Harvester Vimek 404 T5	Forwarder 606 TT
Weight, kg	3800	2960
Length, m	3.35	6.20
Width, m	1.84	1.80
Engine	CAT 2.2T	Kubota D 902-E
Engine power, kW	44.7	18.0
Crane/length, m	Mowi 2046/4.6 m	Mowi 2046/4.6 m
Head/max. diameter	Keto Forst Silver/30 cm	-
Lifting capacity, kg	-	300
Loading capacity, kg	-	3000
Tyres, front	Mitas 405/70-24	Mitas 405/70-24
Tyres, rear	Mitas 405/70-24	Mitas 400/60-15.5

Harvesting was carried out by a 43-year-old operator, with two years of experience working on harvesters. The forwarder operator was 27 years old and had two years' work experience. He also carried out the manual felling of trees, of which he had six years' experience.

Both the operators were instructed before thinning. Thinnings with midfield (TM1 and TM2) were new to the harvester operator, therefore several trees were felled by chainsaw and later pulled and processed by the harvester in order to learn the process before the research study began.

The length of all the logs was 2.55 m, and two types of assortments were prepared: pulp wood, thicker logs from the bottom parts of trees, and energy wood, thinner (maximum 7 cm over bark) from the tops of trees. The harvested logs were processed at the side of a strip road for further forwarding. Each pile of logs consisted of only one type of assortment. Bottom logs—pulp wood—from a few trees were put in one pile, top logs—energy wood—also from a few trees were disposed of in another pile. This was done in order to boost forwarder efficiency. Consequently, the volume and productivity of each thinning operation was calculated as a mean for each machine. After forwarding, separate piles of timber were made for each operation. The width, height and length of the piles were measured and suitable official conversion factors were used to calculate the timber volume under bark.

The volume of merchantable timber of the standing trees was also not used in the further calculation of productivity curves, as the harvester usually finished processing at a smaller diameter than 7 cm over bark. Therefore, it was decided that the mean harvester and forwarder productivities should be specified, and these were then used for the cost calculations.

All costs linked to machine use were obtained from the machine owner as they were recorded in the company's books (Table 4). The diesel consumption of the harvester and forwarder was measured at the end of each strip road. The tanks were full before the operation started and after each strip road was finished, both the harvester and forwarder were refuelled. The volume of added diesel was measured using a vessel with a volume accuracy of 0.1 L. The chainsaw fuel and oil tanks were refilled when necessary (when the petrol finished), for which a vessel with an accuracy of 0.1 L was used.

Table 4. General data for cost calculations.

Feature	Harvester	Forwarder	Husqvarna 346 XP
Initial investment, €	229,848	125,134	476
Economic life, years	10	10	3
Salvage value (10%), €	22,985	12,513	48
Fuel price, € l^{-1}	1.09	1.09	1.20
Number of working days per year	242	242	242
Number of working hours per day	8	8	8
Machine utilisation time, h $year^{-1}$	1800	1600	-
Machine utilisation time after reduction, h $year^{-1}$	1500	1400	-
Machine utilisation	0.775	0.723	0.813
Monthly:			
Insurance, €	83.12	49.57	-
Garage, €	3.46	3.46	-
Lubricants, €	119.04	17.89	54,56
Repair and spare parts, €	215.51	75.51	14.43
Relocation costs, €	63.90	63.90	-
Labour costs (wages, incl. all costs), €	2311.69	1728.57	1330.00

Stand damage was measured on 10 sample plots in each thinning operation. Stand damage (expressed in %) was measured on the 10 sample plots used in each thinning operation. Data collection was carried out according to a design-based approach and a statistical method in order to overcome possible pseudoreplication problems [6]. The number of trees with damage (partial bark removal) as proposed by Picchio et al. [7], was recorded as the number of trees with damage against all the trees on the sample plots.

The strip road area was calculated per 1 ha. The strip road width multiplied by total length on 1 ha was calculated and cover was expressed in % of area excluded from forest production.

To compare the data sets obtained from each thinning operation, the Mann-Whitney U test was applied with $p < 0.05$ or Kruskal-Wallis test (for more than two independent measures) with $p < 0.05$.

3. Results

3.1. Productivity and Costs

The highest mean thinning operation productivity, 4.83 m^3 PMH^{-1}, was in TM1, where the distance between the strip roads was ca 20 m and one harvester pass was needed (Table 5). In the TM2 and TWM operations, the mean productivity for both machines, harvester and forwarder, was 4.55 and 4.50 m^3 PMH^{-1}, respectively. The harvester was the most efficient in the operation without midfield, where the distance between the strip roads was only ca 10 m. In contrast, forwarder productivity in TWM was the lowest and amounted to 4.33 m^3 PMH^{-1}. Forwarder productivity in the operations with midfield was similar: 5.25 and 5.35 m^3 PMH^{-1} in TM1 and TM2, respectively. Unfortunately, TM1 and TM2 were 13% and 23% more expensive (respectively) than the TWM operation.

The cost of chainsaw use made the operations with midfield more expensive. Fuel consumption (diesel) also had an impact on the lowest cost of harvester use in the operation without midfield, where only 0.46 L per 1 m^3 of harvested timber was used (Table 6).

Table 5. Productivity and cost of thinning operations.

Feature	TM1	TM2	TWM
Harvester productivity, m^3 PMH^{-1}	4.42	3.74	4.67
Forwarder productivity, m^3 PMH^{-1}	5.25	5.35	4.33
Productivity of thinning, m^3 PMH^{-1}	4.83	4.55	4.50
Harvester costs, € m^{-3}	8.83	10.34	7.89
Forwarder costs, € m^{-3}	4.80	4.69	5.84
Chainsaw felling costs, € m^{-3}	1.88	1.88	-
Costs of thinning, € m^{-3}	15.51	16.91	13.73

Table 6. Fuel consumption by Vimek 404 T5 harvester, Vimek 606 TT forwarder, Husqvarna 346 XP.

Feature	TM1	TM2	TWM
Harvester diesel consumption, L m^{-3}	0.94	1.02	0.46
Forwarder diesel consumption, L m^{-3}	0.31	0.30	0.40
Chainsaw petrol consumption, L m^{-3}	0.11	0.08	-

Diesel consumption by the forwarder was much lower in the operations with midfield in comparison with the harvester. It was also recorded that fuel use in the forwarding without midfield was ca 33% bigger than in the operations with midfield. Petrol consumption by chainsaw was low, ca 0.1 L per m^3 of felled trees, as delimbing and bucking were carried out by the harvester.

3.2. Harvesting Time Consumption

Considering effective time without delays, more time was needed for one tree in operations with midfield: 43 and 48 s in TM1 and TM2 respectively. The shortest time was observed in TWM: 38 s per tree (Table 7).

Table 7. Harvester time consumption per one tree; results of Mann-Whitney U test and descriptive statistics of time categories.

Feature	TM1		TM2		TWM	
Time category	A	B	A	B	A	B
Mean (s)	26^{a}	17^{a}	30^{b}	$18^{a,b}$	23^{a}	15^{b}
Min. (s)	7	2	6	5	9	3
Max. (s)	72	81	77	92	57	69
SD	13	12	14	13	10	10
N	94	94	97	97	104	105

Different letters next to mean values indicate statistically significant differences. TM1—thinning operation with midfield and one harvester pass; TM2—thinning operation with midfield and two harvester passes; TWM—thinning without midfield. Time categories: A: Preparation (tidying residues), machine moving, crane out, head positioning, felling; B: Delimbing and bucking.

Processing time (B: Delimbing and bucking) was similar in all operations, although there was a statistically significant difference between the trees processed in TM1 and TWM. There was also the expected difference between the operations in time of tree preparation (A): More time was needed in operations with midfield in comparison with TWM (Table 7).

Tree chainsaw felling was more time consuming in the operation with one harvester pass, where 43 s were needed for one tree. In TM2, trees were felled after the first harvester pass, and only 24 s per tree was needed on average (Table 8). The time spent by the chainsaw operator moving from one tree to another was twice as long in TM1 as in TM2. These differences were statistically significant. The mean time of felling per tree was 64% longer in TM1 than in TM2 (Table 8), and that difference was also statistically significant.

Table 8. Chainsaw operator time consumption per tree; Mann-Whitney U test and descriptive statistics of time categories.

Feature	Chainsaw Operator, TM1			Chainsaw Operator, TM2		
Time Category	A, s	B, s	C, min.s	A, s	B, s	C, min.s
Mean	20 [a]	23 [a]	20.11	10 [b]	14 [b]	18.12
Minimum	3	4	13.41	3	3	12.08
Maximum	60	70	26.40	64	67	24.15
Standard deviation	15	16	-	10	12	-
N	43	43	2	50	50	2

Different letters next to mean values indicate statistically significant differences. TM1—thinning operation with midfield and one harvester pass; TM2—thinning operation with midfield and two harvester passes. Time categories: A: preparation (tidying residues), machine moving, crane out, head positioning, felling; B: delimbing and bucking; C: delays.

Delay times occurred at regular intervals and they were linked with chainsaw refuelling. Each break was ca 20 min and there were two breaks in each operation with midfield.

3.3. Damage to the Remaining Stand

Recorded damage was very low. On the 10 sample plots in each operation, approximately 2% of the trees had partial bark removal (Table 9). The least frequent damage, 1.5% of trees, was observed in TWM, where a chainsaw was not used and the trees were not pulled from midfield by a harvester.

Table 9. Damage to remaining stand.

Feature	TM1	TM2	TWM
Number of trees with damage	3	3	2
Total number of trees	142	138	130
Share of trees with damage, %	2.1	2.2	1.5

A similar level of damage (2%) was found in the thinning operations with midfield. Damage was observed only on a few sample plots.

3.4. Strip Road Area

As initially designed, TM1 and TM2 had a thinner network of strip roads, the distance between them being 20 m. In such a configuration, the area needed for strip roads only amounted to 15% (Table 10). Twice as many strip roads were needed when TWM was proposed, and the total area excluded from production was 30%.

Table 10. Strip road area, calculations per one hectare.

Strip Road Detail	TM1	TM2	TWM
Number of strip roads	5	5	10
Width, m	3	3	3
Distance between strip roads, m	20	20	10
Total length, m	500	500	1000
Area, %	15	15	30

4. Discussion

Obtaining satisfactory thinning operation production rates in young stands is difficult. Small tree sizes have the biggest impact on lower productivity [8,9], which also leads to higher costs for the whole operation [5]. In the presented research, the configuration of the strip road network was considered, which is mainly dependent on machine parameters and harvester crane reach. Smaller machines are expected to be used in younger stands, where thinning is necessary [10]. Smaller machines are also cheaper and cost less to use due to their lower fuel consumption [5]. The option presented of the Vimek harvester and forwarder was a good economic solution for small-size tree harvesting.

The economic assessment of the three different thinning operations showed that the cheapest option required a dense network of strip roads in the stand. However, the cheapest option contrasts with the long-term economic benefits from stand utilisation. Excluding 30% of the land from further stand production is another cost to be borne by the forest owner.

Thinning with midfield is more difficult to manage. The chainsaw operator has to control the width of the midfield zone between two strip roads, which creates delays in tree felling. Additionally, the trees need to be felled in the direction of the nearest strip road. To avoid delays due to chainsaw operator hesitations, a thinning operation with two harvester passes was proposed. In TM2, after the first harvester pass and the felling of the trees next to the strip road, the midfield zone was visible for the chainsaw operator, and it was also easier to fell trees into the area next to the strip road, where thinning had already been carried out. However, both harvester passes cost 23% more than the cheapest option, TWM.

Timber harvesting with midfield is a known solution, described earlier in pine stands by Mederski [5]. In that study from the previous decade, a midfield was used in older stands of 3rd and 4th age classes, and thinning with midfield was the most economical. The productivities achieved at that time, more than a decade ago, were generally lower when compared to the harvesting nowadays, where in thinning operations approximately 20 m^3 PMH^{-1} is achieved [8,9]. In this context of highly productive thinning operations, the use of a chainsaw is an unwanted necessity. The proposed thinning operations with midfield make it possible to keep a standard distance between the strip roads thanks to the additional use of the chainsaw. However, it would be more convenient if small harvesters were developed with longer cranes for early thinnings.

A forest owner is interested in the best possible land use for production, and from the three different harvester operations presented, the one with midfield (TM1) would be in their best interest. In TM1, only 15% of the area was used for strip roads. The cost of the whole TM1 operation was 13% higher, although more trees (15% extra forest area) were saved in the stand for future production. If this slightly higher cost of the thinning operation can be accepted, strip roads of regular density (20 m apart) will stay for the next thinning, when in the future a bigger harvester with a 10 m-long crane will probably be used for late thinning.

In general, the productivities and costs achieved for these small machines should be considered satisfactory and similar to those achieved in other studies [11,12]. It was expected that, in particular, the forwarder productivity would be higher in the midfield operations due to a higher timber concentration [13] in comparison with TWM. Higher productivities in thinnings with midfield were achieved in previous studies using a Vimek forwarder [5], but these were in older pine stands. Bigger trees positively impact the productivities of both harvesters and forwarders [8,9,13,14], although fork trees [15], tree sizes which are too large [16], or bigger slopes [17] lower the productivity of a harvester considerably.

A wider distance between strip roads allows for a bigger concentration of timber, which impacts positively on forwarding efficiency [5,13]. It was also expected that harvester productivity would be the highest in TWM: in this operation, the time for tree pulling from the stand was excluded, which made harvesting more efficient. However, it has to be pointed out that thinning with midfield was provided as an experimental case study and the harvester operator did not have experience in that type of operation (except a short training session on a few trees). Therefore, it may be expected that there is still potential to obtain higher productivities in this operation.

Harvester time consumption was the shortest when TWM was applied. This was expected as the time-consuming process of pulling trees from midfield was only in TM1 and TM2. Considering the processing time (delimbing and bucking), it was similar in all operations, although there was a statistically significant difference between TM2 and TWM. This was rather accidental, as there was no reason why those times should be different. In contrast, the time of preparation (A) was the shortest in

TWM, which was expected as extra time was needed for tree pulling in the thinning with midfield. Additionally, the difference in time A was statistically significant when comparing TM2 and TWM (but again not in a comparison between TM1 and TWM).

Chainsaw felling time per tree was much shorter in TM2 due to the open space next to the strip roads after the first harvester pass (fewer hanging trees). Additionally, the time spent by the chainsaw operator moving from one tree to another (time A) was shorter in TM2. This was also expected as after the first harvester pass, the chainsaw operator had a clear view of which trees belonged to the midfield area.

An important aspect of the presented research is that the different thinning options of the Vimek harvester and forwarder are proposed with economic and environmental consequences—the cost and stand preparation with strip roads. Indeed, the cheapest harvesting option in this case was not the best one, because in comparison with TM1, a further 15% of forest area (another tree row at this stage of stand development) needed to be excluded from production to create strip roads (Table 8). The loss in the future production of wood in that removed tree row is in fact much more expensive than the saved ca €2 in one thinning operation. Strip roads every 10 m in the stand would also be difficult to accept in relation to ecological aspects. It is also worth mentioning that in the next thinning (older stand), a bigger harvester may be used with a 10-m-long crane and strip roads every 10 m may not be necessary any longer.

When a mechanised thinning operation is used, the network of strip roads has a direct impact on soil compaction. TWM had double the area used for strip roads (every 10 m) in comparison with TM1 or TM2 (every 20 m). Consequently, a bigger area in TWM (30% of the stand area) was under the pressure of machine wheels. This is more than the area affected by the skidder, where ca. 18% of the stand area can be disturbed [18], unless winching is used with a tractor in a coppice stand, which may affect a very small area of the stand (3–4%) [19].

Soil compaction caused by a forwarder may be more severe [20], which may lead to a slower early growth of seedlings [21]. Forest machines, and especially fully-loaded forwarders, cause unwanted soil disturbances [22], which is usually more intensive in older stands, where trees of bigger dimensions are extracted [23].

Damage to the remaining stand was very low—only 1 to 2% of the trees were injured. This was comparable to the low-level damage characteristic for winching (especially with a snatch block) [24]. Otherwise, when the short wood system is used, a higher level of damage can be observed [25,26]. It was expected that in the thinning operations with midfield (with a larger distance between the strip roads), the level of damage would be higher, and it was. However, overall, only a few trees with bark damage were found on the sample plots. In operations with midfield, trees cut with a chainsaw have to be pulled afterwards by a harvester for processing—pulling trees creates additional opportunities for the remaining stands to be damaged. However, in 31-year-old pine stands, trees are of small sizes (Table 1), and even though they are pulled between other trees, it does not necessarily mean that damage will be done. The extraction of bigger trees in older stands may lead to a higher probability of damage to the remaining trees, but not necessarily in the short wood system [27]. In the presented study, only a few trees had damage, and besides, there is still the chance of the healing and overgrowth of wounds [28].

An interesting aspect of the presented research was fuel consumption. When the harvester was used in operations with midfield, a greater use of diesel was recorded. This was a natural consequence of the additional work—pulling the felled trees for processing. At the same time, a higher diesel consumption was noted when the forwarding was carried out in TWM. Again, the smaller timber concentration next to the strip road led to lower forwarder productivity and eventually to higher fuel use.

Considering more aspects (apart from the economic one) in thinning operations is vital in order to select the right forest operation method. Current trends show that combining economic and ecological issues is not enough, and more aspects should be considered: ergonomic, social and product quality [29]. This is very important considering the recent dynamic growth in mechanised forest operations [30–32].

5. Conclusions

The use of the Vimek 404 harvester and Vimek 606 TT forwarder in the 31-year-old pine stand had satisfactory productivity and acceptable costs. Three different thinning operations were proposed for these machines requiring different networks of strip roads. Taking into consideration all the aspects presented here (costs, damage to the remaining stand and density of strip roads), the best solution for the forest owner is to use the thinning operation with midfield and one harvester pass. Although this operation may require higher costs of ca 13% for the whole operation, a reasonable strip road density will exist in the stand. In addition, with this operation, 15% of the stand area was saved for timber production for the next decades. Overall, the level of damage was very low in all cases. However, when thinning with midfield was applied, about 2% of the trees were damaged.

Author Contributions: P.S.M., M.B. and Z.K. conceived and designed the experiments; P.S.M., M.S. and M.R. performed the experiments; M.S. and P.S.M. analysed the data; P.S.M. wrote the paper, R.V., Z.P. and I.L. contributed to the writing of the paper.

Funding: The research was completed with financial support from Ministry of Science and Higher Education, granted to the Department of Forest Utilisation (PULS) for the year 2015.

Acknowledgments: We would like to thank Adam Potocki, the manager of Bobolice Forest District for support during the research and for access to the selected stand. Thank you to Waldemar Winkowski, forest entrepreneur, who put his machines at our disposal, and allowed us access to his financial records. Finally, we would like to thank Fabian Winkowski, harvester operator, and Przemysław Kowalczyk, forwarder and chainsaw operator, for their good cooperation in completing the thinning operations.

Conflicts of Interest: The authors declare no conflict of interest. The founding sponsors had no role in the design of the study; in the collection, analyses, or interpretation of data; in the writing of the manuscript, and in the decision to publish the results.

References

1. Emer, B.; Grigolato, S.; Lubello, D.; Cavalli, R. Comparison of biomass feedstock supply and demand in Northeast Italy. *Biomass Bioenergy.* **2011**, *35*, 3309–3317. [CrossRef]
2. Marchi, M.; Paletto, A.; Cantiani, P.; Bianchetto, E.; De Meo, I. Comparing thinning systems effects on ecosystem services provision in black pine (*Pinus nigra* J.F. Arnold) artificial forests. *Forests* **2018**, *9*, 188. [CrossRef]
3. Marchi, E.; Picchio, R.; Spinelli, R.; Verani, S.; Venanzi, R.; Certini, G. Environmental impact assessment of different logging methods in pine forests thinning. *Ecol. Eng.* **2014**, *70*, 429–436. [CrossRef]
4. Picchio, R.; Neri, F.; Petrini, E.; Verani, S.; Marchi, E.; Certini, G. Machinery-induced soil compaction in thinning two pine stands in central Italy. *For. Ecol. Manag.* **2012**, *285*, 38–43. [CrossRef]
5. Mederski, P.S. A comparison of harvesting productivity and costs in thinning operations with and without midfield. *For. Ecol. Manag.* **2006**, *224*, 286–296. [CrossRef]
6. Hurlbert, S.H. Pseudoreplication and the design of ecological field experiments. *Ecol. Monogr.* **1984**, *54*, 187–211. [CrossRef]
7. Picchio, R.; Neri, F.; Maesano, M.; Savelli, S.; Sirna, A.; Blasi, S.; Baldini, S.; Marchi, E. Growth effects of thinning damage in a Corsican pine (*Pinus laricio* Poiret) stand in central Italy. *For. Ecol. Manag.* **2011**, *262*, 237–243. [CrossRef]
8. Mederski, P.S.; Bembenek, M.; Karaszewski, Z.; Łacka, A.; Szczepańska-Álvarez, A.; Rosińska, M. Estimating and modelling harvester productivity in pine stands of different ages, densities and thinning intensities. *Croat. J. For. Eng.* **2016**, *37*, 27–36.
9. Labelle, E.R.; Bergen, M.; Windisch, J. The effect of quality bucking and automatic bucking on harvesting productivity and product recovery in a pine-dominated stand. *Eur. J. For. Res.* **2017**, *136*, 639–652. [CrossRef]
10. Proto, A.R.; Macrì, G.; Visser, R.; Harrill, H.; Russo, D.; Zimbalatti, G. Factors affecting forwarder productivity. *Eur. J. For. Res.* **2018**, *137*, 143–151. [CrossRef]
11. Spinelli, R.; Magagnotti, N. Performance and cost of a new mini-forwarder for use in thinning operations. *J. For. Res.* **2010**, *15*, 358–364. [CrossRef]
12. Lazdiņš, A.; Prindulis, U.; Kalēja, S.; Daugaviete, M.; Zimelis, A. Productivity of Vimek 404 T5 harvester and Vimek 610 forwarder in early thinning. *Agron. Res* **2016**, *14*, 475–484.

13. Manner, J.; Nordfjell, T.; Lindroos, O. Effects of the number of assortments and log concentration on time consumption for forwarding. *Silva Fenn.* **2013**, *47*. [CrossRef]
14. Spinelli, R.; Cacot, E.; Michelic, M.; Nestorovski, L.; Mederski, P.; Tolosana, E. Techniques and productivity of coppice harvesting operations in Europe: A meta-analysis of available data. *Ann. For. Sci.* **2016**, *73*, 1125–1139. [CrossRef]
15. Labelle, E.R.; Soucy, M.; Cyr, A.; Pelletier, G. Effect of tree form on the productivity of a cut-to-length harvester in a hardwood dominated stand. *Croat. J. For. Eng.* **2016**, *37*, 175–183.
16. Visser, R.; Spinelli, R. Determining the shape of the productivity function for mechanized felling and felling-processing. *J. For. Res.* **2012**, *17*, 397–402. [CrossRef]
17. Ackerman, P.; Martin, C.; Brewer, J.; Ackerman, S. Effect of slope on productivity and cost of Eucalyptus pulpwood harvesting using single-grip purpose-built and excavator-based harvesters. *Int. J. For. Eng.* **2018**, *30*, 1–9. [CrossRef]
18. Kulak, D.; Barszcz, A. The impact of selected skidding techniques on soil damage in beech-fir stands. *Sylwan* **2008**, *12*, 20–28.
19. Marchi, E.; Picchio, R.; Mederski, P.S.; Vusić, D.; Perugini, M.; Venanzi, R. Impact of silvicultural treatment and forest operation on soil and regeneration in Mediterranean Turkey oak (*Quercus cerris* L.) coppice with standards. *Ecol. Eng.* **2016**, *95*, 475–484. [CrossRef]
20. Cambi, M.; Giannetti, F.; Bottalico, F.; Travaglini, D.; Nordfjell, T.; Chirici, G.; Marchi, E. Estimating machine impact on strip roads via close-range photogrammetry and soil, parameters: A case study in central Italy. *iForest* **2018**, *11*, 148–154. [CrossRef]
21. Cambi, M.; Mariotti, B.; Fabiano, F.; Maltoni, A.; Tani, A.; Foderi, C.; Laschi, A.; Marchi, E. Early response of *Quercus robur* seedlings to soil compaction following germination. *Land Degrad. Dev.* **2018**, *29*, 916–925. [CrossRef]
22. Kulak, D.; Stańczykiewicz, A.; Szewczyk, G.; Lubera, A.; Strojny, T. Factors affecting the changes in penetration resistance of forest soils during timber harvesting. *Sylwan* **2015**, *159*, 318–325.
23. Sowa, J.; Kulak, D. Probability of occurrence of soil disturbances during timber harvesting. *Croat. J. For. Eng.* **2008**, *29*, 29–39.
24. Picchio, R.; Magagnotti, N.; Sirna, A.; Spinelli, R. Improved winching technique to reduce logging damage. *Ecol. Eng.* **2012**, *47*, 83–86. [CrossRef]
25. Bembenek, M.; Giefing, D.F.; Karaszewski, Z.; Mederski, P.S.; Szczepańska-Álvarez, A. Tree damage in lowland spruce stands caused by early thinnings. *Sylwan* **2013**, *157*, 747–753.
26. Bembenek, M.; Giefing, D.F.; Karaszewski, Z.; Mederski, P.S.; Szczepańska-Álvarez, A. Tree damage in lowland spruce stands because of late thinning. *Sylwan* **2013**, *157*, 892–898.
27. Bakinowska, E.; Mederski, P.S.; Szczepańska-Álvarez, A.; Karaszewski, Z.; Bembenek, M. The parallel application of two probability models, logit and probit, for the accurate analysis of spruce timber damage due to thinning operations. *Drewno* **2016**, *59*, 49–59.
28. Tavankar, F.; Picchio, R.; Nikooy, M.; Lo Monaco, A.; Venanzi, R.; Iranparast Bodaghi, A. Healing rate of logging wounds on broadleaf trees in Hyrcanian forest with some technological implications. *Drewno* **2017**, *60*, 65–80.
29. Marchi, E.; Chung, W.; Visser, R.; Abbas, D.; Nordfjell, T.; Mederski, P.S.; McEwan, A.; Brink, M.; Laschi, A. Sustainable Forest Operations (SFO): A new paradigm in a changing world and climate. *Sci. Total Environ.* **2018**, *634*, 1385–1397. [CrossRef] [PubMed]
30. Mederski, P.S.; Karaszewski, Z.; Rosińska, M.; Bembenek, M. Dynamics of harvester fleet change in Poland and factors determining machine occurrence. *Sylwan* **2016**, *160*, 795–804.
31. Moskalik, T.; Borz, S.A.; Dvořák, J.; Ferencik, M.; Glushkov, S.; Muiste, P.; Lazdiņš, A.; Styranivsky, O. Timber harvesting methods in Eastern European countries: A review. *Croat. J. For. Eng.* **2017**, *38*, 231–241.
32. Spinelli, R.; Magagnotti, N.; Schweier, J. Trends and perspectives in coppice harvesting. *Croat. J. For. Eng.* **2017**, *38*, 219–230.

forests

MDPI

Article

Species Identification of *Pinus* Pollen Found in Belukha Glacier, Russian Altai Mountains, Using a Whole-Genome Amplification Method

Fumio Nakazawa [1,2,*], Yoshihisa Suyama [3], Satoshi Imura [1,2] and Hideaki Motoyama [1,2]

1. National Institute of Polar Research, 10-3 Midori-cho, Tachikawa, Tokyo 190-8518, Japan; imura@nipr.ac.jp (S.I.); motoyama@nipr.ac.jp (H.M.)
2. SOKENDAI (The Graduate University of Advanced Studies), Shonan Village, Hayama, Kanagawa 240-0193, Japan
3. Field Science Center, Graduate School of Agricultural Science, Tohoku University, 232-3 Yomogida, Naruko-onsen, Osaki, Miyagi 989-6711, Japan; suyama@tohoku.ac.jp

* Correspondence: nakazawa@nipr.ac.jp; Tel.: +81-42-512-0761

Received: 20 April 2018; Accepted: 28 June 2018; Published: 24 July 2018

Abstract: Pollen taxa in sediment samples can be identified based on morphology. However, closely related species do not differ substantially in pollen morphology, and accurate identification is generally limited to genera or families. Because many pollen grains in glaciers contain protoplasm, genetic information obtained from pollen grains should enable the identification of plant taxa at the species level. In the present study, species identification of *Pinus* pollen grains was attempted using whole-genome amplification (WGA). We used pollen grains extracted from surface snow (depth, 1.8–1.9 m) from the Belukha glacier in the summer of 2003. WGA was performed using a single pollen grain. Some regions of the chloroplast genome were amplified by PCR, and the DNA products were sequenced to identify the pollen grain. *Pinus* includes approximately 111 recognized species in two subgenera, four sections, and 11 subsections. The tree species *Pinus sibirica* and *P. sylvestris* are currently found at the periphery of the glacier. We identified the pollen grains from the Belukha glacier to the level of section or subsection to which *P. sibirica* and *P. sylvestris* belong. Moreover, we specifically identified two pollen grains as *P. sibirica* or *P. cembra*. Fifteen species, including *P. sibirica*, were candidates for the remaining pollen grain.

Keywords: pollen, DNA; glacier; *Pinus*; Altai; pollen source

1. Introduction

The natural range of the genus *Pinus* is confined to the Northern Hemisphere, although some species have been introduced to the Southern Hemisphere. The genus is currently a dominant component in forests [1]. The pollen grains of *Pinus* have two sacs and the winged pollen grains can be transported long distances by wind. In fact, *Pinus* pollen grains can reach remote areas, such as mountain glaciers in the Northern Hemisphere, Arctic glaciers, and the Greenland ice sheet, and are found in snow and ice as a predominant pollen type [2–9]. Many pollen grains in glaciers are expected to contain protoplasm, and their maintenance at below 0 °C is favorable for DNA preservation [10]. This property is characteristic of glacial pollen; protoplasm is rarely seen in pollen found in other sediment types, such as peat and lacustrine deposits. Therefore, pollen grains in glaciers are advantageous for obtaining genetic information, which should enable identification to the species level. Modern pollen identification focuses on the morphological characteristics of the pollen wall, but this approach is generally limited to the identification of plant genera or families. In the case of *Pinus*, identification to the genus level is typically possible, although the haploxylon and diploxylon types are sometimes distinguished for *Pinus* pollen grains on the basis of vesicle morphology and other characters. Hence, alternative techniques are needed for species identification, such as DNA analyses of pollen grains. If *Pinus* pollen grains in glaciers are identified to the species level, it may be possible to investigate the provenance and transportation routes from source plants. Genetic data also provide valuable information related to physiological ecology, gene flow, and population dynamics [11,12]. In addition, if genetic analysis is applied to pollen from ice cores, which are cylindrical samples of ice drilled from glaciers and are used to reconstruct past climate conditions and the environmental history of a particular area, then the abovementioned provenance and ecological studies can be extended to trace back into the past.

Nakazawa et al. [10] analyzed the DNA of *Pinus* pollen grains collected from subsurface snow layers on the Belukha glacier in the Altai Mountains of Russia in the summer of 2003. They identified *Pinus* pollen grains to the section level using a PCR-based method. However, it is difficult to achieve species-level identification using this approach. The sequences provide limited information owing to their short lengths, meaning that there is the potential to obtain more DNA information by improving on PCR-based methods.

To identify the grains at the species level, an optimized whole-genome amplification (WGA) approach combined with multiplex PCR was developed in this study. Multiplex PCR can be used to amplify multiple targets in a single PCR experiment. Additionally, DNA barcoding is used to identify materials from known species based on short DNA sequences of standard genomic regions (i.e., DNA barcodes). In general, chloroplast DNA in land plants has a low nucleotide substitution rate, on the order of 10^{-9} per site per year [13]. Therefore, few mutations are expected within a short period of time, and the most promising DNA barcoding loci for plants are chloroplast genes. DNA barcoding technologies are being developed for applications in palynology (pollen DNA barcoding) [14], and these studies have demonstrated that both chloroplast and nuclear barcoding markers can be amplified from pollen. Unfortunately, most plastid candidate barcodes lack species-level resolution [15]. Additionally, DNA barcoding markers with universal primers used in previous pollen DNA barcoding studies [14] provide insufficient information on the species-level taxonomy of *Pinus*. Species-specific primers are generally used for the precise identification of samples [16,17]. Moreover, PCR amplification using limited amounts of DNA template, such as from a single pollen grain, has high risk of contamination, biased amplification, and product redundancy [18,19]. Thus, in this study, we designed primers for species-specific DNA barcoding with high resolution at low taxonomic levels to reduce these risks and increase precision.

2. Materials and Methods

2.1. Study Area and Pollen Samples

The Belukha glacier (49°49′ N, 86°34′ E; 4,110 m a.s.l.) is located on the western side of Mt. Belukha (4500 m a.s.l.) in the Russian Altai Mountains and is situated in the border region between Russia, Mongolia, China, and Kazakhstan (Figure 1). In the summer of 2003, a 4-m-deep pit on the plateau of the glacier was examined (4100 m a.s.l.) [20]. *Pinus* pollen grains in a snow sample were obtained at a depth of 1.8–1.9 m in the pit and used for the DNA analysis. The *Pinus* pollen concentration in the sample was 34,900 grains L^{-1}. The sample was dated to the summer of 2002 by counting the seasonal distribution of pollen [20]. The sample was obtained from the same pit as in our previous study [10], but the pollen grains were previously collected from a depth of 0.4–0.5 m and dated to the summer of 2003. The sample was kept in a frozen state until it was analyzed.

Pinus species surrounding the Belukha glacier are *P. sibirica*, which is distributed between approximately 1000 and 2000 m a.s.l., and *P. sylvestris*, which typically occurs below the *P. sibirica* stands [21]. For a detailed description of the vegetation, see Nakazawa et al. [10].

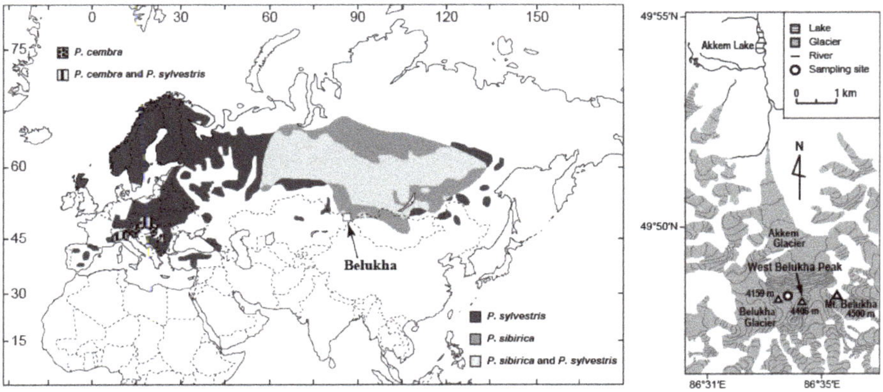

Figure 1. Location of the Belukha glacier in Russia's Altai Republic, and distributions of extant *P. cembra* and two *Pinus* species (*P. sibirica* and *P. sylvestris*) found in the region surrounding the glacier. The sampling site was located on the western side of Mt. Belukha. The distribution map was compiled based on the maps of Farjon [22]. The figure is from our previous study [10].

2.2. DNA Extraction from a Single Pollen Grain

DNA was extracted using a modified version of the extraction method described by Nakazawa et al. [10]. A flow chart of the experimental procedure is shown in Figure 2. Melted snow and ice samples were first filtered through a hydrophilic PTFE membrane filter with a pore size of 10 μm. Next, pollen grains that showed no structural damage were selected from the filter using a micromanipulator (MM-88; Narishige, Tokyo, Japan) under a microscope. Each selected pollen grain was placed onto a new hydrophilic PTFE membrane filter with a pore size of 5 μm, and was then washed by suction filtration with 1 mL of nuclease-free water (Ambion Life Technologies, Foster City, CA, USA). The filter trapping a pollen grain was transferred to a sterile Petri dish. The washed grain was then transferred to the inner side of the lid of a DNA-free PCR tube containing 0.5 μL of water using a pipette.

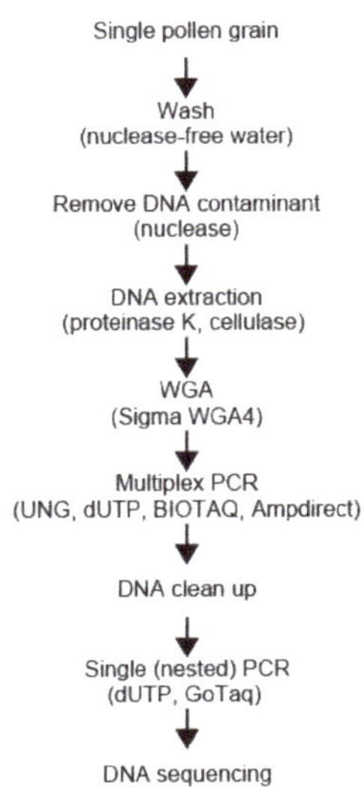

Figure 2. Flow chart of the experimental procedure.

The pollen grain in the lid was treated with endonuclease and exonuclease to eliminate potential contaminants from DNA fragments attached to the surface of the grain. One microliter of reaction mixture containing 0.48 μL of water, 0.2 μL of 10× Exonuclease III Buffer (TaKaRa Biotechnology Co., Ltd., Dalian, Japan), 0.1 μL of Exonuclease I (20 U/μL, Epicentre, Madison, WI, USA), 0.02 μL of Exonuclease III (200 U/μL, TaKaRa Biotechnology), and 0.2 μL of DNase I (1 U/μL, Sigma-Aldrich Co., St. Louis, MO, USA) was added to the lid. The mixture was incubated at 37 °C for 3–5 h, then at 98 °C for 10 min.

The treated grain was crushed directly in the lid of the tube using a sterile plastic pipette tip. For crushing, we used an electric toothbrush with an attached tip-rounded pipette tip, which was made by heating with a gas burner in advance. The vibration of the electronic toothbrush facilitated crushing. One microliter of extraction mixture containing 0.7 μL of Tris-HCl (pH 8.0, 20 mM, Nacalai Tesque, Kyoto, Japan), 0.2 μL of proteinase K (1 μg/μL, TaKaRa Biotechnology), and 0.1 μL of cellulase (Sigma-Aldrich) was added to the lid of each sample and spun down for collection at the bottom of the tube. The mixture was incubated at 50 °C for 6 h, at 95 °C for 10 min, and then used as a template.

2.3. Whole-Genome Amplification

WGA from a pollen grain was performed using the GenomePlex Single Cell Whole Genome Amplification Kit (WGA4; Sigma-Aldrich) according to the manufacturer's instructions, with slight modifications. After the lysis procedure, 1.3 μL of fragmentation solution, including 0.3 μL of fragmentation buffer and 1 μL of water, was added and heated in a thermal cycler (Bioer Technology Co. Ltd., Hangzhou, China) at 99 °C for 3 min. The PCR tubes were then cooled on a cooling rack (Nippon Genetics Co., Ltd., Tokyo, Japan). For library preparation, 1 μL of the reaction solution including 0.7 μL of Library Preparation Buffer and 0.3 μL of Library Stabilization Solution was added to each sample and was placed in the thermal cycler at 95 °C for 2 min. The samples were cooled on the cooling rack. Next, 1.0 μL of enzyme solution including 0.7 μL of water and 0.3 μL of the Library Preparation Enzyme solution was added to each sample. The samples were placed in the thermal cycler. The reaction time and temperature were based on the instructions provided.

For the amplification, 14.6 μL of the reaction mixture, including 11.0 μL of water, 2.0 μL of 10× Amplification Master Mix, 1.3 μL of WGA DNA Polymerase, 0.2 μL of uracil-*N*-glycosylase (UNG; TaKaRa Biotechnology), and 0.1 μL of BIOTAQ HS DNA polymerase (Bioline, London, UK), was added to the sample. UNG was used to degrade uracil-containing PCR contaminants from previous PCR prior to the amplification reaction. This is explained in more detail in the next section. Note that the PCR products included dUTP, instead of dTTP. Therefore, the treatment with UNG should allow the selective removal of carryover PCR products. Samples were first incubated at 25 °C for 10 min, and then amplified using an initial denaturation of 95 °C for 10 min followed by 40 cycles each consisting of a denaturation step at 94 °C for 30 s, an annealing step at 52 °C for 1 min, and an extension step at 72 °C for 1 min, and a final extension at 72 °C for 7 min. The WGA DNA in the reaction mixture was stored at −20 °C until further use, without a DNA purification step.

2.4. UNG Treatment and Multiplex PCR Amplification

The quality of the WGA amplification was evaluated using a multiplex PCR approach. The DNA specimens generated by WGA were subjected to various analyses for chloroplast DNA. Instead of single PCR, a multiplex PCR step was introduced in this study to use the specimens effectively. To preclude carryover contamination of amplification products from previous PCR, the reaction with UNG, an enzyme that degrades uracil-containing DNA, was carried out. Each multiplex PCR amplification was performed with dUTP instead of dTTP. Thus, the PCR products should be obtained from only thymine-containing templates amplified by WGA.

For the multiplex PCR assay, 2–7 primer pairs were designed. PCR was carried out with a 10-μL reaction mixture containing 1 μL of template WGA products. The primers are listed in Table 1. To identify samples at the section or subsection level, a total of 10 primer pairs were used. Seven or 14 primer pairs were selected to narrow the candidates within each subsection. To evaluate the primer sets, the performance of primers was examined using single pollen grains of *P. resinosa* belonging to subsection *Pinus* and needles collected from *P. pumila*, *P. strobus*, *P. taeda*, *P. jeffreyi*, and *P. monophylla*, which belong to subsections *Strobus*, *Strobus*, *Australes*, *Ponderosae* and *Cembroides*, and *P. resinosa*, respectively. From the results, some primers were screened out. The primers presented in Table 1 were effective for sequencing. In addition, the number of cycles of multiplex PCR in this study was minimized to avoid the introduction of significant PCR bias.

Nine microliters of the reaction mixture for UNG and multiplex PCR, containing 1.75 μL of water, 5.1 μL of 5× PCR buffer, 2.0 μL of a mix of primers (2.5 μM each primer), 0.05 μL of BIOTAQ HS DNA polymerase, and 0.1 μL of UNG, was added to a PCR tube. The 5× PCR buffer was prepared by mixing 2 μL of 5× Ampdirect-D (Shimadzu Biotech, Kyoto, Japan), 2 μL of 5× AmpAddition-3 (Shimadzu Biotech, Kyoto, Japan), 0.3 μL of 25 mM $MgCl_2$, and 0.8 μL of dU plus dNTP Mixture (TaKaRa Biotechnology).

Table 1. Primers for the target fragments used in the present study. The primers for nested PCR for Pns1–7 are not shown because the multiplex PCR products showed no visible band by agarose gel electrophoresis, and an additional round of nested PCR was not performed.

Mixture for Multiplex PCR	Primer ID	Region	Primers for Multiplex PCR	Sequence (5′-3′)	Product Size (bp)	Primers for Nested PCR	Sequence (5′-3′)	Product Size (bp)
Mixture 1 (Matsu1-Matsu5)	Matsu1	*clpP*	Matsu m1F	CAACTTGGGTCGACTTATACAACC	185	Matsu n1F	CGACTTATACAACCGACTTTATCG	174
			Matsu m1R	ACCTACCCCTGGTATTACTGATCC		Matsu n1R	CCTACCCCTGGTATTACTGATCC	
	Matsu2	*rpoA*	Matsu m2F	CTGGGTCCAACAATATAAATAGAAGC	174	Matsu n2F	CAATATAAATAGAAGCTTCTCGGATTCC	164
			Matsu m2R	AGTAGAAGGAACATGTATCACACG		Matsu n2R	same as Matsu m2R	
	Matsu3	*atpB*	Matsu m3F	GGAGAACCTGTCGATAATTTGGG	130	Matsu n3F	CCTGTCGATAATTTGGGTCCTG	115
			Matsu m3R	GATCTACTACTTTAATGCCTGTTTCG		Matsu n3R	CTTTAATGCCTGTTTCGAAGATGG	
	Matsu4	*rpoC1*	Matsu m4F	GCCTAGTAAATTTTCACGAAATCTTCCC	134	Matsu n4F	TTCCCTCTTTGCCTTCGATCAC	111
			Matsu m4R	AGGAAGCCGTAGATGCACTTC		Matsu n4R	same as Matsu m4R	
	Matsu5	*ycf3*	Matsu m5F	CTGGAGATAGAACAATTCCTTCTGTC	143	Matsu n5F	GTATCCCCGGTCAATGCAC	115
			Matsu m5R	CCTATCAAATAGGTTCAACTATACAAGC		Matsu n5R	same as Matsu m5R	
Mixture 2 (Matsu6-Matsu10)	Matsu6	*ycf4*	Matsu m6F	GGTTCACATTATCACCAGTACGAG	160	Matsu n6F	CACATTATCACCAGTACGAGTTAAAGG	157
			Matsu m6R	CCCGGAATAAATCGTCGTATTTTC		Matsu n6R	same as Matsu m6R	
	Matsu7	*rpoB*	Matsu m7F	ACTCCAGAATGGCTTTTTCC	133	Matsu n7F	AATGGCTTTTTCCCTCGAC	109
			Matsu m7R	TCAGSTATGGGTTTAAATCTCG		Matsu n7R	TCTCGAAGAGATTCTGGACAATAC	
	Matsu8	*rpoC1*	Matsu m8F	AGGCATAAGACCATCCATTTTGG	186	Matsu n8F	CCATCCATTTTGGTTCTATATTTGTTCGG	177
			Matsu m8R	ACATATGGAATGGAAGAACTTGGTG		Matsu n8R	CACATATGGAATGGAAGAACTTGGTG	
	Matsu9	*ycf3*	Matsu m9F	GATTAATCCCCGGAGAATACAGG	147	Matsu n9F	GGAGAATACAGGGCGTTAAGAAC	136
			Matsu m9R	ATTAAMAGGGGCTAGTGTATTTCC		Matsu n9R	same as Matsu m9R	
	Matsu10	*ycf4*	Matsu m10F	AGTATTTGCTGAGATGACAATAGGG	142	Matsu n10F	same as Matsu m10F	118
			Matsu m10R	AACCTATAACAGGGTCTCGAAAAAG		Matsu n10R	GAAGTAATTTCTTTTGGGCTTGTATCC	
Mixture 3 (Strbs1, 3, 5, 7, 9, 11)	Strbs1	*ycf1*	Strbs m1F	TTCGGATCGAGTGAAAGCTC	146	Strbs n1F	GAAAGCTCTAAGCCATGGATCTC	113
			Strbs m1R	GGTCTATTGTTCCACGCAATG		Strbs n1R	CCCCATTAAGCAATGGATCATAC	
	Strbs3	*ycf1*	Strbs m3F	CTGAGCATTGGCAGGAATTG	148	Strbs n3F	CAGGAATTGGAACACAAAAGC	125
			Strbs m3R	GCTAATGGATAAARCCGTTTCG		Strbs n3R	AGCCGTTTCGAAATAGGTTC	
	Strbs5	*ycf1*	Strbs m5F	GGAAATGCAGATCCAAGAAATC	182	Strbs n5F	same as Strbs m5F	148
			Strbs m5R	CAACGTTTCTARCATCAATTCG		Strbs n5R	CACTAAAGAGTTTGTGTAGATCCGTTC	
	Strbs7	*ycf1*	Strbs m7F	GGATGATTCAACGCAAACG	199	Strbs n7F	CGGATGATTCAACGCAAAC	180 (186)
			Strbs m7R	TTTGACCTTTCTTGTACGAATCC		Strbs n7R	TCCTACTCGTTGATATTTGAATTGG	
	Strbs9	*ycf1*	Strbs m9.1F	CCTAAAGATTATTATGACACGTTCG	194	Strbs n9.1F	CGATTCTGGTAGAGTGAATCAGG	169
			Strbs m9.1R	TTTGATTGAGCCACTAATATGAGAC		Strbs n9.1R	same as Strbs m9.2R	
			Strbs m9.2F	TTATGACACGTTCGATTCTGG	181			
			Strbs m9.2R	TGATTGAGCCACTAATATGAGACC				
	Strbs11	*ycf1*	Strbs m11F	CGTTTGAAGCCTTGGCATAG	211	Strbs n11F	same as Strbs m11F	177
			Strbs m11R	CTCTCTTCAATCCTTTCTTCAATCC		Strbs n11R	TCCTTTCTTCAATCCTTTCTTCAC	

Table 1. *Cont.*

Mixture for Multiplex PCR	Primer ID	Region	Primers for Multiplex PCR	Sequence (5′-3′)	Product Size (bp)	Primers for Nested PCR	Sequence (5′-3′)	Product Size (bp)
Mixture 4 (Strbs2, 4, 6, 8, 10, 12)	Strbs2	*ycf1*	Strbs m2.1F	GGCATTGCGTGGAACAATAG	178	Strb n2.1F	same as Strb m2.1F	129
			Strbs m2.1R	CAATTCCTGCCAATGCTCAG		Strb n2.1R	TTCGGAAATCCCTCTTTACAGTC	
			Strbs m2.2F	GGGGATTCTTTTACTGAAATGTATG	169			
			Strbs m2.2R	TCGGAAATCCCTCTTTACAGTC				
	Strbs4	*ycf1*	Strbs m4F	GTTCCGAGCACTAAATCATCG	196	Strb n4F	CCAAAAGAGGAAAAGTTGAACC	177
			Strbs m4R	GGATCTCCATTTCCAACAAATC		Strb n4R	same as Strb m4R	
	Strbs6	*ycf1*	Strbs m6F	ACCGAATTGATGGTAGAAACG	167	Strb n6F	CCGAATTGATGGTAGAAACG	137
			Strbs m6R	TTGCGTTGAATCATCCGTAG		Strb n6R	CGCGACAATTTCGTAGTATTG	
	Strbs8	*ycf1*	Strbs m8F	CATGCCGAGTCAGATTATCGTC	178	Strb n8F	same as Strb m8F	140
			Strbs m8R	TCACTCTACCAGAATCGAACGTG		Strb n8R	GGATAAGTGGGTATTTCCATTCTTCTC	
	Strbs10	*ycf1*	Strbs m10F	GAGGTCTCATATTAGTGGCTCAATC	256	Strb n10F	same as Strb m10F	228
			Strbs m10R	CAAGGCTTCAAACGAAAAGG		Strb n10R	GCTTTATCTGCATACCATATTTGTACC	
	Strbs12	*ycf1*	Strbs m12F	GGATTGAAGAAAGGATTGAAGAGAG	174	Strb n12F	GGATTGAAGAAAGGATTGAAGAG	135
			Strbs m12R	ACCCATAGGGGTAGTCTCAGTCTC		Strb n12R	TTGTATCCGGTCATTAAGTTCAC	
Mixture 5 (Pns1-Pns5)	Pns1	*ycf1*	Pns m1F	TTTCGGATCGAGTGAAAGCTC	150			
				GGTTTATTGTTCCACGGAATGC				
	Pns2	*ycf1*	Pns m2F	GGTCAAGTAGAAGATCAACAAACTG	149			
	Pns3	*ycf1*	Pns m3F	CCAACCATATCGTTTATCAAGC	213			
				TCTCTACGACGTTTTGGAAGC				
	Pns4	*rbcL*	Pns m4F	TTGTACACAAGCTTCTAGAGCAACC	190			
				AGATTGGGTATCTATGCCAGGTG				
	Pns5	*rpl20-rps18*	Pns m5F	AGAGGCAGTTGCTTCCAAATC	161			
				TCCGGGAGAATCTGTTCTATCC				
Mixture 6 (Pns6-Pns7)	Pns6	*ycf1*	Pns m6F	TTGCTCTTCAGAGGAATGTTCG	126			
				TATACATCAGGAATTGGTCATCCAC				
	Pns7	*ycf1*	Pns m7F	CTCGGCAATAATGAGCCAAAG	149			
				GGGACATTATTTGAATGCTACTGC				

2.5. Nested PCR

The secondary amplification for each strand, which was run with a nested set of primers, was performed using a 0.5-μL aliquot of the purified first PCR products. A new PCR mixture of 10 μL was prepared, containing 0.5 μL of the template, 4.75 μL of water, 2.0 μL of 5× Green GoTaq Flexi Buffer (Promega Co., Madison, WI, USA), 0.9 μL of 25 mM $MgCl_2$, 0.8 μL of dU plus dNTP Mixture, 1.0 μL of 5.0 μM primers, and 0.05 μL of GoTaq Hot Start Polymerase (Promega). The primers used for nested PCR are listed in Table 1. The amplification was performed using a thermal cycler (GeneAmp PCR System 9700; Applied Biosystems, Foster City, CA, USA) under the following conditions: initial activation at 95 °C for 2 min, 20 cycles of denaturation at 95 °C for 30 s, annealing at 52 °C for 30 s, and extension at 72 °C for 30 s, followed by a final incubation at 72 °C for 5 min. Amplified PCR products were then sequenced using the BigDye Terminator v.3.1 Sequencing Kit (Applied Biosystems, Foster City, CA, USA) and an ABI 3130xl Genetic Analyzer (Applied Biosystems).

2.6. Identification of Individual *Pinus* Pollen Grains

Each *Pinus* pollen grain was identified based on parsimony-informative characters in various regions of the chloroplast genome, as shown in Table 1. The genus *Pinus* has approximately 111 recognized species in two subgenera, four sections, and 17 subsections (Table 2). Sequence data were collected for these regions from GenBank, which are available for almost all *Pinus* species. In addition, the aligned DNA sequences for each locus were used to determine parsimony-informative sites in MEGA ver. 5 [23] (see Supplementary data). Additionally, a small insertion of 3 bp was considered for the Matsu5 (*ycf3*) and Matsu7 (*rpoB*) regions. Genus-level classification was based on Gernandt et al. [24]. Their classification, based on a chloroplast DNA phylogeny, was a modification of (1) the influential classification of Little and Critchfield [25], which is based primarily on morphological characters and data from interspecific crosses, and (2) the classification of Price et al. [26], which incorporates additional recently described species.

Table 2. Classification system for the genus *Pinus*. The classification system is based on Gernandt et al. [24]. The colors indicating various taxa correspond to those in Table 3.

Genus	Subgenus	Section	Subsection
Pinus	*Strobus*	*Parrya*	*Cembroides*
			Nelsoniae
			Balfourianae
		Quinquefoliae	*Strobus*
			Krempfianae
			Gerardianae
	Pinus	*Trifoliae*	*Australes*
			Ponderosae
			Contortae
		Pinus	*Pinus*
			Pinaster

Table 3. Identification of pollen grains by multiplex PCR. Each pollen grain was identified at the subgenus, section, or subsection level based on regions that were positively amplified. The S, Q, P, and St indicate the subgenus *Strobus*, section *Quinquefoliae*, section *Pinus*, and subsection *Strobus*, respectively. Dashes indicate a lack of amplification. For the sequence data and identification procedures, see Supplementary data.

Primer ID	Region	Belukha 1	Belukha 2	Belukha 3	Belukha 4	Belukha 5
Matsu1	*clpP*	Q	-	-	-	-
Matsu2	*rpoA*	Q	Q	Q	-	-
Matsu3	*atpB*	Q	Q	Q	Q	P
Matsu4	*rpoC1*	S	S	S	-	-
Matsu5	*ycf3*	Q	Q	Q	-	-
Matsu6	*Ycf4*	St	-	-	-	-
Matsu7	*rpoB*	-	S	Q	-	-
Matsu8	*rpoC1*	St	St	-	-	-
Matsu9	*ycf3*	Q	-	Q	-	-
Matsu10	*Ycf4*	Q	-	-	-	-

3. Results and Discussion

3.1. *Pinus* Pollen Identification at the Section or Subsection Level

We analyzed 21 pollen grains, and five samples showed positive amplification from at least one locus in the multiplex PCR with the primer sets Matsu1–10 (Table 3). DNA fragmentation and degradation, particularly in ancient samples, make it difficult to amplify long fragments from a single pollen grain in sediment samples [27,28]. This problem is alleviated by the higher amplification efficiency of short fragments (<200 bp) [29]. To increase the probability of amplification, our primer sets were designed to yield fragments of around 200 bp, even though our samples were not ancient. Although previous pollen DNA barcoding studies have focused on fragments of longer than 300 bp [14], we believe that fragments of around 200 bp may be sufficient to identify pollen at the section or subsection level when the target pollen type is specified and sequence data are available, as demonstrated in the present study.

The controls in our PCRs followed the experimental methodology of Parducci et al. [29]. A positive control was not used owing to the high contamination risk. As a negative control, contamination by exogenous chloroplast DNA from *Pinus* species in the reagents was monitored using a PCR blank that included all reagents. If no band was visible by agarose gel electrophoresis after multiplex PCRs, we concluded that the samples were not contaminated. Samples with bands on the agarose gel were selected for an additional nested PCR amplification step with each single primer set.

Based on the sequence data collected for the *clpP*, *rpoA*, *atpB*, *rpoC1*, *ycf3*, *ycf4*, and *rpoB* regions by nested PCR, the pollen grains were identified to the section or subsection level (Table 3). We identified four out of five pollen grains as members of the subsection *Strobus*. One belonged to the section *Quinquefoliae*, and the remaining grain was part of the section *Pinus*.

For the four grains in the subsection *Strobus*, we successfully obtained sequence data for multiple loci (Table 3). Based on these sequence data, we identified each locus as belonging to the same phylogenetic clade, namely, subsection *Strobus* of the section *Quinquefoliae* in the subgenus *Strobus* (Table 2). This consistency confirms that the WGA reactions could increase the amount of accurate DNA data that can be obtained from a single pollen grain.

3.2. *Species Identification of Pinus Pollen Grains*

To further narrow down the candidate species in the subsection *Strobus*, we used the WGA products for multiplex PCRs with primers to amplify short fragments on the plastid gene *ycf1*. We designed various primer sets targeting regions in the *ycf1* gene (Table 1), which is a highly variable locus that can be used for *Pinus* species identification [30]. The multiplex PCR products were subjected to an additional round of nested PCR.

We successfully amplified target regions for three out of four pollen grains in the subsection *Strobus*. Regions that were successfully amplified are shown in Table S1. In the sequence analysis based on variable base positions, we identified two pollen grains as *P. cembra* or *P. sibirica* (samples No. 1 and 3). Although we examined various regions in the chloroplast genome, to positively identify pollen grains as *P. sibirica* or *P. cembra*, it was sufficient to examine only three regions, namely, Strbs3, Strbs7, and Matsu6 (or Matsu8). For the remaining grain, we determined the following 15 candidate species: *P. albicaulis*, *P. armandii*, *P. bhutanica*, *P. cembra*, *P. dalatensis*, *P. fenzeliana*, *P. lambertiana*, *P. monticola*, *P. morrisonicola*, *P. parviflora*, *P. peuce*, *P. pumila*, *P. sibirica*, *P. wallichiana*, and *P. wangii*.

We inferred that the three grains were P. sibirica. Gugerli et al. [31] observed highly similar chloroplast and mitochondrial DNA sequences between *P. cembra* and *P. sibirica*. This similarity suggests a relatively recent evolutionary separation of the species, despite their currently disjunct distributions (Figure 1). *P. sibirica* appears in the Urals and Siberia of Russia, in eastern Kazakhstan, in northern Mongolia, and in Xingjiang, Nei Mongo, and Heilongjiang of China. *P. cembra* occurs in the Swiss Alps, in the Tirol of Austria, in the High Tatra between Poland and the Slovak Republic, and in the eastern Carpathians of Romania and Ukrine [22]. Heinze and Holzer [32] verified that a nearly complete *P. cembra* chloroplast genome sequence in GenBank (Accession No. FJ899574) is identical to a *P. sibirica* sequence (Accession No. FJ899558). Although the FJ899574 sequence lacks part of the *ycf1* region targeted in this study, the *ycf1* sequence identities for both species were validated by a comparative analysis using FJ899558 and KP128626 of *P. cembra*. Accordingly, the two species likely cannot be discriminated based on comparative sequence analyses of the chloroplast genome. *P. sibirica* is an extant species and is the only member of the subsection *Strobus* that is found near the glacier. This species was a major candidate in our study, suggesting that the pollen grains in the glacier originated from *P. sibirica* trees found in the immediate surroundings. This consistency between the results of our genetic analysis and species distribution data suggests that our method to identify pollen species was reliable.

We were not able to obtain sequence data for three grains. For the pollen grain identified as belonging to the section *Pinus*, the WGA products were used as templates in the multiplex PCRs, with primer sets for the *ycf1*, *rbcL*, and *rpl20-rps18* chloroplast regions (Table 1). We could not identify those grains at a lower taxonomic level. However, the grains appeared to be *P. sibirica* in subsection *Strobus*, section *Quinquefoliae*, and *P. sylvestris*, which belongs to subsection *Pinus* in section *Pinus* based on the consistent results at the subsection or section levels. To obtain accurate sequences and facilitate a more detailed taxonomic identification, additional PCRs with newly designed primer sets may be effective. Although we were able to generate sufficient DNA specimens from single pollen grains for additional PCRs using the WGA technique, we did not perform subsequent analyses owing to a lack of time and resources and an expectation of unremarkable results.

3.3. Potential Use of Pollen Grains as a Tracer for Emission Sources

We believe that our method for *Pinus* pollen identification is suitable for further work aimed at a more detailed characterization of the provenance of aerosols, particularly in Arctic glaciers and the Greenland ice sheet. Pollen is a type of bio-aerosol; *Pinus* pollen as well as other types of pollen travel long distances. This pollen is regarded as exotic, and many studies have investigated its source area and long-distance transportation by trajectory analyses [33–37]. For aerosols reaching ice sheets and mountain glaciers, dust has been used as a tracer for emission sources as well as large-scale atmospheric circulation [38–40]. Clay mineralogy and Sr-Nd isotopic and elemental compositions have suggested that East Asia (i.e., the Gobi Desert, northern Chinese deserts, and the Taklamakan Desert) is the main source of dust arriving in Greenland, both at present and during the last glacial period [38]. Generally, dust seems to originate mainly from arid regions. In contrast, pollen sources are restricted to vegetated areas. Therefore, pollen can be used as another tracer, and a combination of both of these tracers should lead to a better understanding of the provenance of solid aerosols and the materials cycle. This approach has not been used, although some pollen grains, including *Pinus* pollen grains, have been found in Greenland [2,35,36].

As mentioned in Section 3.2, our analysis strongly suggested that two *Pinus* pollen grains found in the Belukha glacier are *P. sibirica*, in consonance with the surrounding *Pinus* vegetation. Therefore, we can assume that the provenance of the *Pinus* pollen is the region that extends from the northwest to the east of the glacier, as shown in Figure 1. A *Pinus* pollen grain is around 50 μm in size; the size of a pollen grain typically ranges from 10 to 200 μm, and the most common size is between 20 and 60 μm [41]. Despite the relatively large size of *Pinus* pollen, the grains are well dispersed in a vesiculate form with two prominent sacci [42]. In addition, *Pinus* is characterized by high pollen grain production. For those reasons, *Pinus* pollen in palaeoecological records are frequently disregarded as evidence for presence or absence in the arctic region with low local production of pollen [1]. However, these properties are favorable for analyses of *Pinus* pollen grains from Arctic glaciers and the Greenland ice sheet by our method to investigate geographic provenances and the materials cycle. Additionally, their large size seems to be beneficial for the treatment of pollen grains in a laboratory.

3.4. Improvement in the Amplification Success Rate

The success rate for obtaining sequence data reported by Nakazawa et al. [10] was 7.6% (n = 105). This was higher than the success rates observed in previous DNA analyses of pollen collected from peat or lacustrine deposits. Suyama et al. [28,43] and Parducci et al. [29] observed success rates ranging from 0 to 3.2%. However, these rates are still insufficient to develop a new field of palynological research based on genetic information. Hence, improving the success rate is a particularly important issue.

In this study, we subjected 21 pollen grains to WGA, and we observed positive amplification from at least one locus for a total of five grains. The success rate for sequence amplification in this study was 24% and exceeded that of Nakazawa et al. [10] who used pollen samples from the same glacier collected from the upper layer of the pit. To improve the success rate of DNA analyses, we introduced multiplex PCR; amplifying multiple loci simultaneously in a single reaction improved the probability that at least one locus was amplified. In addition, the WGA technique enabled an increased quantity of DNA to be obtained from samples with limited DNA content. Since we collected samples from a different layer from that of previous studies, we cannot make a simple comparison of the success rates between studies. However, we were able to obtain sequence data from multiple loci, and this method appeared to be more effective with respect to success rate.

4. Conclusions

We described an initial attempt to identify pollen grains from a glacier at the species level based on chloroplast DNA sequences. For precise identification, we applied an optimized WGA technique for single pollen grains. We subjected WGA products to an additional round of multiplex PCRs. The combined DNA sequence data obtained from a single pollen grain suggested that identification at or near the species level is possible.

We analyzed 21 pollen grains, of which five exhibited positive amplification for at least one locus in the multiplex PCRs. One grain appeared to belong to the section *Quinquefoliae*, four grains were in the subsection *Strobus* of section *Quinquefoliae*, and the remaining grain was identified as a member of the section *Pinus*. In addition, for three out of the four grains in the subsection *Strobus*, the candidates were narrowed down to two species; 15 candidates remained for the other grain. Owing to the identical *P. cembra* and *P. sibirica* chloroplast genome sequences, it was not possible to differentiate between the species using sequence data. However, we inferred that both grains were *P. sibirica* because it is an extant species that is currently distributed around the glacier. Meanwhile, *P. cembra* is distributed in some high mountains in Switzerland, Austria, Poland, the Slovak Republic, Romania, and Ukraine [22]. Moreover, we could assume the grains traveled from the region that expands from the northwest to the east of the glacier based on the forest distribution of *P. sibirica*.

Similarly, the remaining grains appeared to be *P. sibirica* or *P. sylvestris*, which is also found around the glacier. Our pollen identifications based on DNA sequence data are supported by the vegetation in the region, suggesting that the method established in this study enables reliable identification at the species level for pollen grains. Additionally, the method should be useful for future studies on the provenance of solid aerosols and the materials cycle in the polar region and high mountain regions in the Northern Hemisphere, where glaciers exist.

The DNA amplification success rate in this study exceeded that of our previous study of samples from the same snow pit. However, the samples were collected from a different layer, preventing a simple comparison of success rates between studies. However, we demonstrated that our method could be used to obtain sequence data from multiple loci and effectively increase the success rate.

Further investigations using older samples from ice cores are necessary to extend the applications of these methods and to accumulate data in the field. The rarity of suitable, well-preserved pollen samples in sediments has limited the broad utility of DNA studies for the taxonomic identification of pollen. However, owing to low-temperature conditions, pollen grains in glaciers are not strongly affected by diagenesis and their DNA is therefore likely to be preserved. Accordingly, pollen samples from glaciers have broad utility for studies of taxonomy, past vegetation, population genetics, and climate and environmental conditions. Our method based on WGA and multiplex PCR techniques may enable the generation of DNA specimens from single pollen grains that can be analyzed by multiple molecular techniques for a range of applications.

Supplementary Materials: The following are available online at http://www.mdpi.com/1999-4907/9//444/s1. S1: Table S1, S2: Sequence data obtained for Matsu1–10 are shown in the sheets of "Matsu No. seq" of the file. Sequence variation for parsimony-informative characters in *Pinus* and *Pinus* pollen from the Belukha glacier are compared in the individual sheet named "Matsu No. id" to identify pollen taxa. Identical sequences with those of the samples for each sheet are colored to identify candidates.

Author Contributions: F.N. carried out the molecular genetic studies and drafted the manuscript. Y.S. helped to evaluate and edit the manuscript. S.I. participated in the coordination of the study and helped in data interpretation and manuscript evaluation. H.M. supervised the development of the work and helped draft the manuscript.

Funding: This research was funded by the "Systematic Analysis for Global Environmental Change and Life on Earth" research project of the Transdisciplinary Research Integration Center; the Oasis Project (Historical evolution of adaptability in an oasis region to water resource changes) and the Ili Project (Historical interactions between multi-cultural societies and the natural environment in a semi-arid region in Central Eurasia) promoted by the Research Institute for Humanity and Nature, Kyoto, Japan; and JSPS KAKENHI Grant Numbers 24710019, 16K00529, 24248025. The production of this paper was supported by an NIPR publication subsidy.

Acknowledgments: We wish to thank all individuals who generously assisted with the study. We also thank Kenichi Watanabe, Ryoko Yamagishi, and Reiko Sakamoto of the National Institute of Polar Research for help with the DNA analysis.

Conflicts of Interest: The authors declare no conflict of interest.

References

1. Richardson, D.M.; Rundel, P.W. Ecology and biogeography of *Pinus*: An introduction. In *Ecology and Biogeography of Pinus*; Richardson, D.M., Ed.; Cambridge University Press: Cambridge, UK, 1998; pp. 3–46. ISBN 0521789109.
2. Fredskild, B.; Wagner, P. Pollen and fragments of plant tissue in core samples from the Greenland Ice Cap. *Boreas* **1974**, *3*, 105–108. [CrossRef]
3. Short, S.K.; Holdsworth, G. Pollen, oxygen isotope content and seasonality in an ice core from the Penny Ice Cap, Baffin island. *Arctic* **1985**, *38*, 214–218. [CrossRef]
4. Bourgeois, J.C. Seasonal and interannual pollen variability in snow layers of arctic ice caps. *Rev. Palaeobot. Palynol.* **2000**, *108*, 17–36. [CrossRef]
5. Nakazawa, F.; Fujita, K.; Uetake, J.; Kohno, M.; Fujiki, T.; Arkhipov, S.M.; Kameda, T.; Suzuki, K.; Fujii, Y. Application of pollen analysis to dating of ice cores from lower-latitude glaciers. *J. Geophys. Res.* **2004**, *109*, F04001. [CrossRef]

6. Nakazawa, F.; Fujita, K.; Takeuchi, N.; Fujiki, T.; Uetake, J.; Aizen, V.; Nakawo, M. Dating of seasonal snow/firn accumulation layers using pollen analysis. *J. Glaciol.* **2005**, *51*, 483–490. [CrossRef]
7. Hicks, S.; Isaksson, E. Assessing source areas of pollutants from studies of fly ash, charcoal, and pollen from Svalbard snow and ice. *J. Geophys. Res.* **2006**, *111*, D02113. [CrossRef]
8. Nakazawa, F.; Konya, K.; Kadota, T.; Ohata, T. Reconstruction of the depositional environment upstream of Potanin Glacier, Mongolian Altai, from pollen analysis. *Environ. Res. Lett.* **2012**, *7*, 035402. [CrossRef]
9. Festi, D.; Kofler, W.; Bucher, E.; Mair, V.; Gabrielli, P.; Carturan, L.; Oeggl, K. A novel pollen-based method to detect seasonality in ice cores: a case study from the Ortles Glacier (South Tyrol, Italy). *J. Glaciol.* **2015**, *61*, 815–824. [CrossRef]
10. Nakazawa, F.; Uetake, J.; Suyama, Y.; Kaneko, R.; Takeuchi, N.; Fujita, K.; Motoyama, H.; Imura, S.; Kanda, H. DNA analysis for section identification of individual *Pinus* pollen grains from Belukha glacier, Altai Mountains, Russia. *Environ. Res. Lett.* **2013**, *8*, 014032. [CrossRef]
11. Yazdani, R.; Lindgren, D.; Stewart, S. Gene dispersion within a population of *Pinus sylvestris*. *Scand. J. For. Res.* **1989**, *4*, 295–306. [CrossRef]
12. Savolainen, O.; Pyhäjarvi, T.; Knürr, T. Gene flow and local adaptation in trees. *Annu. Rev. Ecol. Evol. Syst.* **2007**, *38*, 595–619. [CrossRef]
13. Wolfe, K.H.; Li, W.H.; Sharp, P.M. Rates of nucleotide substitution vary greatly among plant mitochondrial, chloroplast, and nuclear DNAs. *Proc. Natl. Acad. Sci. USA* **1987**, *84*, 9054–9058. [CrossRef] [PubMed]
14. Bell, K.L.; de Vere, N.; Keller, A.; Richardson, R.T.; Gous, A.; Burgess, K.S.; Brosi, B.J. Pollen DNA barcoding: current applications and future. *Genome* **2016**, *59*, 629–640. [CrossRef] [PubMed]
15. Dong, W.; Xu, C.; Li, C.; Sun, J.; Zuo, Y.; Shi, S.; Cheng, T.; Guo, J.; Zhou, S. *ycf1*, the most promising plastid DNA barcode of land plants. *Sci. Rep.* **2015**, *5*, 8348. [CrossRef] [PubMed]
16. Nachtigall, P.G.; Rodrigues-Filho, L.F.S.; Sodré, D.C.A.; Vallinoto, M.; Pinhal, D. A multiplex PCR approach for the molecular identification and conservation of the Critically Endangered daggernose shark. *Endanger. Spec. Res.* **2017**, *32*, 169–175. [CrossRef]
17. Kim, J.; Hong, J.; Lim, J.A.; Heu, S.; Roh, E. Improved multiplex PCR primers for rapid identifcation of coagulase-negative staphylococci. *Arch. Microbiol.* **2018**, *200*, 73–83. [CrossRef] [PubMed]
18. Miner, B.E.; Stöger, R.J.; Burden, A.F.; Laird, C.D.; Hansen, R.S. Molecular barcodes detect redundancy and contamination in hairpin-bisulfite PCR. *Nucleic Acids Res.* **2004**, *32*, e135. [CrossRef] [PubMed]
19. McCloskey, M.L.; Stöger, R.; Hansen, R.S.; Laird, C.D. Encoding PCR products with batch-stamps and barcodes. *Biochem. Genet.* **2007**, *45*, 761–767. [CrossRef] [PubMed]
20. Nakazawa, F.; Miyake, T.; Fujita, K.; Takeuchi, N.; Uetake, J.; Fujiki, T.; Aizen, V.B.; Nakawo, M. Establishing the timing of chemical deposition events on Belukha glacier, Altai Mountains, Russia, using pollen analysis. *Arctic Antarct. Alp. Res.* **2011**, *43*, 66–72. [CrossRef]
21. Eichler, A.; Tinner, W.; Brütsch, S.; Olivier, S.; Papina, T.; Schwikowski, M. An ice-core based history of Siberian forest fires since AD 1250. *Quat. Sci. Rev.* **2011**, *30*, 1027–1034. [CrossRef]
22. Farjon, A. *Pines: Drawings and Descriptions of the Genus Pinus*, 2nd ed.; Brill: Leiden, The Netherlands, 2005; p. 236. ISBN 9004139168.
23. Tamura, K.; Peterson, D.; Peterson, N.; Stecher, G.; Nei, M.; Kumar, S. MEGA5: Molecular Evolutionary Genetics Analysis using Maximum Likelihood, Evolutionary Distance, and Maximum Parsimony Methods. *Mol. Biol. Evol.* **2011**, *28*, 2731–2739. [CrossRef] [PubMed]
24. Gernandt, D.S.; GeadaLópez, G.; Garcia, S.O.; Liston, A. Phylogeny and classification of *Pinus*. *Taxon* **2005**, *54*, 29–42. [CrossRef]
25. Little, E.L., Jr.; Critchfield, W.B. *Subdivisions of the Genus Pinus (Pines)*; Miscellaneous Publications 1144; U.S. Department of Agriculture: Washington, DC, USA, 1969; p. 51.
26. Price, R.A.; Liston, A.; Strauss, S.H. Phylogeny and systematics of *Pinus*. In *Ecology and Biogeography of Pinus*; Richardson, D.M., Ed.; Cambridge University Press: Cambridge, UK, 1988; pp. 49–68. ISBN 0521789109.
27. Pääbo, S. Ancient DNA: Extraction, characterization, molecular cloning, and enzymatic amplification. *Proc. Natl. Acad. Sci. USA* **1989**, *86*, 1939–1943. [CrossRef] [PubMed]
28. Suyama, Y.; Kawamuro, K.; Kinoshita, I.; Yoshimura, K.; Tsumura, Y.; Takahara, H. DNA sequence from a fossil pollen of *Abies* spp. from Pleistocene peat. *Genes Genet. Syst.* **1996**, *71*, 145–149. [CrossRef] [PubMed]
29. Parducci, L.; Suyama, Y.; Lascoux, M.; Bennett, K.D. Ancient DNA from pollen: a genetic record of plant population history. *Mol. Ecol.* **2005**, *14*, 2873–2882. [CrossRef] [PubMed]

30. Parks, M.; Cronn, R.; Liston, A. Increasing phylogenetic resolution at low taxonomic levels using massively parallel sequencing of chloroplast genomes. *BMC Biol.* **2009**, *7*, 84. [CrossRef] [PubMed]
31. Gugerli, F.; Senn, J.; Anzidei, M.; Madaghiele, A.; Büchler, U.; Sperisen, C.; Vendramin, G.G. Chloroplast microsatellites and mitochondrial nad1 intron 2 sequences indicate congruent phylogenetic relationships among Swiss stone pine (*Pinus cembra*), Siberian stone pine (*Pinus sibirica*), and Siberian dwarf pine (*Pinus pumila*). *Mol. Ecol.* **2001**, *10*, 1489–1497. [CrossRef] [PubMed]
32. Heinze, B.; Holzer, K. A review of research on Pinus cembra in Austria, with special reference to the conservation of genetic resources. In Proceedings of the 5th Symposium for Research in Protected Areas, Mittersill, Austria, 10–12 June 2013; pp. 279–284.
33. Bourgeois, J.C.; Koerner, R.M.; Alt, B.T. Airborne pollen: a unique mass tracer, its influx to the Canadian High Arctic. *Ann. Glaciol.* **1985**, *7*, 109–116. [CrossRef]
34. Rogers, C.A.; Levetin, E. Evidence of long-distance transport of mountain cedar pollen into Tulsa, Oklahoma. *Int. J. Biometeorol.* **1998**, *42*, 65–72. [CrossRef]
35. Rousseau, D.D.; Duzer, D.; Cambon, G.; Jolly, D.; Poulsen, U.; Ferrier, J.; Schevin, P.; Gros, R. Long distance transport of pollen to Greenland. *Geophys. Res. Lett.* **2003**, *30*, 1765. [CrossRef]
36. Rousseau, D.D.; Schevin, P.; Duzer, D.; Cambon, G.; Ferrier, J.; Jolly, D.; Poulsen, U. New evidence of long distance pollen transport to southern Greenland in late spring. *Rev. Palaeobot. Palynol.* **2006**, *141*, 277–286. [CrossRef]
37. Mohanty, R.P.; Buchheim, M.A.; Anderson, J.; Levetin, E. Molecular analysis confirms the long-distance transport of Juniperus ashei pollen. *PLoS ONE* **2017**, *12*, e0173465. [CrossRef] [PubMed]
38. Bory, A.J.M. A 10,000 km dust highway between the Taklamakan Desert and Greenland. In *Past Global Changes Magazine*; Merkel, U., Rousseau, D.D., Stuut, J.B., Winckler, G., von Gunten, L., Kiefer, T., Eds.; PAGES International Project Office: Bern, Switzerland, 2014; Volume 22, pp. 72–73.
39. Nagatsuka, N.; Takeuchi, N.; Nakano, T.; Shin, K.; Kokado, E. Geographical variations in Sr and Nd isotopic ratios of cryoconite on Asian glaciers. *Environ. Res. Lett.* **2014**, *9*, 045007. [CrossRef]
40. Vallelonga, P. The enigma of dust provenance: Where else does Antarctic dust come from? In *Past Global Changes Magazine*; Merkel, U., Rousseau, D.D., Stuut, J.B., Winckler, G., von Gunten, L., Kiefer, T., Eds.; PAGES International Project Office: Bern, Switzerland, 2014; Volume 22, pp. 74–75.
41. Matsushita, M. *Pollen Analysis and Archaeology*; Douseisha: Tokyo, Japan, 2004; p. 135. ISBN 4886213030. (In Japanese)
42. Willis, K.J.; Bennett, K.D.; Birks, H.J.B. The late Quaternary dynamics of Pines in Europe. In *Ecology and Biogeography of Pinus*; Richardson, D.M., Ed.; Cambridge University Press: Cambridge, UK, 1998; pp. 107–121. ISBN 0521789109.
43. Suyama, Y.; Kawamuro, K.; Takahara, H.; Shichi, K.; Yoshimaru, H.; Kinoshita, I.; Yoshimura, K.; Tsumura, Y. Study on vegetation dynamics and biological evolution from DNA analyses of ancient pollen. *Chikyu Mon. Symp.* **2003**, *42*, 187–192. (In Japanese)

forests

MDPI

Article

Corsican Pine (*Pinus laricio* Poiret) Stand Management: Medium and Long Lasting Effects of Thinning on Biomass Growth

Rodolfo Picchio [1,*], Rachele Venanzi [1], Francesco Latterini [1], Enrico Marchi [2], Andrea Laschi [2] and Angela Lo Monaco [1]

1 Department of Agriculture and Forest Sciences (DAFNE), Tuscia University, 01100 Viterbo, Italy; venanzi@unitus.it (R.V.); francesco.latterini@libero.it (F.L.); lomonaco@unitus.it (A.L.M.)

2 Department of Agriculture (GESAAF), Florence University, 50144 Florence, Italy; enrico.marchi@unifi.it (E.M.); andrea.laschi@unifi.it (A.L.)

* Correspondence: r.picchio@unitus.it; Tel.: +39-0761-357-400

Received: 10 April 2018; Accepted: 7 May 2018; Published: 10 May 2018

Abstract: With the aim of acquiring better comprehension of the ecological and productive aspects of the management of pine forests, we monitored logging damage and evaluated the effects of thinning on stand growth 20 years after the treatment in a *Pinus laricio* Poiret stand in central Italy. The objectives of the present study were to estimate the injury levels to the remaining trees after thinning; to assess logging damage in the long-term by monitoring residual trees at the end of thinning; to evaluate the effect of damage on the radial growth of trees; to assess the stand dynamics in relation to injury levels and the treatment applied in a twenty-year range; to understand a possible treatment return time; and to evaluate the existence of the "thinning shock". The results were that 20 years after treatment, the stand dynamics showed a complete recovery; logging damage did not affect the radial growth of *P. laricio* over time; a second treatment seem to be sustainable starting from the fifteenth year after the previous treatment; and the thinning shock can be clearly evaluated in the first six to seven years after the treatment.

Keywords: stand growing; thinning shock; forest logging; tree wounds; tree ring width

1. Introduction

The functionality and resilience of forest ecosystems are strictly related to forest structure [1–3], which represents the distribution of biomass, i.e., the vertical and horizontal spatial arrangement of plant species, tree sizes, age distribution, and tree canopy layering. Stand structure attributes are increasingly recognized to be of theoretical and practical importance in analyzing the management quality in forest ecosystems. In general, the characteristics of forest cover are closely related to stand density, stand age, and species composition and are altered by natural disturbances or anthropogenic interference such as forest management practices. Among the various forest management practices, the thinning or removal of some trees from the forest is widely conducted to produce more valuable and large-diameter timber. It reduces competition among the remaining trees, reduces the risk of fire [4], and helps to maintain a healthy forest [5].

Considering that in the last fifty years, Mediterranean forests have experienced a rapid expansion of pioneer conifer stands driven by natural colonization after land abandonment in SE France [6] or afforestation efforts in SE Spain [7] and in Italy [8], forest fires and pest outbreaks are predicted to increase in the future, also due to the impact of climate changes on drought and extreme weather events [9,10]. As already mentioned, the appropriate silvicultural treatments play a key role in increasing the resistance and resilience of Mediterranean forests to environmental perturbations [11,12].

Structural changes in the forest canopy due to thinning alter the microclimatic conditions within the forest [13–17]. These microclimatic changes could improve the growth conditions for the remaining trees and understory species [13,18]. However, a short- and medium-term decline in the growth of remaining trees, called "thinning shock," has also been reported for some forest stands and species [19–21].

Furthermore, the positive changes due to thinning could be offset by high vulnerability to insect damage and mortality, which might be caused by trunk or root damage during the thinning procedure [22,23]. For these reasons, the methodologies of felling, processing, and extraction need to be planned on a larger scale; they cannot be sporadic events, not connected to the social, environmental, and economic contexts. The forest operations have to be planned during the implementation and execution of the working phases [24].

Research on the damage caused by forest operations to the remaining trees and/or to regeneration in forest stands started at the beginning of the twentieth century and is more important as the use of mechanized wood harvesting has been increasing [21,25,26]. Logging injury to the remaining trees may lead to serious economic losses in terms of timber quality at the final harvest, wood losses in the damaged trees, and tree growth reduction [27]. Long term effects may significantly reduce the benefits of commercial thinning [21,25]. Other forest ecosystem services, e.g., carbon sequestration, soil protection, and water control, may also be affected.

The effect on tree growth of logging mechanical damage to the stems remains unclear, although some authors have conducted detailed studies [21,28]. Moreover, for Mediterranean pine plantations, there are some knowledge gaps regarding the late first thinning effect on tree growth. Studies on this topic are still limited and the long-term effects of damage are not well known [29].

Analysis of the annual growth rings can be used as an "environmental archive" [30,31] and as an indicator of growth changes showing the effect of silvicultural treatments.

With the aim of acquiring a better comprehension of some ecological and productive aspects of the management of pine forests, logging damage was monitored and the effects of thinning on stand growth 20 years after the treatment were evaluated in a *Pinus laricio* Poiret stand in central Italy. Specifically, the aims of this study were an extension of previous work [21] developed ten years ago, with the following objectives: (1) to estimate the injury levels to the remaining trees after thinning; (2) to assess logging damage in the long-term by monitoring residual trees at the end of thinning; and (3) to evaluate the effect of damage on the radial growth of trees. In the present study, further investigations were carried out on the same stand in order to assess the stand dynamics in relation to the injury levels and the treatment applied in a twenty-year range. This continuing research was aimed at assessing the effect of damage on the radial growth of trees; at understanding a possible treatment return time; and evaluating the existence of "thinning shock".

Understanding all of these aspects is a key factor in the ecological management of Corsican pine forests located in the Mediterranean area and is an important factor in supporting the decision making of forest managers.

2. Materials and Methods

2.1. Area Description

The study was carried out in a forest located in Grottone (Viterbo—Central Italy-42°23′35.58″ N, 12°10′25.52″ E 720 m a.s.l.). The stand was an even-age, *P. laricio* (Corsican pine) forest, thinned in 1998. This forest originated from a plantation established in the 1970s on a mostly flat terrain (maximum slope gradient 10%). Seedlings were two to five years old at planting time and the plantation scheme was rectangular (2 m × 2.5 m), with a density of about 2000 trees/ha.

The climate of the area is moist temperate, with a mean annual precipitation of 1180 mm, mean annual temperature of 12.8 °C, and mean summer precipitation of 55 mm; the driest month is July (30 mm), the temperature of the warmest month is 22.0 °C, and the temperature of the coldest month is 4.2 °C. Data were the averages over the period 1916–2010 at the San Martino al Cimino weather station (510 m a.s.l., less than 5 km from the study area). The soil, formed on a volcanic parent material, is sandy loam, slightly acid, and showed a well-structured, humus-rich A horizon.

As described in a previous paper [21], the silvicultural treatment used was selective thinning from below, carried out in the spring of 1998 on about 32 ha. For research purposes, an adjacent control area of about 6 ha was not treated. Logging operations were carried out with a semi-mechanized logging system, felling and processing by a chainsaw, and bunching and extraction by skidding with a tracked agricultural tractor (crawler). The skidding performed was inter-row and only one of two inter-rows was used as a skid trail. Reference [21] described the detailed working cycle: felling of two or more trees; bunching by the tractor (average distance 33 m); partial delimbing (on about 3/4 of the tree circumference); topping; and skidding to the landing (average distance was about 73 m; average load for skidding trip was 0.87 t, for an average skidding speed of about 1.4 km/h).

2.2. Data Collection

Data collection was carried out for the first time in 1998 through a design-based approach, a statistical approach that establishes the choosing methods and using of the sites, allowing possible pseudoreplication problems to be overcome [32]. The subsequent data records were made before and after thinning, in 2008 and in 2017, in 25 circular plots of 1256 m^2 (dendrometric plots DP) in the treated area (total surface 31,400 m^2) and in five DPs in the control area (total surface 6280 m^2). Plots were randomly selected, at a >30 m distance from the edge of the treated and control areas. No thinning or other important disturbance were carried out in the time between 1998 and 2017.

For the three temporal field analyses (1998–2008–2017), after the measurements of the main dendrometric characteristics, the growing stock was estimated with a two-way table [33], developed for *Pinus laricio* of Tuscany and the slenderness ratio (average value of H/DBH of each trees) was calculated as an index of stand stability [34].

Tree damage was detected in four rectangular plots (10 m × 250 m wound plots WP). In 1998, these WP and all damaged trees were numbered and marked. The field analysis repetitions were done in 2008 and 2017. The following parameters were recorded: DBH of all the trees; crown class of all the trees; type of damaged tissues; cause of damage; location of wounds, in meters above ground level; and size of wounds, by measuring maximum length and width by a ruler (1 mm accuracy) and calculating the ellipse surface area. Type and size of damage were used as the classifications in two damage classes, i.e., severe (wood wound >200 cm^2 size) and light (bark/bast wound <200 cm^2 size), according to [21,35]. Trees damaged in several parts were classified on the basis of the most severe damage. The agent of damage (felling, skidding) was classified in 1998.

In May–July 2008 and 2017, stem core samples were collected at breast height on the north side of two trees per DP in each of the seven 5-cm diameter classes (from 15 to 50 cm) by using a Pressler increment borer. In the thinned area, stem core samples were collected from one damaged and one undamaged tree in each DP. Ring width was measured by using Aniol Dendroware, with an accuracy of 0.01 mm. In order to check the differences in growth among the damaged, undamaged, and control trees, different numbers of rings were considered: all rings before thinning (BT); ten rings before thinning (TBT); post thinning 2008 (ten rings TPT1); and post thinning 2017 (twenty rings TPT2). In order to evaluate the differences at the stand level, the yearly weighted ring width (YWW) was calculated as a mean of yearly ring width weighted on the basis of the frequency distribution of trees per diameter class in 2008 and in 2017.

2.3. Statistical Analysis

Statistical analyses were carried out using the Statistica 7.1 (2007) software (StatSoft Inc., Tulsa, OK, USA). After checking for normality (Kolmogorov-Smirnov test) and homogeneity of variance (Levene test), the *t*-test was applied to all dendrometric parameters. Two-way ANOVA was applied to DBH and the number of trees to test the effect of year and the factors "damaged and undamaged", "severity", or "agent". The post-hoc Tukey-HSD test was used to compare the significance of differences among the means. Factorial ANOVA was applied to the ring width to test the effect of the number of rings (BT, TBT, TPT1, TPT2), diameter class, and the thinning and damage effect (trees in control area, undamaged, and damaged trees in thinned area). The post-hoc Tukey HSD test per unequal N was used to compare the significance of differences among the means. The Kruskal–Wallis non-parametric multiple-comparison test was used to test the effect of the factor "position" on DBH and the number of damaged trees in every year (1998 and 2008).

In order to determine the relationship between DBH, damage agent (felling, skidding), position (roots or stem at different levels above ground), and severity (light, severe), a nonparametric correlation analysis (Spearman correlation matrix) was applied. Nonlinear regression analysis between ring width (dependent variable) and year (independent variable), for damaged (TD) and undamaged (TU) trees and the trees from the control area (C), was applied. The Wilcoxon non-parametric test was used to test the differences in YWW before and after thinning between the control and damaged; control and undamaged; and damaged and undamaged trees. Non-metric multidimensional scaling (NMDS) was used to show the differences in the average ring width per diameter class for the trees in the control area (C), for damaged trees (TD), and for undamaged trees (TU) in the thinned area.

3. Results

As reported in the previous step of this research by [21], about 20% of the volume, 38% of trees, and 26% of the basal area were removed by the treatment; the basal area before thinning was in fact 56.5 m^2 ha^{-1}, and after thinning, it was 41.9 m^2 ha^{-1}. Regarding the dendrometric situation before the treatment, no statistically significant difference was recorded between the thinned areas and control (Table 1). The treatment effect in the year of the thinning (1998) led to some significant changes in most of the dendrometric parameters. DBH and tree height were significantly increased (Table 1). Tree density, basal area, growing stock, and slenderness ratio were significantly decreased (Table 1). In detail, the changes due to the thinning and their development during the last two decades are shown in Table 1. The number of trees recorded in 1998 decreased in the thinned area by about 23% and 26% in 2008 and 2017, respectively. In the control area, the reduction was 13% in 2008 and 14% in 2017. The reduction in the number of trees was mainly due to trees that were uprooted or broken off by wind or snow. The tree reduction recorded in about twenty years in the thinned area was mainly due to the death of undamaged (84.8% of dead trees) rather than damaged trees (15.2%). The DBH increase recorded in 2017 was about +54% higher than that in 1998 after thinning, thus maintaining the positive trend shown in 2008. The decrease of basal area recorded in the thinned area in 1998 and 2008 (26% and 1%, respectively), changed trend completely in 2017, showing an increase of 28.5%, similar to that recorded for the control area (31.9%), although the highest value was in the thinned area (Table 1). The average tree height and growing stock showed an important increase. Tree height showed no statistical difference between the thinned and control areas in 2017; growing stock increased similarly to that of the control area over time, showing a lower statistical difference. The slenderness ratio in 2017 significantly decreased (15%) (Table 1) in the thinned area, while it increased in the control area (2%) relative to 1998 (Table 1).

From 1998 to 2017, 270 trees died in the WPs (wound plots) (Table 2), i.e., a further 50 trees in comparison with 2008. The dead trees consisted of 230 undamaged trees (26.2% of the total undamaged trees) and 40 damaged trees (29% of the total damaged trees).

Immediately after thinning, 13.6% of the standing trees showed damage (Table 2) and 36.2% of the damaged trees showed severe injuries (Table 3). As found ten years after thinning, no wound was completely healed in 2017. In 2017, the percentage of damaged trees was about 17.9%, and out of this, 87.2% showed severe wounds (Table 3). These results were partially concordant with the data of 2008 (17.0% of damaged trees and 86.7% with severe wounds).

Table 1. Dendrometric parameters before and after thinning (1998; 2008; 2017) (mean ± SD).

Dendrometric Parameters	Before Thinning 1998			After Thinning							Control Area	
	Control	Thinned	A	1998	B	2008	C	2017	D	E	2017	F
Growth rings at breast height (N.)	22 ± 1.7	22 ± 1.9	ns	22 ± 1.9	ns	32 ± 2.3	***	40 ± 0.5	***	**	40 ± 0.6	ns
Trees (N. ha^{-1})	1585 ± 7.5	1600 ± 11.3	ns	988 ± 12.9	***	764 ± 19.6	***	726 ± 10.2	**	*	1368 ± 15.7	***
DBH [a] (cm)	20.8 ± 0.6	21.2 ± 0.9	ns	23.2 ± 1.1	**	30.5 ± 1.8	**	35.7 ± 0.8	***	*	25.7 ± 0.9	**
Basal area (m^2 ha^{-1})	53.9 ± 1.1	56.5 ± 0.9	ns	41.9 ± 2.1	**	55.8 ± 3.6	**	72.6 ± 4.1	***	**	71.1 ± 10.2	*
Tree height (m)	14.1 ± 0.7	14.4 ± 1.2	ns	15.0 ± 0.7	*	20.6 ± 1.3	**	24.4 ± 1.1	***	**	24.3 ± 0.6	ns
Growing stock (m^3 ha^{-1})	395 ± 10.5	410 ± 8.2	ns	329 ± 5.4	*	610 ± 6.1	**	827 ± 4.1	***	**	797 ± 6.5	*
Slenderness ratio [b]	87 ± 1.6	86 ± 0.9	ns	83 ± 1.8	*	78 ± 0.7	**	73 ± 0.9	**	*	89 ± 2.6	***

t-test applied to: A, thinned (25 DP) vs. control areas (5 DP) in 1998; B, thinned areas before (25 DP) and after thinning (25 DP) in 1998; C, thinned areas in 1998 (25 DP) and 2008 (25 DP); D, thinned areas (25 DP) and 2017 (25 DP); E, thinned areas in 2008 (25 DP) and 2017 (25 DP); F, thinned (25 DP) and control (5 DP) areas in 2017. [a] calculated by means of the average value of the basal area of each tree; [b] average value of H/DBH of each tree. * $p \leq 0.05$; ** $p \leq 0.01$; *** $p \leq 0.001$; ns = not significant.

The DBH of the damaged trees was significantly higher in 1998 than the DBH of undamaged trees, but for this parameter in 2017, no statistical difference was found (Table 2).

Table 2. Damage to trees and diameter at breast height (±SD) after thinning in 1998, 2008, and 2017.

Tree Condition	Year	DBH (cm)	Trees (N. ha^{-1})	Dead Trees (N. ha^{-1})
Undamaged	1998	20.8 a ± 0.1	877 a ± 30	
Damaged	1998	22.1 b ± 0.5	138 b ± 7	
Undamaged	2008	30.9 c ± 0.2	660 c ± 20	182 ± 50.4
Damaged	2008	29.8 c ± 0.6	135 b ± 19	38 ± 8.3
Undamaged	2017	35.7 d ± 0.3	612 c ± 12	230 ± 25.1
Damaged	2017	34.9 d ± 0.9	133 b ± 11	40 ± 6.1
ANOVA *p*-value				
Damage		0.902	<0.001	
Year		<0.001	<0.001	
Damage × Year		<0.01	<0.05	

Different letters show significant differences among values in a column (Tukey test, N = 4 wound plots).

The DBH of trees with severe or light damage did not show statistical differences (Table 3).

Some logging damage developed over time, and therefore delayed damage due to undetectable injuries was only recognized sometime after the thinning. However, the disclosure of "hidden" wounds, due to impacts, as shown in Table 3, is closely linked to the first ten years after the damaging event.

Significant differences were recorded by [25] in the number of damaged trees in relation to the damage agent (Table 4). In particular, damage to standing trees was mainly caused by skidding and bunching (81.2% of damaged trees); about 14.5% of trees were damaged by felling and only 4.3% showed injuries due to both felling and bunching/skidding. The damage agent had no significant incidence on tree death from 1998 to 2017; as shown in Table 4, the death rate of 25% was similar between the felling and skidding damage agents, and only multiple damage was strictly related to tree death (100%). Tree mortality was greater in the first 10 years of the survey and then decreased.

Table 3. Damage severity in trees and diameter at breast height (±SD) after thinning in 1998, 2008, and 2017.

Damage Severity	Year	DBH (cm)	Trees (N. ha^{-1})	Dead Trees (N. ha^{-1})	Variation 1998–2017 (N. ha^{-1})
Light damage	1998	22.0 a ± 0.6	88 a ± 13		
Severe damage	1998	22.0 a ± 0.5	50 b ± 8		
Light damage	2008	30.7 b ± 1.1	18 c ± 11	28 ± 8.6	−42 ± 5.1
Severe damage	2008	29.7 b ± 0.5	117 d ± 11	10 ± 2.3	+77 ± 4.8
Light damage	2017	36.1 c ± 1.4	17 c ± 11	29 ± 8.4	−42 ± 5.1
Severe damage	2017	35.1 c ± 0.6	116 d ± 11	11 ± 1.6	+77 ± 4.8
ANOVA *p*-value					
Severity		0.122	<0.001		
Year		<0.001	0.420		
Severity × Year		0.081	<0.001		

Different letters show significant differences among values in a column (Tukey test, N = 4 wound plots).

At the time of thinning (Table 5), about 40.6% of the damage was located at the stem base, 39.1% between 0.3 m a.g.l. and 1 m a.g.l. (above ground level), and 10.9% at more than 1 m a.g.l. Only 2.9% of trees showed recognizable root damage and 6.5% showed damage in several parts. In 2017, all trees with damage to roots or several parts died, 56.4% showed damage at the stem base, 36.1% between 0.3 m a.g.l. and 1 m a.g.l., and only 7.5% showed damage at more than 1 m a.g.l. Damage usually occur right after the treatment, but sometimes developed over time. In this study, 35 trees classified as undamaged in 1998 showed severe damage in 2008. The other 42 trees classified as lightly damaged in 1998 showed severe wounds in 2008. These variations may be explained by a late damage effect, i.e., some trees were hit during logging but did not show clear and recognizable damage in 1998.

Table 4. Damage agent in trees and diameter at breast height (±SD) after thinning in 1998 and dead trees in 2017.

Damage Agent	Year	DBH (cm)	Trees (N. ha^{-1})	Dead Trees in Year 2017 (N. ha^{-1})
Felling	1998	21.2 ± 1.3	20 b ± 11	5 ± 2.3
Skidding	1998	22.1 ± 0.6	112 a ± 9	28 ± 3.1
Both	1998	21.8 ± 4.2	6 c ± 2	7 ± 1.1
ANOVA *p*-value				
Agent		0.389	<0.001	

Different letters show significant differences among values in a column (Tukey test, N = 4 wound plots).

Table 5. Damage position in trees and DBH (±SD) after thinning in 1998, 2008, and 2017.

Damage Position	Year	DBH (cm)	Trees (N. ha^{-1})	Dead Trees (N. ha^{-1})
Roots	1998	23.8 ± 5.4	4 a ± 3	
Stem base	1998	22.2 ± 0.3	56 b ± 11	
0.3–1 m a.g.l.	1998	21.6 ± 0.6	54 b ± 35	
>1 m a.g.l.	1998	21.0 ± 1.0	15 c ± 11	
Several parts	1998	21.6 ± 4.2	9 c ± 5	
Kruskal-Wallis *p* [a]		0.366 df(4, N. = 17)	0.018 df(4, N. = 17)	
Roots	2008	n.a.	0	4 ± 3.3
Stem base	2008	30.1 ± 0.3	77 a ± 23	0
0.3–1 m a.g.l.	2008	29.6 ± 1.5	48 a,b ± 23	11 ± 3.8
>1 m a.g.l.	2008	29.9 ± 1.9	10 b ± 5	14 ± 5.2
Several parts	2008	n.a.	0	9 ± 2.0
Kruskal-Wallis *p*		0.874 H(2, N. = 12)	0.014 H(2, N. = 12)	
Roots	2017	n.a.	0	4 ± 3.3
Stem base	2017	35.4 ± 0.6	75 a ± 15	2 ± 1.2
0.3–1 m a.g.l.	2017	36.1 ± 0.5	48 b ± 12	11 ± 3.8
>1 m a.g.l.	2017	35.9 ± 0.9	10 c ± 3	14 ± 5.2
Several parts	2017	n.a.	0	9 ± 2.0
Kruskal-Wallis *p*		0.771 df(2, N. = 12)	0.021 df(2, N. = 12)	

[a] Damage to roots and several positions was only found in two and three plots, respectively, in 1998. Different letters show significant differences among positions according to the K–W test. Data were recorded in each of the four wound plots. n.a., not available because the trees with damage to either roots or several positions were dead.

Some statistically significant correlations were found in 1998 (agent vs. severity, position vs. severity and agent) and in 2017 (severity vs DBH), but all were characterized by a low value of R, which was always less than 0.4 (Table 6).

The average ring width analysis showed a significant statistical difference among the different periods (Table 7). Data showed that the mean ring width before thinning (BT) was the largest and the mean ring width in the ten years after thinning (TPT1) was the smallest.

Significant differences in the average ring width were found per diameter class, as well as a positive relation between the DBH classes and average ring width. No statistical differences in the average ring width were found for the periods BT and 10 years before thinning (TBT), between trees of the control area, damaged trees, and undamaged trees.

In contrast, significant statistical differences in the average ring width were found for the periods TPT1 and TPT2 in the trees of the control area, with respect to the damaged and undamaged trees.

Table 6. Correlation matrix among diameter at breast height, damage severity, agent, and position (Spearmann R, $p < 0.05$ in bold) after thinning in 1998 (N. = 135), 2008 (N. = 138) and 2017 (N. = 140). Severity, agent, and position were ranked as in Tables 3–5.

	DBH	Severity	Agent
1998			
Severity	R = 0.042	-	
Agent	R = 0.120	R = −0.397	-
Position	R = −0.160	R = 0.242	R = −0.220
2008 [a]			
Severity	R = −0.189	-	-
Agent			
Position	R = −0.088	R = 0.035	-
2017 [a]			
Severity	R = −0.191	-	-
Position	R = −0.082	R = 0.023	-

[a] In 2008 and 2017, it was not possible to determine the agent of damage in trees that showed late damage. Thus, results refer to severity and position only.

The trends of the yearly average ring width (YWW) for the control and damaged and undamaged trees in the thinned area are shown in Figures 1 and 2. The YWW gradually decreased starting from the 1980s, with a similar trend in the thinned (both damaged and undamaged) and control areas. From 2005 to 2006, a clear change in the trend was shown, with differentiation in the control with respect to undamaged and damaged trees. From 2006, the tree YWW of the control area stopped with a negative trend (Table 8 and Figure 1) but showed a low value and without any major changes until 2017, as also demonstrated from the results of the regression analysis (Figure 2).

The damaged and undamaged trees showed a negative trend until 2004, at which point there was then a change in the trend from 2005, as also demonstrated from the results of the regression analysis (Table 8 and Figure 2).

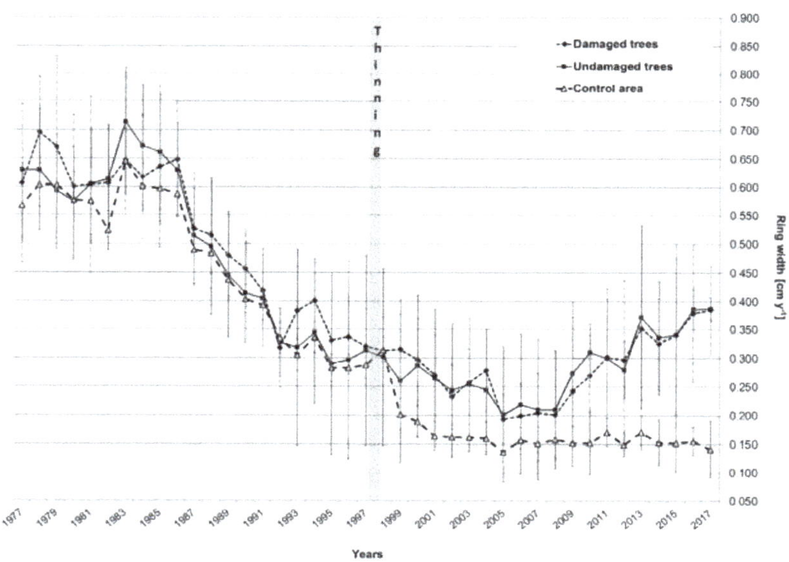

Figure 1. Trend of weighted yearly ring width (YWW) (i.e., the mean of yearly ring width weighted on the basis of the frequency distribution of trees per each diameter class in 2017) from '70s to 2017.

Table 7. Average ring width per diameter class in control area C (N. = 5 plots, 7 diameter classes, 2 stem core samples = 70) and damaged TD (N. = 25 plots, 7 diameter classes, 1 stem core samples = 175) and undamaged trees TU (N. = 25 plots, 7 diameter classes, 1 stem core samples = 175) in thinned area over different time periods (mean ± SD).

Time Periods	Treatment	Ring Width (cm year^{-1}) Per Diameter Class						
		15–20	**20–25**	**25–30**	**30–35**	**35–40**	**40–45**	**45–50**
BT	TU	0.241 ± 0.06 a	0.281 ± 0.05 d	0.377 ± 0.04 f	0.451 ± 0.04 g	0.482 ± 0.07 g	0.595 ± 0.09 h	0.628 ± 0.08 h,i
	TD	0.255 ± 0.09 a	0.301 ± 0.02 d	0.352 ± 0.05 f	0.455 ± 0.05 g	0.507 ± 0.08 g	0.577 ± 0.09 h	0.651 ± 0.09 i
	C	0.228 ± 0.04 a	0.284 ± 0.04 d	0.369 ± 0.06 f	0.437 ± 0.06 g	0.465 ± 0.08 g	0.505 ± 0.10 g	0.608 ± 0.10 h
TBT	TU	0.198 ± 0.02 b	0.222 ± 0.03 a	0.302 ± 0.02 d	0.364 ± 0.02 f	0.455 ± 0.05 g	0.505 ± 0.08 g	0.514 ± 0.08 g
	TD	0.192 ± 0.05 b	0.251 ± 0.06 a	0.305 ± 0.05 d	0.384 ± 0.04 f	0.465 ± 0.02 g	0.511 ± 0.08 g	0.525 ± 0.08 g
	C	0.201 ± 0.08 b	0.212 ± 0.08 a	0.283 ± 0.09 d	0.314 ± 0.08 d	0.386 ± 0.11 f	0.475 ± 0.10 g	0.490 ± 0.12 g
TPT1	TU	0.141 ± 0.09 c	0.168 ± 0.08 c	0.211 ± 0.09 a	0.225 ± 0.03 a	0.278 ± 0.05 d	0.315 ± 0.06 d	0.335 ± 0.05 d
	TD	0.150 ± 0.04 c	0.151 ± 0.07 c	0.205 ± 0.07 a,b	0.214 ± 0.05 a	0.280 ± 0.03 d	0.327 ± 0.07 d,f	0.348 ± 0.07 d,f
	C	0.128 ± 0.10 c	0.140 ± 0.07 c	0.175 ± 0.07 c	0.195 ± 0.09 b	0.252 ± 0.07 a	0.291 ± 0.09 d	0.318 ± 0.10 d
TPT2	TU	0.132 ± 0.10 c	0.156 ± 0.09 c	0.210 ± 0.11 a	0.238 ± 0.09 a	0.298 ± 0.10 d	0.336 ± 0.16 f	0.351 ± 0.07 d,f
	TD	0.135 ± 0.06 c	0.141 ± 0.09 c	0.211 ± 0.09 a	0.229 ± 0.08 a	0.295 ± 0.13 d	0.331 ± 0.08 d,f	0.350 ± 0.09 d,f
	C	0.118 ± 0.15 e	0.114 ± 0.05 e	0.155 ± 0.09 c	0.184 ± 0.09 b	0.241 ± 0.08 a	0.294 ± 0.11 d	0.328 ± 0.11 d
ANOVA					*p*-level			
Time period					**0.019**			
Treatment					**0.048**			
Diameter class					**0.001**			
Time period × Treatment					0.085			
Time period × Diameter class					**0.032**			
Treatment × Diameter class					**0.021**			
Time period × Treatment × Diameter class					**0.040**			

BT: all rings before thinning; TBT: ten rings before thinning; post thinning 2008 (ten rings—TPT1); post thinning 2017 (twenty rings—TPT2). Different letters show significant differences among time periods and diameter classes.

The Wilcoxon non-parametric test showed significant differences after thinning, for the YWW between trees of the control area and damaged trees of the thinned area (p-level 0.008). Similar results were obtained for trees of the control area and undamaged trees of the thinned area (p-level 0.006). Furthermore, no statistically significant difference was exhibited after thinning between damaged and undamaged trees of the thinned area (p-level 0.095).

Table 8. Non-linear regression analysis among ring width (dependent variable) and year (independent variable), for damaged (TD) and undamaged (TU) trees and the trees from the control area (C).

Typologies	Regression Results	Parameters	Values	p-Level
TD	R = 0.929; R^2 = 0.865; R^2 adj = 0.862; df(2,120); $p < 0.001$; Std. Err.: 0.057	Intercept year year^2	2223.65906 −2.21614 0.00055	<0.001 <0.001 <0.001
TU	R = 0.911; R^2 = 0.829; R^2 adj = 0.826; df(2,120); $p < 0.001$; Std. Err.: 0.064	Intercept year year^2	2422.75717 −2.41617 0.00060	<0.001 <0.001 <0.001
C	R = 0.957; R^2 = 0.916; R^2 adj = 0.915; df(2,120); $p < 0.001$; Std. Err.: 0.058	Intercept year year^2	1439.80071 −1.42622 0.00035	<0.001 <0.001 <0.001

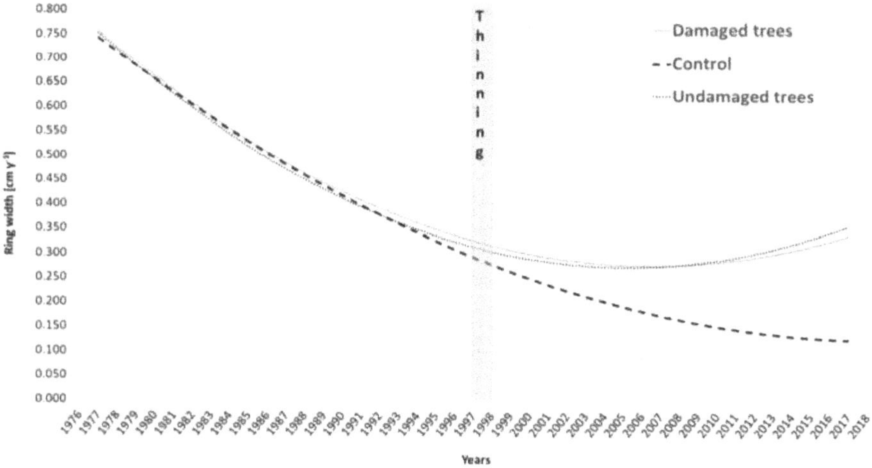

Figure 2. Polynomial regression of weighted yearly ring width (YWW) (i.e., the mean of yearly ring width weighted on the basis of the frequency distribution of trees per each diameter class in 2017) from '70s to 2017. The non-linear regression analysis results are shown in Table 8.

Before thinning, no significant difference was found between the treatments (control, damaged, and undamaged), both in the BT and TBT. After thinning, no significant differences of YWW were found in the thinned area between the damaged and undamaged trees, while the control and thinned area differed significantly, both for damaged (p-level 0.008) and undamaged (p-level 0.006) trees. The non-metric multidimensional scaling (NMDS, Figure 3) of the average ring width per diameter class for the 1998 to 2008 (TPT1) interval highlighted that the C area showed the lowest average ring width, while TD and TU showed a slightly higher average ring width, mainly due to the higher DBH classes. The NMDS of the average ring width per diameter class for C, TD, and TU, focusing on the post thinning period 1998–2017 (TPT2-Figure 4), showed that C had a low average ring width in comparison with TU and TD, while TD with respect to TU showed a similar average ring width. From Figures 3 and 4, the increasing difference over time is clearly visible, between C, and TU and TD. Between TD and TU, no difference was shown over time, and they both showed a higher average ring width of the higher DBH classes (>35 cm) than the trees of the control area.

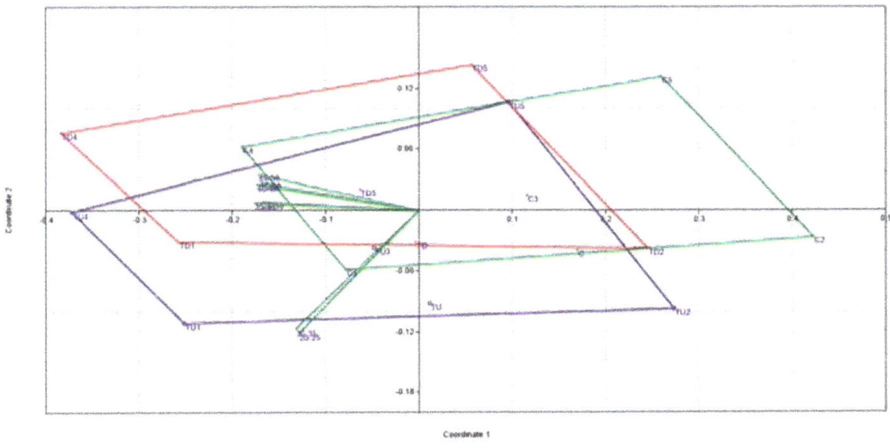

Figure 3. Non-metric multidimensional scaling (NMDS) of the average ring width per diameter class for the trees in the control area C (green polygon, N. = 5 plots, 7 diameter classes, 2 stem core samples = 70), for damaged trees TD (violet polygon, N. = 25 plots, 7 diameter classes, 1 stem core samples = 175), and for undamaged trees TU (red polygon, N. = 25 plots, 7 diameter classes, 1 stem core samples = 175) in the thinned area. The analysis is focused on a specific time period post thinning, from 1998 to 2008 (ten rings—TPT1).

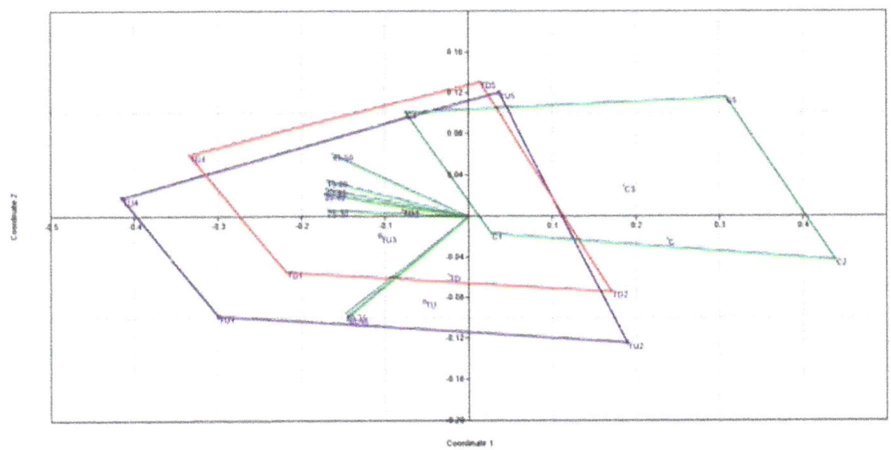

Figure 4. Non-metric multidimensional scaling (NMDS) of the average ring width per diameter class for the trees in the control area C (green polygon, N. = 5 plots, 7 diameter classes, 2 stem core samples = 70), for damaged trees TD (violet polygon, N. = 25 plots, 7 diameter classes, 1 stem core samples = 175), and for undamaged trees TU (red polygon, N. = 25 plots, 7 diameter classes, 1 stem core samples = 175) in the thinned area. The analysis is focused on a specific time period post thinning, from 1998 to 2017 (twenty rings—TPT2).

4. Discussion

This study is a prosecution of a previous one [21] carried out to obtain a long-term analysis of some of the ecological and productive aspects related to thinning in a *Pinus laricio* Poiret stand in Italy. It also takes into consideration the effects of logging damage and of thinning on stand growth 20 years after the treatment, with an intermediate survey in the year 2008.

The decrease in the basal area recorded in the thinned area in 1998 and 2008 completely changed its trend in 2017. In the same period, the average DBH and height in the treated area were greater than in the control area. The effect of thinning was recorded on each dendrometric parameter as a general rule. Thinning from below forced an increase in mean diameter, directly caused by the removal of the trees with smaller diameters [36]. Ten years after thinning, only a light growing stock increase was shown with respect to the control area. Twenty year later, thinning increased the growing stock significantly. In thinned stands, the tree to tree competition was reduced. The resource redistribution to a lesser number of trees had consequences on the diametric increase of the residual trees, affecting the growing stock [37–39]. The reductions in stand density through thinning had no effect on tree height. As other authors have observed, height growth was less reactive to thinning than diameter [37,40–42]. Twenty years after the thinning, the height of the thinned area and the control showed no statistically significant differences, probably due to the high natural fertility of the site. As demonstrated by some authors, species with higher values of height are associated with the greatest resource availability due to high metabolic demands and biomass accumulation [43,44].

Thinning modified both the stand density and the tree shape. Immediately after thinning, only a slightly lower slenderness ratio was obtained (Table 1), similar to that before thinning, showing a stand stability improvement, but again was very low [34].

Ten years after thinning, the ratio value decreased, suggesting a higher stand stability in comparison to the untreated area. In 2017, this trend was confirmed with a further decrease of the ratio value in comparison to an increase in the untreated area. The slenderness ratio in the thinned area was lower than that found for black pine in central Italy by [45]. The increased tree stability was due to the greater average DBH in the thinned stand rather than height. The age of these stands was the same. The thinning favored larger spaces for the trees, lowering the competition, allowing an increase in diameter both for the technical treatment (thinning from below), and for better allocation of resources. Reference [46] observed that denser stands were more vulnerable to stem breakage due to the high slenderness ratio. The improvement in stability progressed over time after thinning, involving not only the diameter increase, but also the development of the crown.

Only multiple damage led to tree death. Damage to the roots cannot be excluded, as detection is difficult, and may have been the cause of death. However, mortality was mainly caused by wind or snow-throw as no effective presence or evidence that showed causes directly associated with logging wound or operations was detected. The stand achieved better stability in the thinned area. The thinning from below seemed to favor the stand stability in cases of snowfall or wind storm, as by removing the less stable trees, the "domino effect" was avoided [47]. The damaged tree rate was similar to that found in other studies on logging damage in the thinning of conifer stands [25,48]. It is interesting to note that the DBH of the damaged trees was significantly higher than the undamaged trees, but only in the first 10 years of the survey.

Lightly damaged trees in 1998 were classified as severely damaged 10 years later, showing a worsening condition that was unchanged in the following period. Moreover, some trees that were classified as undamaged in 1998 were classified as severely damaged in 2017, as non-evident injuries occurred over time [21]. This evidence was apparently conflicting as some trees were impacted during logging operations, but this event did not give rise to injuries or to scrapes on bark. Only sometime after the thinning were undetectable injuries recognized [49].

As shown in [21], the extraction was the main damage agent, as evidenced by the applied equipment (skidding) and logging system (tree length system). In fact, ground-based extraction methods, if not supported by adequate technologies and planning, cause higher damage than aerial-based methods (i.e., cable yarder) [24,35,48,50], and the shorter the logs, the lower the damage [51,52].

As found from the analysis up to 2017, no wound was completely healed, while partially healed wounds comprised about 30% of the damaged trees. As found by [27], the wound healing rate is influenced by tree species, tree growth and vigor, and wound characteristics. The authors in [25] suggested a 25–50 year healing period for a 10 cm wide wound in Norway spruce. Reference [53] indicated a 15-year period for the complete healing of <60 cm^2 wounds in Sitka spruce. The commercial value of these pine plantations is not the main objective today, but pinewood has a role in ecosystem services such as soil protection, landscape, and social services. However, injuries can affect the potential forest efficiency. Diseases and insect infestations may afflict trees in an unhealthy physiological condition. Logging injuries to the roots or at the lower part of the bole are easy access points for biological attacks [27,54–56], and in our case, death causing agents.

This result suggests that logging damage does not affect the radial growth of *P. laricio* over time, at least up to 40 years of age. However, some species seems to be more susceptible to logging damage. In Norway spruce, a growth reduction of about 10–30% was observed [57–59], even in height. Scots pine suffered a mean diameter increment reduction of 12.3% [60]. In lime trees, diameter growth was dramatically affected by wounds, decreasing as much as 43%, thus increasing injury severity [61]. In Alder, wounds significantly reduced diameter growth [62]. Other authors have highlighted the lack of difference in damaged or undamaged trees. Reference [63] underlined that ponderosa and lodgepole pine, three years after artificial wounding, showed higher healing rates in the thinned area, but no significant difference in diameter growth was observed between trees in the thinned and control areas. They observed that no decay occurred in wounded pine, while it frequently occurred in grand fir. On *Pinus pinaster* trees, scarred by debris transported during flash floods, [64] noticed that ring widths of the disturbed trees decreased dramatically near the injuries, but on the opposite site, ring widths were not significantly different.

Damage to roots is difficult to detect since soil modification can be arduous to quantify in relation to single trees. Root injuries can reduce the radial and height increment of Norway spruce more than stem injuries [59].

However, there are many factors that may explain the difference between our results and those of most of the other previous studies. First of all, the species and geographic area, the Mediterranean area in our study, and North-Europe in most of the other studies, may affect the results. In a recent study on Norway spruce and Scots pine, [32] did not find any relation between artificial damage and radial, height, or volume increment.

All treatments (damaged, undamaged, and control trees) showed a reduction over time of average ring width, in the periods BT and TPT1. However, the different trend in the yearly average ring width (YWW) after thinning highlighted that at the stand level, the growth was higher in the thinned area than in the control area. The delay of about seven years in the growth response after thinning was also found by other authors [19,20] and corresponded to a short-term decline in the growth of remaining trees, called "thinning shock". After this phenomenon, a clear recovery of growth started.

5. Conclusions

The results of the first step of this investigation [21] suggested that more specific studies on this topic were needed, as well as an effective assessment of logging damage to be made over long time periods. In this study, logging damage was monitored and the effects of thinning on stand growth were evaluated in the 20 years after the treatment.

These analyses were done for a better comprehension of the ecological and productive aspects due to the management of pine forests, in a *Pinus laricio* Poiret stand in Central Italy. In detail, the aims were:

- to assess the stand dynamics in relation to the injury levels and the treatment applied in a twenty-year range: in about twenty years after the treatment, the stand dynamics showed a complete recovery, and associated implementations, of growth and stability characteristics, independent from the logging injury;
- to evaluate the effect of damage on the radial growth of trees: the results found suggested that logging damage did not affect the radial growth of *P. laricio* over time;
- to try to understand a possible treatment return time and evaluate the existence of the "thinning shock" situation: in consideration of what was found, a second treatment seemed to be sustainable starting from the fifteenth year after the previous treatment. Regarding thinning shock, this can be clearly evaluated in the tree ring width in the first six to seven years after the treatment.

The findings are key factors in the ecological management of Corsican pine forest located in the Mediterranean area and an important reading vital to supporting the decision making of forest managers. In addition, from this case study, some suggestions are proposed to reduce the impact during logging in the thinning of pine stands: (i) promote environmentally responsible and locally acceptable forest mechanization levels; (ii) invest in workforce training that improves not only operational skills, but also an awareness of ecological and environmental issues; (iii) the selection of an adequate timber extraction system during silvicultural planning; (iv) pursue detailed operational planning to reduce potential negative impacts and improve their economic efficiency and safety; and (v) make a multitasking plan based on a close dialogue between sustainable forest management and sustainable forest operations.

Furthermore, for proper ecological management, as well as for an active forest management of pine forests, pragmatic harvesting guidelines and best management practices will be needed. This approach could minimize negative impacts related to active management practices.

Author Contributions: Conceptualization: R.P. and E.M.; Methodology: R.P., R.V. and A.L.M.; Validation: R.P., R.V. and F.L.; Investigation: R.P., R.V., F.L. and A.L.; Data Curation: R.P., R.V., A.L.M.; Writing-Original Draft Preparation: R.P., R.V., E.M., A.L. and A.L.M.; Writing-Review & Editing: R.P. and E.M.; Supervision: R.P., E.M. and A.L.M.

Conflicts of Interest: The authors declare no conflict of interest.

References

1. Kint, V.; Mohren, G.M.J.; Geudens, G.; De Wulf, R.; Lust, N. Pathways of stand development in ageing Pinus sylvestris forests. *J. Veg. Sci.* **2004**, *15*, 549–560. [CrossRef]
2. Zenner, E.K.; Hibbs, D.E. A new method for modeling the heterogeneity of forest structure. *For. Ecol. Manag.* **2000**, *129*, 75–87. [CrossRef]
3. Chiavetta, U.; Skudnik, M.; Becagli, C.; Bertini, G.; Ferretti, F.; Cantiani, P.; Di Salvatore, U.; Fabbio, G. Diversity of structure through silviculture. *Ital. J. Agron.* **2016**, *11*, 1–175.
4. Corona, P.; Ascoli, D.; Barbati, A.; Bovio, G.; Colangelo, G.; Elia, M.; Garfi, V.; Iovino, F.; Lafortezza, R.; Leone, V.; et al. Integrated forest management to prevent wildfires under Mediterranean environments. *Ann. Silvic. Res.* **2015**, *39*, 1–22.
5. Kerr, G.; Hauf, J. Thinning Practice. A Silvicultural Guide. Version 1.0. January 2011. Forestry Commission. Available online: http://www.forestry.gov.uk/pdf/Silviculture_Thinning_Guide_v1_Jan2011.pdf/$FILE/Silviculture_Thinning_Guide_v1_Jan2011.pdf (accessed on 1 February 2011).
6. Barbero, M.; Bonin, G.; Loisel, R.; Quézel, P. Changes and disturbances of forest ecosystems caused by human activities in the western part of the Mediterranean Basin. *Vegetatio* **1990**, *87*, 151–173. [CrossRef]
7. Ortuño, F. El plan para la repoblación forestal de España del año 1939. Análisis y comentarios. *Ecol. Fuera Ser.* **1990**, *1*, 373–392.
8. Marchi, M.; Paletto, A.; Cantiani, P.; Bianchetto, E.; De Meo, I. Comparing Thinning Systems Effects on Ecosystem 2 Services Provision in Black Pine (*Pinus nigra* J.F. Arnold) Artificial Forests. *Forests* **2018**, *9*, 188. [CrossRef]

9. Kovats, R.S.; Valentini, R.; Brower, L.M.; Georgopoulou, E.; Jacob, D.; Martin, E.; Rounsevell, M.; Soussana, J.-F. *Climate Change 2014: Impacts, Adaptation, and Vulnerability. Part B: Regional Aspects*; Contribution of Working Group II to the Fifth Assessment Report of the Intergovernmental Panel on Climate Change; Cambridge University Press: Cambridge, UK; New York, NY, USA, 2014; pp. 1267–1326.
10. Moriondo, M.; Good, P.; Durao, R.; Bindi, M.; Giannakopoulos, C.; Corte Real, J. Potential impact of climate change on fire risk in the Mediterranean area. *Clim. Res.* **2006**, *31*, 85–95. [CrossRef]
11. Lindner, M.; Garcia-Gonzalo, J.; Kolström, M.; Green, T.; Reguera, R.; Maroschek, M.; Seidl, R.; Lexer, M.J.; Netherer, S.; Schopf, A.; et al. Impacts of Climate Change on European Forests and Options for Adaptation. Report to the European Commission Directorate-General for Agriculture and Rural Development. 2008. Available online: https://ec.europa.eu/agriculture/external-studies/euro-forests_en (accessed on 20 January 2018).
12. Gavinet, J.; Vilagrosa, A.; Chirino, E.; Granados, M.E.; Vallejo, V.R.; Prévosto, B. Hardwood seedling establishment below Aleppo pine depends on thinning intensity in two Mediterranean sites. *Ann. For. Sci.* **2015**, *72*, 999–1008. [CrossRef]
13. Park, J.; Kim, T.; Moon, M.; Cho, S.; Ryu, D.; Seok Kim, H. Effects of thinning intensities on tree water use, growth, and resultant water use efficiency of 50-year-old *Pinus koraiensis* forest over four years. *For. Ecol. Manag.* **2018**, *408*, 121–128. [CrossRef]
14. Breda, N.; Granier, A.; Aussenac, G. Effects of thinning on soil and tree water relations, transpiration and growth in an oak forest (*Quercus petraea* (Matt.) Liebl.). *Tree Physiol.* **1995**, *15*, 295–306. [CrossRef] [PubMed]
15. Aussenac, G. Interactions between forest stands and microclimate: Ecophysiological aspects and consequences for silviculture. *Ann. For. Sci.* **2000**, *57*, 287–301. [CrossRef]
16. Ganatsios, H.P.; Tsioras, P.A.; Pavlidis, T. Water yield changes as a result of silvicultural treatments in an oak ecosystem. *For. Ecol. Manag.* **2010**, *260*, 1367–1374. [CrossRef]
17. Jutras, S.; Plamondon, A.P.; Hokka, H.; Begin, J. Water table changes following precommercial thinning on post-harvest drained wetlands. *For. Ecol. Manag.* **2006**, *235*, 252–259. [CrossRef]
18. Gebhardt, T.; Häberle, K.-H.; Matyssek, R.; Schulz, C.; Ammer, C. The more, the better? Water relations of Norway spruce stands after progressive thinning. *Agric. For. Meteorol.* **2014**, *197*, 235–243. [CrossRef]
19. Harrington, C.A.; Reukema, D.L. Initial shock and long-term stand development following thinning in a Douglas-fir plantation. *For. Sci.* **1983**, *29*, 33–46.
20. Dore, S.; Montes-Helu, M.; Hart, S.C.; Hungate, B.A.; Koch, G.W.; Moon, J.B.; Finkral, A.J.; Kolb, T.E. Recovery of ponderosa pine ecosystem carbon and water fluxes from thinning and stand-replacing fire. *Glob. Chang. Biol.* **2012**, *18*, 3171–3185. [CrossRef] [PubMed]
21. Picchio, R.; Neri, F.; Maesano, M.; Savelli, S.; Sirna, A.; Blasi, S.; Baldini, S.; Marchi, E. Growth effects of thinning damage in a Corsican pine (*Pinus laricio* Poiret) stand in central Italy. *For. Ecol. Manag.* **2011**, *262*, 237–243. [CrossRef]
22. Zhang, Q.; Alfaro, R.I.; Hebda, R.J. Dendroecological studies of tree growth, climate and spruce beetle outbreaks in Central British Columbia, Canada. *For. Ecol. Manag.* **1999**, *121*, 215–225. [CrossRef]
23. Larson, A.J.; Franklin, J.F. The tree mortality regime in temperate old-growth coniferous forests: The role of physical damage. *Can. J. For. Res.* **2010**, *40*, 2091–2103. [CrossRef]
24. Picchio, R.; Spina, R.; Calienno, L.; Venanzi, R.; Lo Monaco, A. Forest operations for implementing silvicultural treatments for multiple purposes. *Ital. J. Agron.* **2016**, *11*, 156–161.
25. Vasiliauskas, R. Damage to trees due to forestry operation and its pathological significance in temperate forests: A literature review. *Forestry* **2001**, *74*, 319–336. [CrossRef]
26. Bertolotto, P.; Calienno, L.; Conforti, M.; D'Andrea, E.; Lo Monaco, A.; Magnani, E.; Marinšek, A.; Micali, M.; Picchio, R.; Sicuriello, F.; et al. Assessing indicators of forest ecosystem health. *Ann. Silvic. Res.* **2016**, *40*, 64–69.
27. Tavankar, F.; Picchio, R.; Nikooy, M.; Lo Monaco, A.; Venanzi, R.; Iranparast Bodaghi, A. Healing rate of logging wounds on broadleaf trees in Hyrcanian forest with some technological implications. *Drewno* **2017**, *60*, 65–80. [CrossRef]
28. Lutz, J.; Halpern, C. Tree Mortality during Early Forest Development: A Long-Term Study of Rates, Causes, and Consequences. *Ecol. Monogr.* **2006**, *76*, 257–275. [CrossRef]
29. Mäkinen, H.; Hallaksela, A.M.; Isomäki, A. Increment and decay in Norway spruce and Scots pine after artificial logging damage. *Can. J. For. Res.* **2007**, *37*, 2130–2141. [CrossRef]

30. Zhang, J.W.; Webster, J.; Young, D.H.; Fiddler, G.O. Effect of thinning and soil treatments on *Pinus ponderosa* plantations: 15-year results. *For. Ecol. Manag.* **2016**, *368*, 123–132. [CrossRef]
31. Fontana, G.; Cherubini, P.; Rigling, D.; Dobbertin, M.; Brang, P.; Innes, J.L. Ricostruzione della storia di un popolamento di pino montano (*Pinus mugo* Turra) nel Parco Nazionale Svizzero: Gli anelli annuali testimoni di azioni di disturbo antropiche e naturali. *Monti e Boschi* **2003**, *6*, 34–39.
32. Hurlbert, S.H. Pseudorepliction and the design of ecological field experiments. *Ecol. Monogr.* **1984**, *54*, 187–211. [CrossRef]
33. Hermanin, L.; Sani, L. Indagini sulla produzione del pino laricio in Toscana. *Annali dell'Istituto Sperimentale per la Selvicoltura* **1989**, *20*, 571–600.
34. La Marca, O. Il problema degli schianti nei boschi. Ricerche sperimentali su alcuni popolamenti di conifere. *Ann. Acad. Ital. Sci. For.* **1983**, *32*, 69–114.
35. Picchio, R.; Neri, F.; Petrini, E.; Verani, S.; Marchi, E.; Certini, G. Machinery-induced soil compaction in thinning two pine stands in central Italy. *For. Ecol. Manag.* **2012**, *285*, 38–43. [CrossRef]
36. Mäkinen, H.; Isomäki, A. Thinning intensity and long-term changes in increment and stem form of Scots pine trees. *For. Ecol. Manag.* **2004**, *203*, 21–34. [CrossRef]
37. Mäkinen, H.; Isomäki, A. Thinning intensity and growth of Scots pine stands in Finland. *For. Ecol. Manag.* **2004**, *201*, 311–325. [CrossRef]
38. Martin-Benito, D.; Kint, V.; del Río, M.; Muys, B.; Cañellas, I. Growth responses of West-Mediterranean Pinus nigra to climate change are modulated by competition and productivity: Past trends and future perspectives. *For. Ecol. Manag.* **2011**, *262*, 1030–1040. [CrossRef]
39. Sohn, J.A.; Gebhardt, T.; Ammer, C.; Bauhus, J.; Häberle, K.-H.; Matyssek, R.; Grams, T.E.E. Mitigation of drought by thinning: Short-term and long-term effects on growth and physiological performance of Norway spruce (*Picea abies*). *Ann. For. Sci.* **2008**, *65*, 308. [CrossRef]
40. Del Río, M.; Calama, R.; Cañellas, I.; Roig, S.; Montero, G. Thinning intensity and growth response in SW-European Scots pine stands. *Ann. For. Sci.* **2008**, *65*, 308. [CrossRef]
41. Peterson, J.A.; Seiler, J.R.; Nowak, J.; Ginn, S.E.; Kreh, R.E. Growth and physiological responses of young loblolly pine stands to thinning. *For. Sci.* **1997**, *43*, 529–534.
42. Oliver, W.W. *Growth after Thinning Ponderosa and Jeffrey Pine Pole Stands in Northeastern California*; USDA Forest Service Research Paper PSW-85; Pacific Southwest Forest and Range Experiment Station: Berkeley, CA, USA, 1972.
43. Oliveira-Filho, A.T.; Curi, N.; Vilela, E.A.; Carvalho, D.A. Variation in tree community composition and structure with changes in soil properties within a fragment of semideciduous forest in South-eastern Brazil. *Edinb. J. Bot.* **2001**, *58*, 139–158. [CrossRef]
44. Coelho, M.S.; Carlos, P.P.; Pinto, V.D.; Meireles, A.; Negreiros, D.; Morellato, L.P.C.; Fernandes, G.W. Connection between tree functional traits and environmental parameters in an archipelago of montane forests surrounded by rupestrian grasslands. Flora: Morphology, Distribution. *Funct. Ecol. Plants* **2018**, *238*, 51–59. [CrossRef]
45. Cantiani, P.; Chiavetta, U. Estimating the mechanical stability of *Pinus nigra* Arn. using an alternative approach across several plantations in central Italy. *iForest* **2015**, *8*, 846–852. [CrossRef]
46. Nykänen, M.-L.; Peltola, H.; Quine, C.P.; Kellomäki, S.; Broadgate, M. Factors affecting snow damage of trees with particular reference to European conditions. *Silva Fenn.* **1997**, *31*, 193–213. [CrossRef]
47. Del Río, M.; Bravo-Oviedo, A.; Pretzsch, H.; Löf, M.; Ruiz-Peinado, R. A review of thinning effects on Scots pine stands: From growth and yield to new challenges under global change. *For. Syst.* **2017**, *26*, 3. [CrossRef]
48. Spinelli, R.; Magagnotti, N.; Nati, C. Benchmarking the impact of traditional small-scale logging systems used in Mediterranean forestry. *For. Ecol. Manag.* **2010**, *260*, 1997–2001. [CrossRef]
49. Legere, G. Reduction of stem damage by integrating skidding with declaiming. *For. Eng. Res. Inst. Can.* **2001**, *2*, 1–19.
50. Marchi, E.; Picchio, R.; Spinelli, R.; Verani, S.; Venanzi, R.; Certini, G. Environmental impact assessment of different logging methods in pine forests thinning. *Ecol. Eng.* **2014**, *70*, 429–436. [CrossRef]
51. Matzka, P.J.; Kellogg, L.D. Harvest System Selection and Design for Damage Reduction in Noble Fir Stands: A Case Study on the Warm Springs Indian Reservation, Oregon. *West. J. Appl. For.* **2003**, *18*, 118–126.

52. Bakinowska, E.; Szczepańska-Álvarez, A.; Mederski, P.S.; Karaszewski, Z.; Bembenek, M. Parallel application of two probability models, logit and probit, for accurate analysis of spruce timber damage due to thinning operations. *Drewno* **2016**, *59*, 49–59. [CrossRef]
53. Welch, D.; Scott, D.; Staines, B.W. Bark stripping damage by red deer in a Sitka spruce forest in Western Scotland. III. Trends in wound condition. *Forestry* **1997**, *70*, 113–120. [CrossRef]
54. Bettinger, P.; Kellogg, L. Residual stand damage from cut-to-length thinning of second growth timber in the Cascade Range of western Oregon. *For. Prod. J.* **1993**, *43*, 59–64.
55. Camp, A. Damage to residual trees by four mechanized harvest systems operating in small diameter, mixed conifer forests and steep slopes in northeastern Washington: A case study. *West. J. Appl. For.* **2002**, *17*, 14–22.
56. Annesi, T.; Calienno, L.; Picchio, R.; De Simone, D.; Lo Monaco, A. Degradation of some technological features in the wood of ornamental species caused by *Inonotus rickii* (Pat.) Reid. *Drewno* **2015**, *58*, 5–18. [CrossRef]
57. Baader, G. Damage by game in Rheinland-Pfalz and the possibilities to minimize it. *Allg. For. Jagdztg.* **1956**, *127*, 190–212.
58. Vanek, J. Orientacni studio nasledku loupani lesnich porostu zveri. *Lesictvi* **1956**, *3*, 59–78.
59. Isomäki, A.; Kallio, T. Consequences of injury caused by timber harvesting machines on the growth and decay of spruce (*Picea abies* (L.) KARST.). *Acta For. Fenn.* **1974**, *136*, 1–24. [CrossRef]
60. Andersson, S.-O Om lövröjning i plant-och ungskogar. *Sveriges Skogsvaardsfoerbunds Tidskrift* **1984**, *82*, 69–95.
61. Tavankar, F.; Lo Monaco, A.; Picchio, R.; Venanzi, R.; Nikooy, M. Healing ability and diameter growth of lime-trees (*Tilia begonifolia* Stev.) following logging wounds. *Eur. J. For. Res.* **2018**, *137*, 1–11. [CrossRef]
62. Tavankar, F.; Nikooy, M.; Picchio, R.; Bonyad, A.; Venanzi, R. Effects of logging wounds on Caucasian Alder trees (*Alnus subcordata* C.A. Mey.) in Iranian Caspian Forests. *Croat. J. For. Eng.* **2017**, *38*, 73–82.
63. Filip, G.M.; Parks, C.A.; Wickman, B.E.; Mitchell, R.G. Tree Wound Dynamics in Thinned and Unthinned Stands of Grand Fir, Ponderosa Pine, and Lodgepole Pine in Eastern Oregon. *Northwest Sci.* **1995**, *69*, 276–283.
64. Ballesteros, J.A.; Stoffel, M.; Bodoque, J.M.; Bollschweiler, M.; Hitz, O.; Díez-Herrero, A. Changes in Wood Anatomy in Tree Rings of *Pinus pinaster* Ait. Following Wounding by Flash Floods. *Tree-Ring Res.* **2010**, *66*, 93–103. [CrossRef]

forests

MDPI

Article

Strip Clear-Cutting Application and Logging Typologies for Renaturalization of Pine Afforestation—A Case Study

Rodolfo Picchio [1,*], Roberto Mercurio [2], Rachele Venanzi [1], Loretta Gratani [3], Tommaso Giallonardo [4], Angela Lo Monaco [1] and Anna Rita Frattaroli [4]

1 Department of Agriculture and Forest Sciences (DAFNE), Tuscia University, Via S. Camillo de Lellis, 01100 Viterbo, Italy; venanzi@unitus.it (R.V.); lomonaco@unitus.it (A.L.M.)

2 Former full professor of silviculture and forest management at Mediterranean University of Reggio Calabria (Italy); rmercurio@unirc.it or rmercurio@inwind.it

3 Department of Environmental Biology, Università degli Studi di Roma La Sapienza, Piazzale Aldo Moro, 5, 00185 Rome, Italy; loretta.gratani@uniroma1.it

4 Department of Life, Health and Environmental Sciences, Università degli Studi dell'Aquila, Piazzale Salvatore Tommasi 1, Blocco 11, Coppito, 67100 L'Aquila, Italy; tommy.giallonardo@libero.it (T.G.); annarita.frattaroli@univaq.it (A.R.F.)

* Correspondence: r.picchio@unitus.it; Tel.: +39-076-135-7400

Received: 18 April 2018; Accepted: 8 June 2018; Published: 19 June 2018

Abstract: Renaturalization treatment in Black pine afforestation is an important topic that should be considered. There is a need to favor the evolution of artificial pine forests toward natural forest systems. Overall, this study focused on pine forests, and suggests one typology of clear-cutting (dismantling cutting) on strips, which is associated with different extraction management techniques. Some ecological and environmental aspects associated with renaturalization treatments that have been applied by different mechanizations in black pine afforestation have been highlighted, as well as how renaturalization and the active ecological management of these stands could affect soil and vegetation. The main objectives of this research were to: (1) analyze the impact of silvicultural treatment and logging activities on forest soil, and (2) assess tree regeneration and floristic biodiversity in an ecological management system, in terms of both quantity and quality characteristics. These analyses were planned to obtain an overview of the environmental impact related to a multifunctional approach to the forest management of black pine afforestation. Essentially, the answers to the main research questions are: (1) less invasive extraction systems seem to use a cable yarder and forest winch; (2) a clear soil recovery trend with good capabilities is visible, in particular for the two extraction systems by cable; however, over a three-year period, only a partial but substantial recovery has been shown; (3) in general, silvicultural treatment showed qualitative and quantitative improvement in terms of tree regeneration; in particular, the extraction systems by forest winch and cable yarder showed better results; (4) silvicultural treatment seems not to have led to improvement at the level of the herbaceous and shrubby layers; however, clear differences are shown among the different harvesting systems. Significant recovery after use of the cable yarder was observed.

Keywords: dismantling cutting; horse skidding; winch skidding; cable yarder; tree regeneration; soil recovery

1. Introduction

Among the different coniferous species, black pine (*Pinus nigra* Arnold) has been successfully used in Italy since the beginning of last century for afforestation, as well as the recolonization of pastures and degraded environments due to its favorable frugal characteristics [1,2]. Black pine is appropriate for afforestation in shallow soils with severe climatic conditions such as prolonged summer drought, due to its root system extension [1]. In 2005, *Pinus nigra* forests covered an area of approximately 240,000 ha (about 3% of the forested area in Italy) [3]. This afforestation provides soil protection and the restarting of dynamic and evolutionary processes of vegetation.

Tonon et al. [4] highlighted that black pine afforestation modified microclimate and soil conditions, making them suitable for more discerning species. Nevertheless, today many of these afforested areas are in a poor physical and biological state with no dynamic processes. This is due to the combination of several factors (i.e., the inadequate treatment and biotic and environmental adversity). Thus, these areas require urgent silvicultural action in the form of renaturalization [5,6] to ensure the permanence of land cover, with the gradual replacement of pine with late successional tree species that are typical of more mature stages of evolution (native broadleaves). Furthermore, the woody material that was required as a source of renewable energy opens up interesting prospects for the use of biomass obtainable by the renaturalization of these types of forests.

The application of silvicultural treatments has important effects on all of the ecological aspects [7–9]. Soil characteristics and tree functioning may be affected by harvesting due to the modified input of light (quality and quantity), temperature, and water availability. Moreover, it is important to highlight that logging activities, with machine and load traffic, may cause soil compaction, soil horizon mixing, and topsoil removal [10]. Compaction reduces soil porosity and the connectivity of pores, thus increasing soil density and shear strength [9,11,12]. Thus, soil degradation could reduce tree growth [13], and carbon dioxide efflux from the soil may change significantly [14]. After a renaturalization treatment, the restocking of the stand is left to two main natural regeneration processes: through seedlings (gamic) and sprouts (agamic). Due to the high sprouting potential of some broadleaves growing under the canopy cover of the pine forest, the dynamics of recruiting new individuals to existing stands has largely been overlooked [15]. This aspect will be of particular interest for these new forests that are composed of a mosaic of artificial and natural stands.

Biodiversity conservation has long been a goal of European conservation policy [16,17]. However, despite more than 25% of European land having been given some level of protection for conservation, biodiversity continues to decline [18]. One factor contributing to this decline may be unsuitable management practices in protected areas, or, more specifically, an insufficient application of evidence-based conservation recommendations [18,19]; for example, a lack of diversity of management systems with a total abandonment of traditional systems, without considering the possibility of improving them. Biodiversity monitoring is essential in order to support management decisions in maintaining multiple forest ecosystem functions [20]. A better understanding of the roles of the components of biological diversity that support the provision of multiple forest ecosystem services is necessary [21,22].

Logging systems may differ depending on silvicultural management and the final products. The technical and economic utilization of forests depend on various factors concerning terrain conditions, transportation networks, and harvesting technologies, as well as silvicultural treatment and forest operation systems [23,24]. Although in recent times there have been significant innovations in forest operations [8,25] in terms of both technology and methodology, logging activities in Italy are still developed by traditional methods, i.e., motor–manual felling with chainsaws and the use of mules and/or agricultural tractors for extraction [25–27].

Actually, studies focusing on the effects of silvicultural treatment and utilization on soil and plant biodiversity are rare, or are overly guided by extreme currents of thought. Without some clear effects and recovery times, it is very difficult to respect the concept of sustainability. Also for these reasons, companies are seldom required to take into account the impact of their operations on the land and on forest sustainability, or, rather, to consider the real application of sustainable forest management as suggested by Forest Europe.

For these reasons as well as the fundamental purposes of renaturalization, it is important to know and assess the possible impacts on the soil and biodiversity of silvicultural treatments and logging operations, considering that one of the most important problems of the forest sector is to minimize ground damage caused by forest operations [28] and biodiversity loss. In general, the effects of harvesting include changes in vegetation, nutrient availability, and soil microclimate and structure, as well as litter quantity and quality [28]. In particular, forest operations, such as forwarding and skidding, have a high potential for soil compaction [29–31]. However, adequately managed forest ecosystems are claimed to be highly resilient in the long term [32].

The modern silvicultural approaches lead to aggregated retention, which is the practice of leaving unlogged patches within logged areas, and is a form of retention forestry that has gained increasing use globally as an alternative to clear-cut logging [33–35]. In contrast to clear-cutting, retention forestry maintains habitats for species affiliated with a closed forest—thus mitigating the negative effects of timber harvest—while also providing a habitat for early seral species [36,37]. These approaches are characterized by a set of fundamental principles, including an avoidance of clear-cutting, an emphasis on structural diversity and small-scale variability, the deployment of mixed species with natural regeneration, and an avoidance of intensive site-preparation methods [38]. However, although these modern approaches find real justification in the management of "natural" forests, they have some complications in the renaturalization practices or management of forest monospecific plantations. In these cases, which are similar to problems with new silvicultural approaches [38], a suite of ecological, economic, logistical, informational, cultural, and historical constraints currently hinder the wider adoption of alternative silvicultural practices. Individual contexts display their own unique combinations and the relative significance of these constraints, and accordingly, targeted efforts, such as regulations and incentives, may help overcome specific challenges. In a broader context, the possibility of clear-cutting treatments on adequate surfaces and with proper methodologies might provide an additional flexibility and facilitate renaturalization within a broader set of ecological circumstances.

In order to improve silvicultural management and logging methods, better knowledge of the long-term impact of forest operations is needed [26,39]. Reducing the negative effects of felling and extraction is one of the main goals of sustainable forest management [40–42].

The main objectives of this research were to (1) analyze the impact of silvicultural treatment and logging activities on forest soil, and (2) assess tree regeneration and floristic biodiversity in an ecological management system, in terms of both quantity and quality characteristics.

To analyze the impact on soil and short-term recovery in particular, in addition to physical and chemical analyses of soil (i.e., pH, organic matter, bulk density, penetration and shear resistance) [30], an innovative arthropod-based Soil Biological Quality index, QBS-ar [43,44], was used. This is a valuable tool in ecosystem restoration programs for monitoring the development of soil functions and biodiversity, and preventing the negative effects of soil compaction that result from logging activities.

Moreover, other ecological and environmental aspects associated with renaturalization treatments were evaluated in black pine afforestation in relation to tree regeneration characteristics and vegetation dynamics. In particular, the most important points in investigating the renaturalization and active ecological management of these *Pinus nigra* forests were as follows:

- Three mainly different logging methodologies and their immediate impact on some soil features
- Possible recovery time of the impact on soil features over a range of three years
- The effect of silvicultural treatment and logging activities on natural tree regeneration
- First analysis of silvicultural treatment and logging activities on plant structure and functioning

These analyses were planned so as to obtain an overview of the environmental impacts related to a multifunctional approach to the forest management of black pine afforestation.

2. Materials and Methods

2.1. Study Site

The studied areas are located near Passo delle Capannelle Municipality of Pizzoli (AQ 42°26′49 N, 13°20′14 E) (Figure 1). Abruzzo is a region in Italy that has a fairly developed forest sector and a huge forest surface (about 41% of the region) accounting for about 4% of the entire Italian forest surface. In this region, there are about 19,158 ha of coniferous plantations, and of these, black pine afforestation covers about 13,004 ha [44]. The pine afforestation that was studied covers about 27 ha along the middle mountain slope. The two blocks that were chosen for the experiment are located on the southeastern slope in the altitudinal range of 1200–1300 m a.s.l. (above sea level), with an average slope of about 50%. On the slope are three different calcareous formations dating back to the Triassic–Jurassic: lower limestone calcareous, ammonite green, limestone, Posidonia marl, and Corniola. In some portions, there is generally stratified and cemented Pleistocene stratum debris consisting of gravels and breccias, predominantly limestone.

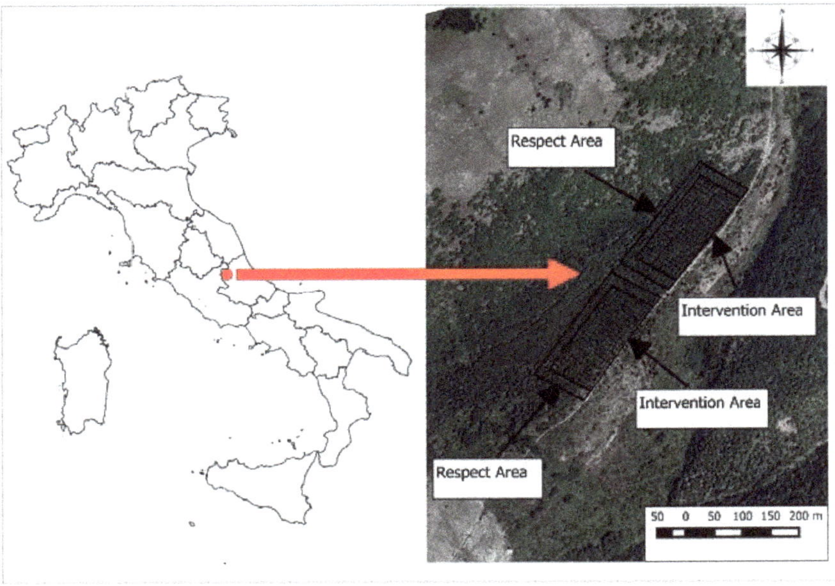

Figure 1. Geographical information and detailed scheme of the treatment area.

A general climatic analysis of the area was conducted on the basis of data available from the meteorological stations of Montereale and Assergi, relating to the years 1950–2000. Rainfall has a bimodal regime, with two annual maximums, in April and November. The driest month is usually July, with about 40 mm of precipitation, and only six rainy days. The winter cold has remarkable peaks, with monthly averages of minimum air temperatures lower than 0 °C for four months; average daily temperatures below 10 °C are exhibited for six months. From the study of the Mitrakos diagrams, it appears that the maximum cold stress was reached only in the month of January, whereas aridity stress was in July. The classification in Rivas-Martínez [45] was used for the bioclimatic description, which defines the considered area as a continental temperate bioclimate, and more precisely as within the attenuated subcontinent subtype. The thermo type was found to be supratemperate lower and ombrotype humid. Additional details regarding temperature and precipitation values throughout the years of the study period are reported in Table 1.

Table 1. Detailed information regarding temperature and precipitation values throughout the years of the study period.

Years	2015	2016	2017	2015	2016	2017
Months	Temperature (°C) (min/mean/max)			Precipitation (mm) Total		
1	−10.0/2.8/14.0	−8.0/4.9/15.0	−3.1/0.1/4.4	69.5	70.1	7.5
2	−5.0/5.2/11.0	−3.0/7.9/18.0	2.5/6.6/11.5	67.6	72.9	0
3	−1.2/10.1/20.1	−0.4/10.8/23.1	2.6/9.0/16.0	67.1	66.8	0
4	1.0/10.8/24.2	−1.1/11.6/27.3	4.3/10.6/17.6	68.9	65.5	0
5	6.1/16.6/31.2	4.1/15.5/30.4	8.6/14.9/21.8	55.1	58.8	3.2
6	10.0/20.6/30.4	9.1/21.4/32.2	14.2/21.5/28.5	45.2	46.6	0.3
7	13.3/23.1/36.1	11.1/22.8/34.0	14.8/22.5/30.1	35.1	36.1	7.4
8	12.3/23.2/33.4	7.1/21.8/32.4	16.0/24.2/32.6	43.2	43.4	0
9	7.1/17.9/33.2	4.1/16.8/30.1	10.1/15.8/21.7	62.1	57.2	0.2
10	4.2/12.1/22.1	2.2/12.6/23.4	5.2/12.6/20.8	89.9	89.1	0
11	−2.0/8.5/21.3	−2.3/8.2/19.4	−0.8/7.3/19.7	85.2	88.8	10.7
12	−5.4/4.5/13.2	−9.1/3.1/12.4	−0.8/3.3/7.9	72.1	92.9	60.5

The black pine (*Pinus nigra* Arnold subsp. *nigra* var. *italica* Villetta Barrea) plantation studied was about 60 years old. The soil preparation was done in steps (called "gradoni", or small terraces) 2 m to 4 m apart, with numerous stone artefacts. The planting was carried out with bare root black pine transplants at a distance of 1 m in the step. It was a pure stand, with poor social differentiation and a high slenderness ratio, which is a clear sign of a lack of thinning. The degree of coverage was high (90–100%) and homogeneous. The distance between the trees was similar to the original planting scheme. The shrubby undergrowth was very poor, while dense *brachypodium* carpets and mosses were intertwined with needles and cones. An analysis conducted on the wooden bores of some trees showed that the annual growth had undergone a significant slowdown in the last eight years (fractions of mm), accentuating the slowdown that occurred 16 years ago, and making it difficult to hypothesize a reaction to any thinning. The limit of 60–65 years represents a threshold beyond which the reaction capacity of the stand to the thinning is greatly reduced [1]. In the considered afforestation, no significant meteoric damage had occurred, there were no obvious signs of fungal and insect attacks, dead wood snags were substantially absent, and logs were not consistent.

2.2. Renaturalization Treatment and Logging Methods

Among the possible renaturalization techniques [46], taking into account the stand age, the need to combine harvesting cost-effectiveness, that it was solid ground despite being on a high slope, and that the black pine forest was sheltered from northern winds, clear-cutting (dismantling cutting) on strips was chosen. This consisted of a series of strips 15 m wide and 100 m long, alternating with uncut strips (to eliminate 50% of the surface of the pine plantation). The width was proportional to the tree height, while the length was proportional to the logging systems and planning. The choice of the two experimental blocks was made through a design-based approach, which is a statistical approach that establishes the methods of choice and use of the sites, allowing possible pseudoreplication problems to be overcome [47]. The two experimental blocks (replicates) were delimited upstream of a forest road, and they differed in altitude location, forest growth conditions, and soil texture. The first experimental block, which was at an altitude of 1200 m a.s.l. (east–southeast), consisted of 12 strips that were 100 m long (according to the lines of maximum slope) and 15 m wide. This block was surrounded on all sides, excluding the track, with a protection buffer that was a minimum of 20 m wide. The second block, with similar characteristics, was realized slightly lower, at an altitude of 1100 m a.s.l. (southeast). The logging treatments within each block were randomly assigned, while the silvicultural treatments were systematically assigned (one uncut strip and one clear-cut strip). Within each block, the silvicultural treatment was applied on six alternating strips (one harvested and one untouched) to have six silvicultural repetitions and six control areas. In the harvested strips, three logging treatments were applied, with two repetitions for every replicate (block). The different extraction methods were characterized by different machines and techniques (detailed in Table 2): (a) animal skidding downhill by TPR (heavy rapid skidding) horse, (b) winching downhill by a forestry-fitted farm tractor using snatch block, and (c) yarding downhill by a cable yarder with automatic carriage. The two blocks (replicates) differed in altitude, soil texture, and dendrometric characteristics.

Table 2. Different extraction methods applied, characterized by different machines and techniques.

Logging Operation	System Detail	Mechanization and Work Team
H	direct skidding of whole tree, on forest soil	Two operators, by "TPR" horse with mass of about 1200 kg
W	indirect skidding of whole tree, from the forest road	Two operators, by forestry-fitted farm tractor using snatch block. Winch of pull force of 70 kN and maximum work distance 100 m, tractor with engine power of 63 kW and mass of about 4 t
C	yarding of whole tree, from the forest road	Two operators, by double drum cable yarder, with automatic carriage with pull force of 20 kN and maximum line work distance 600 m

H: animal skidding; W: winching downhill by a forestry-fitted farm tractor using snatch block; C: yarding downhill by a cable yarder with automatic carriage.

2.3. Analytical Methods

A dendrometric analysis was performed by measuring the diameters at breast height (DBH) of all of the trees, and the plant height of 10% of the trees, which allowed constructing the hypsometric curve and then determining the growing stock by applying the model tree method.

In order to determine the soil particle size distribution for the studied areas, six soil samples in each block were randomly sampled from the top 30 cm of the mineral soil, which was considered to be a crucial indicator of vulnerability to soil compaction [48]. Rock fragments (particles >2 mm in diameter) were removed from the air-dried samples for particle size distribution. Afterwards, three sand fractions—2.00–0.50 mm (coarse), 0.50–0.25 mm (medium), and 0.25–0.05 mm (fine)—were separated by wet sieving. Finally, silt and clay were determined using a hydrometer [31].

Three linear transects, consisting of 1 m × 30 m rectangles for every harvested strip, were tracked to estimate the tree regeneration community composition and perform quantitative analysis. A systematic sampling method was applied as shown in Figure 2. These post-operation analyses were conducted using research methods based on internationally shared protocols, elaborated, and adapted to this context of study as proposed in [8,9,25,26,49].

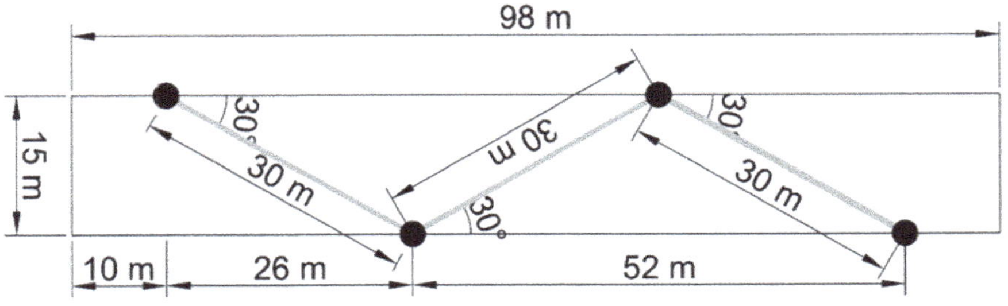

Figure 2. Scheme of the systematic sampling method applied in every harvested strip. The black points are only the starting points of every linear transect (grey lines).

The vegetation study was set up according to the phytosociological method [44]. The surveys were performed in summer, when most of the species were flowering. The analysis was carried out in the entire slope of the forest to investigate the potential vegetation and identify the relationships between plant associations and topography that were established between the different vegetation series. Vascular species were recorded with the percentage of coverage-abundance for each area. The phytosociological surveys carried out in the reforestation area allowed us to characterize the current state of the underbrush. The analysis of the groups of species related to the phytosociological syntaxa allowed us to interpret the current phase of reforestation dynamics and its location in the vegetation series that characterizes the area in which the pinewood was located. The floristic study took place through sample collections before cutting. The samples were determined using the main national [50–54] and European flora [54,55].

The tree composition of natural regeneration was analyzed by the Species Importance Value (SIV) index, which was calculated for each species, as reported in the literature [48,56] (Table 3). The regeneration species diversity index was computed using the Shannon–Wiener information function (Table 3).

Table 3. Detailed formulas and descriptions of the Species Importance Value (SIV) index, Regeneration species diversity index and structural evenness index.

Description	Formula
Species Importance Value (SIV) index	Relative density (RDe) + Relative frequency (RF) + Relative dominance (RDo)
Relative density (RDe)	(Number of individuals of a species × 100)/total number of individuals of all species
Relative frequency (RF)	(Number of plots containing a species × 100)/sum of frequencies of all species
Relative dominance (RDo)	(Sum of the height of a species × 100)/sum of total height of all species
Regeneration species diversity index (Shannon diversity index)	$H' = -\Sigma(ni/n) \cdot Ln\,(ni/n)$ ni = denotes the SIV of a species and n = denotes the sum of the total SIV of all the species
Structural evenness index	$J = H'/Hmax$ H′ is the number derived from the Shannon diversity index and Hmax is the maximum value of H′

The natural regeneration and floristic biodiversity of tree composition were also assessed using species richness. The Shannon index and the evenness index were used for floristic biodiversity, without considering the tree layer before cutting. Biodiversity indices were compared by year and type of intervention. Species richness was defined as the total number of species sampled. The Shannon index [57] is based on information theory and the degree of difficulty in accurately predicting the next species sampled. This diversity index is sensitive to changes in rare species, has good discriminant ability, and is well represented in the literature [58]. Biodiversity indices were compared by year and type of intervention. The structural evenness index was calculated as reported in [59] (Table 2). This index varies between 0 and 1, where a value of 1 symbolizes an exact uniform distribution. The Shannon index is a model that measures species diversity and the degree of homogeneity in species abundance. One of its applications is to correctly estimate the anthropogenic impact on ecological systems.

The impact on soil due to silvicultural treatment and forest operations was assessed by a systematic sampling method selecting three sample plots (SPs) for each harvested strip, one every 30 m along the strip axis, starting at 15 m from the lower border. Each SP consisted of a circular area 12 m in diameter. In each SP, three tests for every parameter were done, which measured: bulk density (BD), pH, organic matter content (OM), inorganic carbon content (IC), penetration resistance (PR), and shear resistance (or strength) (SR), and QBS-ar index as detailed in [43–47]. For the two blocks, two different control areas were selected at the edge of the blocks, which were at least 30 m from the edges. In each area, three SPs were randomly selected, and three tests for all of the parameters were done on each of them.

Natural regeneration analyses were done in 2016 and 2017, which were one and two years after harvesting, respectively. Soil impact analyses were done in 2015, immediately after the logging operations, and in 2016 and 2017, one and two years after harvesting, respectively. All of the analyses were replicated in the same months.

2.4. Statistics

Statistical analyses were carried out using Statistica 7.1 (2007) software (StatSoft Inc., Tulsa, OK, USA). As a first step, data distribution was plotted and checked for normality (Lilliefors) and homogeneity of variance (Levene test). All of the data points then underwent a *t*-test, an ANOVA, or a MANOVA to test the effects of the different treatments. In order to determine the relation between QBS-ar, BD, PR, and SR, a non-parametric correlation analysis (Spearman correlation matrix) was applied. The data, which were not normally distributed and had insufficient homogeneity of variance, were statistically processed using the non-parametric ANOVA Kruskal–Wallis test. Principal component analysis (PCA) was applied to investigate any linear correlations between the expressed soil conditions of the main six characteristics studied, and also between the tree regeneration situations of the main four characteristics studied. To minimize the scaling effect due to the different measurement units, the data corresponding to each independent variable were standardized using Box–Cox lambda.

3. Results

3.1. Silvicultural Treatment and Dendrometric Analysis

The silvicultural treatment applied was clear-cutting (dismantling cutting) on strips, with harvesting of 50% of the surface of the pine plantation. The strip dimensions were proportioned to the tree height and length to the logging system and planning. This treatment aimed to achieve the renaturalization of these forests within a sustainable management system.

The dendrometric data collected before the harvesting of the two blocks showed average values, which were statistically similar for trees age, DBH tree height, basal area, tree density, and aboveground biomass stock (Table 4).

Table 4. Main dendrometric characteristics of the pine forest before cutting (ANOVA test, df 3, 40; average ± SD). DBH: diameters at breast height.

Treatment	Block	Age (Year)	DBH (m)	Height (m)	Density (Trees/ ha)	Basal Area (m^2/ha)	Above-Ground Biomass Stock (m^3/ha)
H	1	60	0.24 ± 0.05 ^{a}	14.4 ± 0.9 ^{a}	728 ± 67 ^{a}	32.9 ± 8.2 ^{a}	308.1 ± 15.2 ^{a}
	2	60	0.19 ± 0.06 ^{b}	13.4 ± 0.6 ^{b}	1220 ± 74 ^{b}	34.6 ± 2.2 ^{b}	301.1 ± 10.5 ^{b}
W	1	60	0.23 ± 0.07 ^{a}	14.3 ± 1.1 ^{a}	720 ± 83 ^{a}	29.9 ± 9.1 ^{c}	277.9 ± 19.3 ^{c}
	2	60	0.18 ± 0.04 ^{b}	13.2 ± 0.8 ^{b}	1232 ± 91 ^{b}	31.3 ± 6.8 ^{a,c}	268.9 ± 19.9 ^{c}
C	1	60	0.22 ± 0.08 ^{a}	14.1 ± 1.8 ^{a}	745 ± 54 ^{a}	28.3 ± 6.7 ^{c}	259.4 ± 21.1 ^{d}
	2	50	0.20 ± 0.08 ^{b}	13.7 ± 0.3 ^{a,b}	1228 ± 47 ^{b}	38.6 ± 5.1 ^{d}	343.4 ± 10.5 ^{e}
Control	1	60	0.22 ± 0.05 ^{a}	14.1 ± 1.0 ^{a}	729 ± 25 ^{a}	27.7 ± 7.1 ^{c}	253.8 ± 11.1 ^{d}
	2	60	0.18 ± 0.03 ^{b}	13.2 ± 0.5 ^{b}	1231 ± 32 ^{b}	31.3 ± 2.9 ^{a,c}	268.6 ± 10.8 ^{c}
p-value			<0.05	<0.05	<0.05	<0.05	<0.05

Different letters show significant differences among values in a column (Tukey test).

3.2. Analysis of Stand Regeneration

No statistical difference was observed between the two blocks, and the nine species were indifferently present ($p > 0.05$). The three logging treatments and the control showed different percentages (ANOVA $p < 0.05$) of species distribution (Figure 3). The tree species most represented were *Sorbus aria* L., *Fraxinus ornus* L., *Quercus pubescens* Willd., and *Quercus cerris* L. According to the Tukey test, the strips extracted by cable yarder were statistically grouped with the control data, while the other two treatments were statistically different. *Pinus nigra* regeneration was largely present in the strips extracted by horse (9%) and winch (7%). There was an exotic tree present in these strips (*Robinia pseudoacacia*).

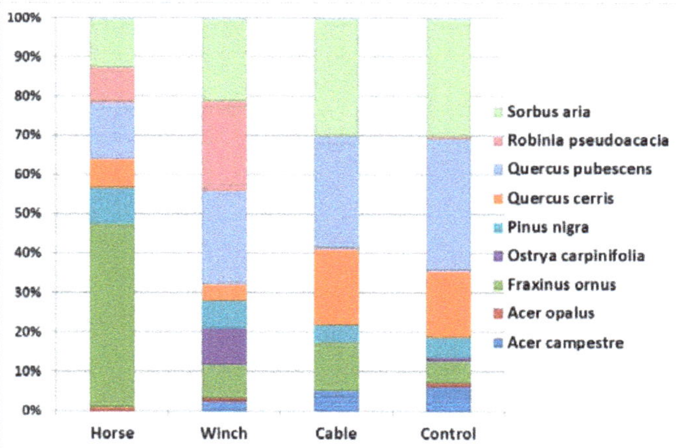

Figure 3. Tree species percentage distribution, average values grouped for logging treatments and control.

One year after harvesting, the highest natural regeneration density was found for the strips extracted by winch and cable yarder: 23,580 trees/ha and 21,150 trees/ha, respectively (Figure 4). The strips extracted by horse showed the lowest value, 6250 trees/ha, and the control had stable values ranging between 10,260–11,150 trees/ha. Two years after harvesting, the highest regeneration density was found for the strips extracted by winch: 20,650 trees/ha, which was slightly lower than in 2016. The strips extracted by horse and cable yarder showed similar values ranging between 18,850–19,170 trees/ha; the first treatment had a considerable increase, while the second one had a slight decrease compared with 2016.

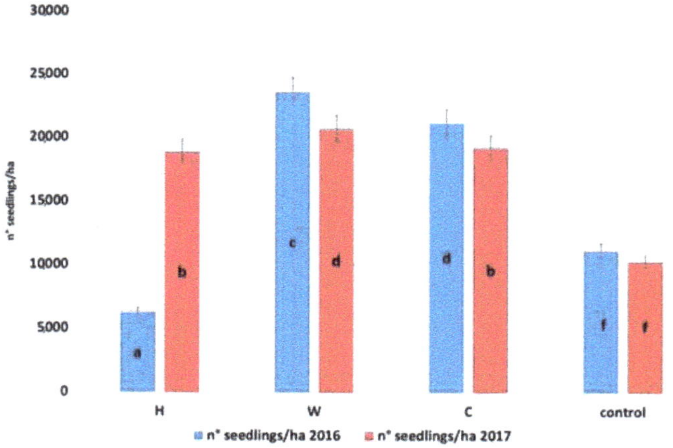

Figure 4. Tree regeneration density, average values with standard deviation for the different treatments, one year after harvesting (2016) and two years after harvesting (2017). ANOVA results shown statistically significant differences, $p < 0.05$ and the Tukey test applied found five groups, showed by different letters.

The SIV detailed analysis for 2017 (Figure 5) shows that the most important tree regeneration was for the species *Q. pubescens* and *Sorbus aria*. The worst SIV was shown by *Q. cerris* and *Fraxinus ornus*. *Fraxinus ornus* only showed an SIV peak in the strips extracted by horse. The analysis of the total SIV (Table 5) showed that the strips harvested by horse and winch increased compared to the control, while for the strips harvested by cable yarder, the trend was positive, but the value was lower than the control.

Tree species diversity, as tested by the Shannon–Wiener and the evenness indices, showed different situations. First, there was a general slightly positive trend from 2016 to 2017 for the control and the strips harvested by winch and cable yarder, while the index was stable for the strips harvested by horse. The highest diversity was found for the strips harvested by cable yarder and winch (in particular for the evenness index, amounting to 0.88 ± 0.07 and 0.85 ± 0.09, respectively, as shown in Table 5), while the lowest diversity was found in the control (Shannon index ranged from 1.68 to 1.70 and the evenness index ranged from 0.73 to 0.77, Table 5).

Table 5. Tree species diversity indexes for the studied areas (Kruskal–Wallis test analysis, df 3, 144; $p < 0.05$; average ± SD).

Treatments	SIV		Shannon Index		Evenness Index	
	2016	2017	2016	2017	2016	2017
H	212 ^{a}	214 ^{a}	1.60 ^{a}	1.58 ^{a}	0.81 ^{a}	0.81 ^{a}
W	205 ^{b}	216 ^{a}	1.79 ^{b}	1.87 ^{c}	0.81 ^{a}	0.85 ^{b}
C	205 ^{b}	208 ^{c}	1.61 ^{a}	1.58 ^{a}	0.82 ^{a}	0.88 ^{b}
Control	215 ^{a}	213 ^{a}	1.68 ^{d}	1.70 ^{d}	0.73 ^{c}	0.77 ^{c}

Different letters show significant differences among values in a column (Tukey test).

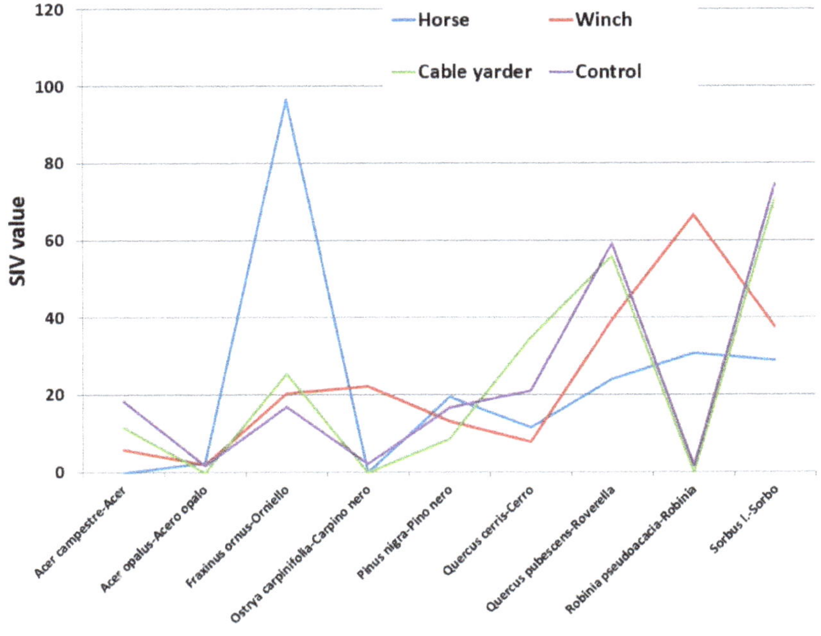

Figure 5. SIV detailed analysis between the treatments and the control for the year 2017.

3.3. Analysis of Floristic Biodiversity

The most consistent group of species, by both qualitative and quantitative analysis, taking into account coverage, was arid secondary grazed meadows (Figure 6). The pool of diagnostic species of the Apennine *Phleo ambigui–Bromion erecti* alliance was highlighted, among which the graminaceous *Brachypodium rupestre* was relevant, forming a fairly uniform mat. It was accompanied by *Cytisus spinescens*, *Asperula purpurea* subsp. *purpurea*, and *Galium lucidum* subsp. *lucidum*. The group of plants belonging to the superior categories of the suborder *Artemisio albae-Bromenalia*, order *Phleo ambigui-Brometalia erecti*, and class *Festuco-Brometea* was very abundant. Among these, the species forming the structure of secondary grasslands in the Apennines was *Bromus erectus*. Even if this vegetation was attributable to the *Phleo ambigui–Bromion erecti* alliance, the difference was the presence in the pinewood of a community dominated by *Brachypodium rupestre*, which was absent in the pasture. This vegetation settled and became dominant in the stations with a more advanced soil type and a higher water content than that in the Apennines. Therefore, it was less xeric, and had a more closed structure compared to the xeric meadows, highlighting the effect of the tree cover.

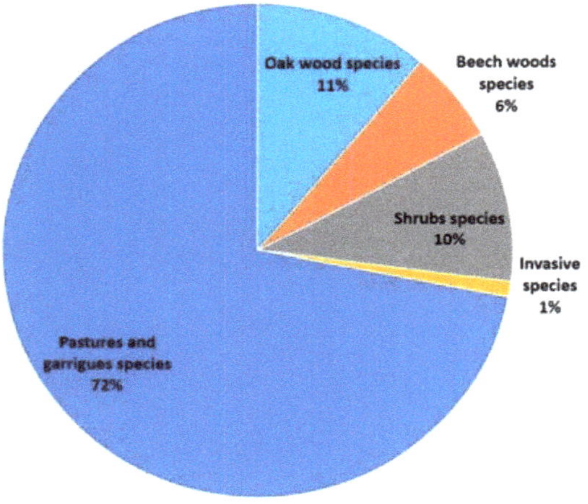

Figure 6. Phytosociological surveys carried out in the afforested area and analysis of the groups of species related to the phytosociological syntaxa.

The results show a constant increase in species richness after cutting (Table 6). This increase does not correspond to an increase in the values of the biodiversity indices, which showed a sharp decline after cutting. However, these values reached higher values after two years before pre-cutting.

A decrease in the values of floristic biodiversity was more evident in the strips harvested by horse, as shown in Figure 7. On the contrary, in the strips where the collection of material was carried out by cable yarder, there was no decrease in Shannon index values (Table 6), but even after the second year, the greatest increases occurred for both indices.

Table 6. Floristic species diversity indexes for the studied areas.

Treatments	Species Richness			Shannon Index			Evenness		
	Before Cutting	1 Year After	2 Years After	Before Cutting	1 Year After	2 Years After	Before Cutting	1 Year After	2 Years After
All	29.92	34.75	36.83	2.03	1.91	2.08	0.6	0.54	0.58
Horses	29.67	32.00	34.67	2.07	1.82	1.90	0.61	0.53	0.54
Winch	25.33	30.00	36.00	1.91	1.91	2.13	0.59	0.56	0.59
Cable Yarder	35.00	45.00	42.00	2.09	2.09	2.39	0.59	0.55	0.64

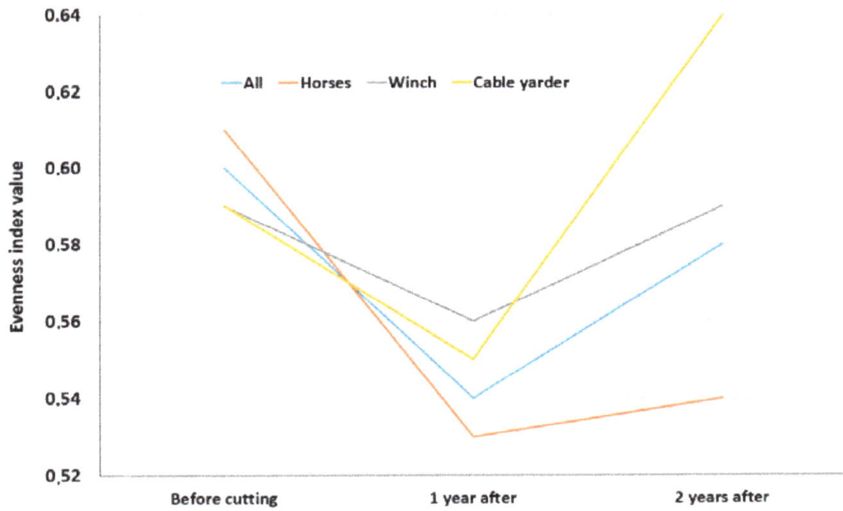

Figure 7. Floristic evenness index trend related to the different harvesting systems.

3.4. Physical and Chemical Analyses of Soil

Soil texture was characterized by a high content of rocky skeleton for both blocks (48.6 ± 17.3%), while sand, silt, and clay content were statistically different between the two blocks ($p < 0.05$). In block 1, sand was 59.2 ± 1.4%, silt was 33.7 ± 1.3%, and the clay was 7.0 ± 1.1%. Thus, the soil can be assigned to the sand loam (SaLo) soil material class. In block 2, the sand was 36.6 ± 3.7%, the silt was 59.4 ± 3.6%, and the clay was 4.0 ± 1.3%. Thus, the soil can be assigned to the silt loam (SiLo) soil material class. Relative to the granulometric values and the soil water method (K. Saxton), the soil field capacity (CC) was calculated at 24% and 27%, respectively.

For each year, sampling was done once in spring, once in summer, and once in autumn. During the sampling periods (2015, 2016, and 2017) soil moisture showed significant differences ($p < 0.05$) between spring and autumn (average moisture 31 ± 5%) and summer (average moisture 22 ± 3%).

Soil BD showed significant differences only among treatments and years (Table 7). In particular, in the strips harvested by horse, BD was higher than the control and the strips harvested by winch and cable yarder. Nevertheless, in 2017, a recovery trend was highlighted. In the strips harvested by winch, the BD was similar to that of strips harvested by cable yarder, but in 2017, no clear recovery trend was shown. In the strips harvested by cable yarder, BD was higher than that of the control, but lower than that of the strips harvested by horse and winch, although in 2017, a complete recovery was shown.

Table 7. Results of the ANOVA and Tukey test for soil bulk density (BD) during the three years after felling (df 3, 144; average ± SD), difference tested between the soil of the harvested area, and control.

Treatments	Bulk Density 2015 (g/cm^3)	Bulk Density 2016 (g/cm^3)	Bulk Density 2017 (g/cm^3)	p-Value
H	1.133 ± 0.08 [a]	1.069 ± 0.09 [d]	0.962 ± 0.03 [b]	<0.05
W	0.946 ± 0.11 [b]	0.945 ± 0.09 [b]	0.928 ± 0.02 [b]	
C	0.947 ± 0.09 [b]	0.913 ± 0.09 [b]	0.867 ± 0.02 [c]	
Control	0.853 ± 0.10 [c]	0.866 ± 0.02 [c]	0.878 ± 0.01 [c]	
p-value	<0.01	<0.01	<0.05	

Different letters show significant differences among values in a column (Tukey test).

Soil PR showed statistically significant differences only among treatments and years (Table 8). In particular, in the strips harvested by horse, PR was higher than that of the control as well as the strips harvested by winch and cable yarder. Nevertheless, in 2017, no clear recovery trend was shown. In the strips harvested by winch, the PR was similar to that of the control from 2015. In the strips harvested by cable yarder, in 2015, the PR was higher than that of the control and strips harvested by winch, but lower than that of the strips harvested by horse, although in 2017, a complete recovery was shown.

Table 8. Results of the ANOVA and Tukey test for soil penetration resistance during the three years after felling (df 3, 144; average ± SD), difference tested between the soil of the harvested area, and control.

Treatments	Penetration Resistance 2015 (MPa)	Penetration Resistance 2016 (MPa)	Penetration Resistance 2017 (MPa)	p-Value
H	0.177 ± 0.06 ^{a}	0.177 ± 0.08 ^{a}	0.175 ± 0.06 ^{a}	<0.05
W	0.161 ± 0.02 ^{b}	0.161 ± 0.03 ^{b}	0.160 ± 0.03 ^{b}	
C	0.169 ± 0.09 ^{c}	0.165 ± 0.07 ^{b,c}	0.162 ± 0.03 ^{b}	
Control	0.160 ± 0.07 ^{b}	0.160 ± 0.05 ^{b}	0.159 ± 0.03 ^{b}	
p-value	<0.05	<0.05	<0.05	

Different letters show significant differences among values in a column (Tukey test).

The soil SR showed statistically significant differences only among treatments and years (Table 9). In particular, in the strips harvested by horse, in 2015 and 2016, SR was higher than that of the control as well as the strips harvested by winch and cable yarder. Nevertheless, in 2017, a complete recovery was shown. In the strips harvested by winch and cable yarder, in 2015, SR was higher than that of the control, although already in 2016, a complete recovery was shown.

Table 9. Results of the ANOVA and Tukey test for soil shear resistance during the three years after felling (df 3, 144; average ± SD), difference tested between soil of harvested area, and control.

Treatments	Shear Resistance 2015 (t/m^2)	Shear Resistance 2016 (t/m^2)	Shear Resistance 2017 (t/m^2)	p-Value
H	2.963 ± 0.11 ^{a}	2.964 ± 0.08 ^{a}	2.089 ± 0.95 ^{c}	<0.05
W	2.294 ± 0.20 ^{b}	2.215 ± 0.14 ^{c}	2.015 ± 0.12 ^{c}	
C	2.347 ± 0.12 ^{b}	2.245 ± 0.14 ^{b,c}	2.016 ± 0.11 ^{c}	
Control	2.185 ± 0.11 ^{c}	2.122 ± 0.32 ^{c}	2.089 ± 0.91 ^{c}	
p-value	<0.05	<0.05	>0.05	

Different letters show significant differences among values in a column (Tukey test).

The pH is a very important parameter for the correct functioning of soil, and its variations influence various soil parameters and processes [60]. The pH values showed statistically significant differences between blocks, treatments, and years. However, this parameter did not seem to be clearly influenced by silvicultural treatment or logging operations (Table 10). The pH variations were recorded only for 2017.

Soil organic matter content (OM) showed significant differences only among treatments and years (Table 11). In particular, in the strips harvested by horse, in 2015, OM was lower than that of the control as well as the strips harvested by winch and cable yarder, and the trends in 2016 and 2017 were negative. In 2015, in the strips harvested by winch and cable yarder, the OM was lower than the control, but for the strips harvested by winch, the trends in 2016 and 2017 were negative. There was a complete recovery for the strips harvested by cable yarder in 2017.

Table 10. Results of the ANOVA and Tukey test for pH (df 3, 72; average ± SD), difference tested between the soil of the harvested area, and control.

Treatments	Block	pH 2015	pH 2016	pH 2017	*p*-Value
H	1	6.5 ± 0.1 [a]	6.4 ± 0.3 [a]	6.1 ± 0.6 [b]	<0.05
	2	6.2 ± 0.2 [b]	6.4 ± 0.1 [a]	6.5 ± 0.3 [a]	
W	1	6.6 ± 0.2 [a]	6.5 ± 0.3 [a]	6.5 ± 0.2 [a]	
	2	6.5 ± 0.3 [a]	6.4 ± 0.2 [a]	6.1 ± 0.1 [b]	
C	1	6.7 ± 0.1 [c]	6.4 ± 0.3 [a]	5.5 ± 0.6 [d]	
	2	6.2 ± 0.1 [b]	6.3 ± 0.1 [a]	6.4 ± 0.3 [a]	
Control	1	6.5 ± 0.2 [a]	6.4 ± 0.7 [a]	6.0 ± 0.6 [b]	
	2	6.9 ± 0.3 [c]	6.5 ± 0.5 [a]	6.5 ± 0.4 [a]	
p-value		<0.05	>0.05	<0.05	

Different letters show significant differences among values in a column (Tukey test).

Table 11. Results of the ANOVA and Tukey test for organic matter (df 3, 72; average ± SD), difference tested between the soil of the harvested area, and control.

Treatments	Organic Matter 2015 (%)	Organic Matter 2016 (%)	Organic Matter 2017 (%)	*p*-Value
H	30.4 ± 2.5 [a]	25.8 ± 2.0 [d]	24.8 ± 1.3 [d]	<0.05
W	34.6 ± 3.2 [b]	25.6 ± 6.1 [d]	17.7 ± 5.1 [e]	
C	33.9 ± 1.2 [b]	35.0 ± 2.1 [b]	35.3 ± 5.3 [b]	
Control	38.2 ± 2.2 [c]	38.1 ± 1.5 [c]	36.6 ± 2.5 [b]	
p-value	<0.05	<0.05	<0.05	

Different letters show significant differences among values in a column (Tukey test).

Soil inorganic carbon content (IC) analysis was done only in 2017. The results showed significant differences only among the treatments and control (Table 12). In particular, in the strips harvested by horse, the IC was lower than that of strips harvested by cable yarder, but it was higher than that of the strips harvested by winch, and similar to the control.

Table 12. Results of the ANOVA and Tukey test for inorganic carbon (df 3, 24; average ± SD), difference tested between soil of harvested area and control.

Treatments	Total Inorganic Carbon 2017 (%)	*p*-Value
H	20.2 ± 3.6 [a]	<0.05
W	13.6 ± 3.3 [b]	
C	26.4 ± 2.1 [c]	
Control	22.4 ± 2.6 [a]	

Different letters show significant differences among values in a column (Tukey test).

3.5. Soil Biodiversity Analysis

The QBS-ar index (Table 13) showed significant differences only among treatments and years. In particular, in the strips harvested by horse and by winch, in 2015, QBS-ar was higher than that of the strips harvested by cable yarder, and lower than that of the control. In 2016, a strong decrease was recorded for the strips harvested by horse, and a slightly decrease was recorded for those harvested by winch compared to the control. Nevertheless, both showed complete recovery in 2017. For the strips harvested by cable yarder, in 2015 and 2016, QBS-ar was unchanged and lower than the control, but in 2017, its value appreciably increased.

Table 13. Results of the Kruskal–Wallis and Tukey test for Soil Biological Quality (QBS-ar) index data (df 3, 144; median), difference tested between disturbed, undisturbed, and control soil.

Treatments	QBS-ar Index 2015	QBS-ar Index 2016	QBS-ar Index 2017	p-Value
H	193 ^{a}	111 ^{d}	228 ^{c}	<0.01
W	195 ^{a}	179 ^{e}	199 ^{a}	
C	143 ^{b}	143 ^{b}	179 ^{e}	
Control	244 ^{c}	199 ^{a}	244 ^{c}	
p-value	<0.05	<0.01	<0.05	

Different letters show significant differences among values in a column (Tukey test).

Microarthropod density (Table 14) showed statistically significant differences among treatments and years. In particular, in the harvested strips, the density values were always lower than the control, but the trends were positive from 2015 to 2017. In 2017, density that was more similar to the control was shown in the strips harvested by winch. More time is necessary for the recovery of the strips harvested by horse and by cable yarder.

Table 14. Results of the Kruskal–Wallis and Tukey test for soil microarthropod density data (df 3, 144; median), difference tested between disturbed, undisturbed, and control soil.

Treatments	Microarthropod Density 2015 (ind/dm^2)	Microarthropod Density 2016 (ind/dm^2)	Microarthropod Density 2017 (ind/dm^2)	p-Value
H	64 ^{a}	81 ^{a}	124 ^{f}	<0.05
W	107 ^{b}	159 ^{d}	161 ^{d}	
C	75 ^{a}	106 ^{b}	100 ^{b}	
Control	175 ^{c}	223 ^{e}	175 ^{c}	
p-value	<0.01	<0.01	<0.01	

Different letters show significant differences among values in a column (Tukey test).

4. Discussion

The management of pine plantations has strong and variable effects on plant species occurrence and diversity due to plantation and treatment operations, and the alteration of ecological processes caused by changes in the landscape and stand structure [61,62]. Plantations contribute to biodiversity conservation in various ways. In Poorbabaei [63], a high similarity in species composition between plantation and the adjacent natural forest, which is the main source of seed in plantations, was considered. Neighboring plantation and natural forest has resulted in the dispersion of hardwood tree seeds within the plantation.

The results highlight that the good density and richness of tree species in this pine plantation indicate the high potential reached by the stand for biodiversity restoration. Similar results are shown in studies where plantations of native and/or exotic tree species increased biodiversity by promoting woody understory regeneration [64,65]. The silvicultural treatment applied in this research showed positive effects on density, richness, and biodiversity of tree species in three years. Logging operations have important effects on regeneration and tree biodiversity, which are the main topics for restoration and renaturalization purposes.

From the results concerning stand regeneration, different taxonomic compositions of the tree forest community among the logging treatments are shown, in particular in the percentage of distribution. A clear simplification with respect to the others is shown in the strips harvested by horse, with the presence of about 45% *Fraxinus ornus*. The general finding shows that ground-based logging systems allowed for the presence of *Robinia pseudoacacia* and only marginal *Pinus nigra* regeneration. In general, felling produced an abundance of light-demanding species due to the increase of solar radiation [48].

In terms of regeneration density, silvicultural treatment showed a positive effect with greater consistency in the strips harvested by winch and cable yarder. When compared to the control, they had increases of 85% and 72%, respectively. The strips harvested by horse showed a positive trend, with an increase of about 69% compared to the control.

In this study, attention was paid to other important aspects, such as the richness and diversity of tree species. In particular, silvicultural treatment more positively influenced the richness (an evenness index increase ranging from 5.2% to 14.3%) and marginally influenced diversity. The strips harvested by winch and by cable yarder had higher richness values than the control and the strips harvested by horse.

Floristic biodiversity, which consists of the numerically and structurally consistent presence of species referable to herbaceous vegetation, and is also present in other pine woods of the Aterno Valley [55], suggested a slow evolution or in some cases no evolution in the undergrowth of artificial pine forests. However, one sign of an ongoing evolutionary process can be found in the analysis of species packages of the shrub layer with the presence of entities referable to mantle vegetation, which are shrub-like formations that set up at the edge of the forest or colonize fields and abandoned pastures. These formations had particular importance in the analysis of the dynamic relationships between the various types of Apennine vegetation. This is confirmed by the significant presence of *Cytisophyllum sessilifolium*, *Rosa canina*, *Amelanchier ovalis* subsp. *ovalis*, *Juniperus communis*, *Juniperus oxycedrus* subsp. *deltoides*, *Viburnum lantana*, and *Prunus spinosa*.

In accordance with results of other similar interventions [66,67] in the years following cutting, we observe an increase in the floristic richness linked to the opening of the tree layer. Increased species richness following canopy disturbance is largely attributed to early seral, shade-intolerant herbs and shrubs invading sites to take advantage of increased light conditions [68,69].

It is important to note that the data presented so far concern a limited period of time that does not allow for more structured statistical analyses. More time is needed to evaluate whether the cutting effect on biodiversity will last long, and observe whether non-forest species are able to regenerate when the crown cover tends to close [70,71]. The future effects on biodiversity of each treatment will also require more time to be assessed.

The findings show that soil BD, PR, and SR were influenced by both silvicultural treatment and logging operations. In particular, silvicultural treatment had an impact mainly in the first year post-harvesting, and after two and three years post-harvesting. Recovery was shown, as also found in other studies [44,48]. Among the strips harvested by winch and by cable yarder, only a few differences were found, and the values after three years were similar to the control. The strips harvested by horse showed an important impact in the first year, and the recovery was much slower. Similar values and situations have been observed in other studies, where logs were skidded and vehicles or animals were moved across the forest soil [9,31].

From the data analysis concerning pH, OM, and IC content, it was possible to note different behaviors among the variables. The pH values, as found in other studies [44,48,72], did not show any statistical relation to treatments, periods, or blocks. OM content was affected by both silvicultural treatment and logging operations. In particular, silvicultural treatment had an impact mainly in the first year post-harvesting only for the strips harvested by cable yarder, and recovery was reached after two years post-harvesting. For the strips harvested by horse and by winch (ground-based extraction), a negative effect was highlighted in the first year post-harvesting. The trends in the following years were negative, which was mainly attributable to the partial remixing of the superficial soil layers. These negative effects in the first year were due to the silvicultural treatment and the logging operations, while in the following years, the ground-based extraction systems negatively affected OM content, due to partial litter removal [73]. IC content seemed to be affected mainly by the logging operations, and ground-based systems in particular. In the short term, the ground-based extraction systems that cause the mixing of the litter with the soil superficial layers (horse and mainly winch) seem to have a negative effect on the IC content. The loss of tree cover, on the contrary, where no litter mixing (cable yarder) is observed, seems instead to increase the IC content in the short term.

The QBS-ar values show a significant positive correlation with soil physical parameters according to the literature [44,74]. The observed variation is explained by the different degrees of soil compaction and the abundance of litter associated with sudden stand removal. This mainly affected the strips harvested by cable yarder. In addition to the QBS-ar index, population density was evaluated during sampling in terms of individuals per dm^2. As can be observed from the data gathered, microarthropod density was inferior in all of the areas involved in the impact caused by the extraction activities. There was a significant difference between the logging operations, for both the QBS-ar index and population density. The strips harvested by winch had higher values than those harvested by horse and cable yarder. In particular, the values recorded in the strips extracted by cable yarder were lower with respect to the control and the other extraction systems. This is mainly due to the concomitant effect of removing the tree cover and maintaining the pine litter in its original condition. In this study, soil compaction and uncovering led to the rarefaction of specialized groups such as Protura, Diplura, and Pauropoda, and between the logging operations, no particular difference is shown. However, regarding microarthropod density, this index did not efficiently describe the conditions in terms of soil biodiversity. The values that it assumes should be interpreted in terms of trend over time.

Two principal component analyses (PCAs) were carried out to investigate any linear correlations between the expressed soil conditions of the main six studied traits, and between the tree regeneration situations of the main four studied traits. Data corresponding to each independent variable were standardized using Box–Cox lambda to minimize the scaling effect due to the different measurement units. For the soil conditions, the principal components, PC1 and PC2, explained 68% and 22% of the total variance, respectively. The PC1 and PC2 scores for the three logging operations and the control are shown in Figure 8. Each logging operation condition is distinct from the other two and from the control based on the score plot, as depicted in Figure 8. In general, a positive trend is shown from 2015 to 2017. In particular, strips harvested by winch and cable yarder showed higher recovery, and for some parameters, the recovery was complete. In detail, the strips harvested by cable yarder had a higher recovery in their physical soil parameters, which implies that these parameters were closely associated with their PC1 scores, according to Marchi [75]. Similarly, the ground-based extraction systems (horse and winch) had a higher recovery of the biological soil parameters that were closely associated with their PC2 scores.

For the tree regeneration situations, the principal components, PC1 and PC2, explained 66% and 30% of the total variance, respectively. The PC1 and PC2 scores for the three different logging operations and the control are shown in Figure 9. The tree regeneration situation was similar among the strips harvested by cable yarder and winch. The situation for the strips harvested by horse was similar to the control in 2016, and showed an implementation in 2017 that approached the values of the other two logging operations based on the score plot (Figure 9). In general, a positive trend can be associated with stand harvesting. In detail, the strips harvested by cable yarder and winch had a higher score for tree regeneration density and in the evenness index. These parameters are closely associated with the PC2 axis. SIV values are always associated with the PC2 axis, but with an inverse relation. The Shannon index is closely associated with PC1 scores, and it only shows the best situation for the strips harvested by winch.

Properly managed and programmed clear-cutting within silvicultural systems, such as the strip clear-cutting described here, can positively influence sustainability. This work indicates that less invasive extraction systems (i.e., cable yarder, forest winch) also inflict the least damage on forest soils.

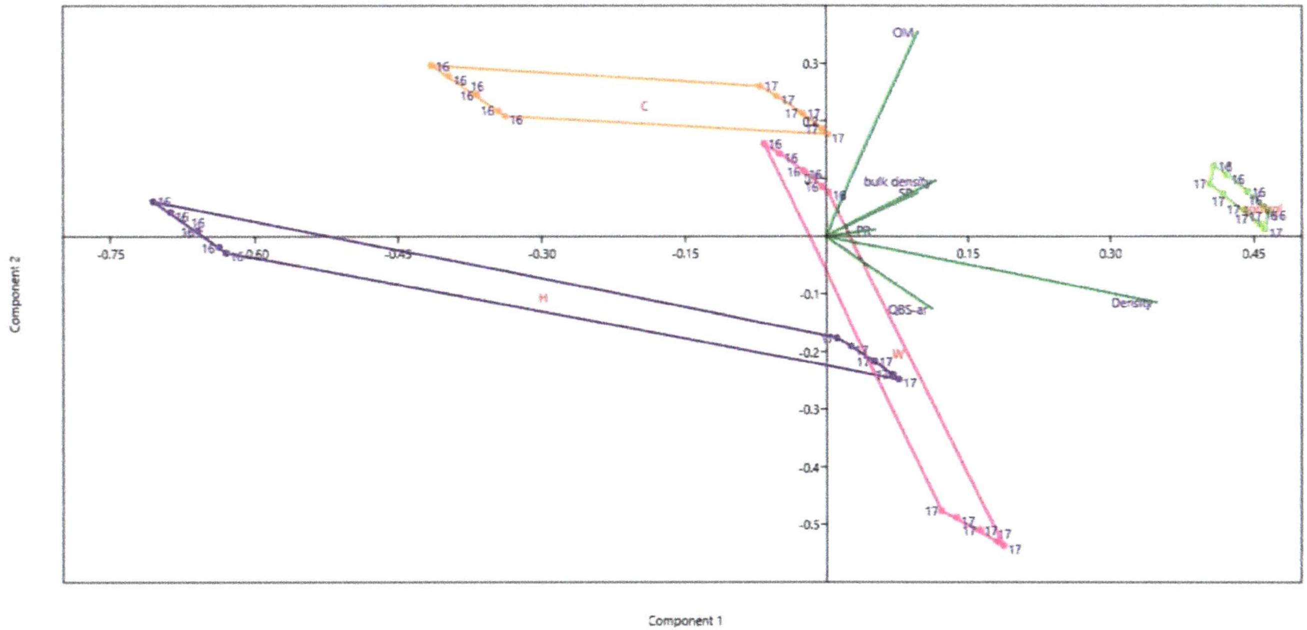

Figure 8. Score plot of the components PC1 and PC2 of the principal component analysis referred to soil characteristics. C (orange): strips harvested by cable yarder; W (pink): strips harvested by winch; H (violet): strips harvested by horse; Control (green): stand not harvested. The numbers 16 and 17 represent the years 2016 and 2017, respectively.

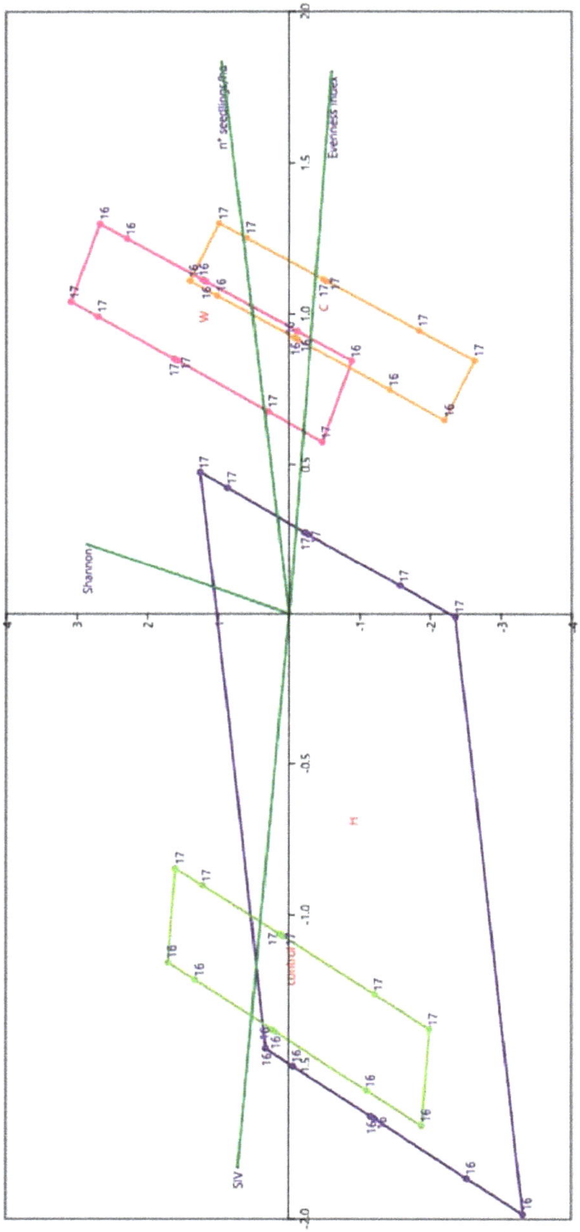

Figure 9. Score plot of the components PC1 and PC2 of the principal component analysis referred to tree regeneration situations. C (orange): strips harvested by cable yarder; W (pink): strips harvested by winch; H (violet): strips harvested by horse; Control (green): stand not harvested. The numbers 16 and 17 represent the years 2016 and 2017 respectively.

5. Conclusions

Renaturalization treatments in black pine afforestation is an important topic that should be considered. There is a need to favor the evolution of artificial pine forests toward natural forest systems. The original main purpose of these plantations was to maintain and improve soil characteristics to encourage more complex forest systems. This shift from even-aged pine monocultures to a mixed forest type through natural regeneration is the final goal of these plantations. In Italy in recent years, this substitution has been planned by means of thinning, whereas clear-cutting has been used less. However, the limit of 60–65 years represents the threshold beyond which the reaction capacity of the stand to thinning is greatly reduced, especially in low-fertility areas.

On the whole, this study focused on pine forests and tried to suggest one typology of clear-cutting (dismantling cutting) on strips associated with different extraction management techniques. Some ecological and environmental aspects associated with renaturalization treatments, including techniques applied by different mechanizations in black pine afforestation, have been highlighted as well as how the renaturalization and the active ecological management of these stands could affect soil and vegetation. An outline of answers to the main research questions is as follows:

- In terms of immediate impact on soil features, the less invasive extraction systems seem to be cable yarder and forest winch.
- A clear soil recovery trend with good capability was visible, in particular for the two extraction systems by cable. However, over a three-year period, only a partial but substantial recovery was shown. The main physical soil characteristics reached quick recovery in particular for the strips harvested by cable yarder and forest winch; the organic matter content highlighted a heavy impact, but only for the strips harvested by horse and forest winch. The biological soil characteristics, in particular arthropod communities, still showed a clear impact, which was more highlighted in the strips harvested by cable yarder
- Silvicultural treatment, particularly logging activities, affected natural tree regeneration in a different way. In general, silvicultural treatment showed qualitative and quantitative improvement in terms of tree regeneration. In particular, the extraction systems by forest winch and cable yarder showed better situations
- Silvicultural treatment seems not to have led to improvements at the level of the herbaceous and shrubby layers. However, clear differences are shown among the different harvesting systems. In particular, the strips harvested by cable yarder showed clear recovery, in terms of positive compositions and quantities, whereas the situation was slightly worse for the strips harvested by forest winch, and definitely worse for the strips harvested by horse.

It is important to highlight that the data concern a limited period of time that does not allow for more structured statistical analyses. More time is needed to evaluate whether cutting has more effects. This first step of the research was planned in order to obtain an overview in terms of the environmental impacts related to multifunctional approaches to the forest management of black pine afforestation.

Author Contributions: Conceptualization, R.P., R.M., L.G. and A.R.F.; Methodology, R.P., R.M., L.G. and A.R.F.; Validation, R.P., R.V. and A.L.M.; Investigation, R.P., R.V., T.G. and A.R.F.; Data Curation, R.P., R.V., A.L.M. and A.R.F.; Writing-Original Draft Preparation, R.P., R.M., L.G., T.G., A.R.F. and A.L.M.; Writing-Review & Editing, R.P., R.V. and A.L.M.; Supervision, R.P., R.M. and A.R.F.

Funding: This research received funding by European Agricultural Fund for Rural Development (PSR-EAFRD 2007–2013) Abruzzo Region, Mes. 124, Project "MORINABIO".

Conflicts of Interest: The authors declare no conflict of interest.

References

1. Del Favero, R. *I Boschi Delle Regioni dell'Italia Centrale. Tipologia, Funzionamento, Selvicoltura (Forests in the Regions of Central Italy)*; CLUEP: Padova, Italy, 2010. (In Italian)
2. Mercurio, R.; Schirone, B. Black Pine Afforestations in Abruzzo (Central Italy): Perspectives and Management. *J. Environ. Sci. Eng.* **2015**, *4*, 494–500. [CrossRef]
3. Italian National Forest Inventory (INFC). *Inventario Nazionale Delle Foreste E Dei Serbatoi Forestali Di Carbonio (Forest Area Estimates 2005—Section 1. MiPAF-Corpo Forestale dello Stato-Ispettorato Generale, CRA-ISAFA)*; Italian National Forest Inventory (INFC): Trento, Italy, 2007; pp. 1–413. (In Italian)
4. Tonon, G.; Panzacchi, P.; Grassi, G.; Gianfranco, M.; Cantoni, L.; Bagnaresi, U. Spatial dynamics of late successional species under *Pinus nigra* stands in the northern Apennines (Italy). *Ann. For. Sci.* **2005**, *62*, 669–679. [CrossRef]
5. Collalti, D.; D'Alessandro, L.; Marchetti, M.; Sebastiani, A. *La Carta Tipologico-Forestale della Regione Abruzzo*; Abruzzo Region, Italy, 2009. Available online: http://geoportale.regione.abruzzo.it/Cartanet/catalogo/agricoltura-uso-del-suolo/carta-tipologico-forestale-della-regione-abruzzo-1 (accessed on 15 February 2018).
6. Muscolo, A.; Settineri, G.; Bagnato, S.; Mercurio, R.; Sidari, M. Use of canopy gap openings to restore coniferous stands in Mediterranean environment. *iForest* **2017**. [CrossRef]
7. Frey, B.; Niklaus, P.A.; Kremer, J.; Lüscher, P.; Zimmermann, S. Heavy-machinery traffic impacts methane emissions as well as methanogen abundance and community structure in oxic forest soils. *Appl. Environ. Microbiol.* **2011**, *77*, 6060–6068. [CrossRef] [PubMed]
8. Picchio, R.; Magagnotti, N.; Sirna, A.; Spinelli, R. Improved winching technique to reduce logging damage. *Ecol. Eng.* **2012**, *47*, 83–86. [CrossRef]
9. Picchio, R.; Neri, F.; Petrini, E.; Verani, S.; Marchi, E.; Certini, G. Machinery-induced soil compaction in thinning two pine stands in central Italy. *For. Ecol. Manag.* **2012**, *285*, 38–43. [CrossRef]
10. Korb, J.E.; Fulé, P.Z.; Gideon, B. Different restoration thinning treatments affect level of soil disturbance in ponderosa pine forests of Northern Arizona, USA. *Ecol. Restor.* **2007**, *25*, 43–49. [CrossRef]
11. Klvač, R.; Vrána, P.; Jiroušek, R. Possibilities of using the portable falling weight deflectometer to measure the bearing capacity and compaction of forest soils. *J. For. Sci.* **2010**, *56*, 130–136. [CrossRef]
12. Williamson, J.R.; Neilsen, W.A. The influence of forest site on rate and extent of soil compaction and profile disturbance of skid trails during ground-based harvesting. *Can. J. For. Res.* **2000**, *30*, 1196–1205. [CrossRef]
13. Grigal, D.F. Effects of extensive forest management on soil productivity. *For. Ecol. Manag.* **2000**, *138*, 167–185. [CrossRef]
14. Olajuyigbe, S.; Tobin, B.; Saunders, M.; Nieuwenhuis, M. Forest thinning and soil respiration in a Sitka spruce forest in Ireland. *Agric. For. Meteorol.* **2012**, *157*, 86–95. [CrossRef]
15. Muscolo, A.; Bagnato, S.; Sidari, M.; Mercurio, R. A review of the roles of forest canopy gaps. *J. For. Res.* **2014**, *25*, 725–736. [CrossRef]
16. Convention on Biological Diversity (CBD). *CBD COP 10 Decision X/2 Strategic Plan for Biodiversity 2011–2020*; CBD: Nagoya, Japan, 2010.
17. CEC European Community Biodiversity Strategy. 1998. Available online: https://eur-lex.europa.eu/legal-content/EN/TXT/?uri=LEGISSUM%3Al28183 (accessed on 15 February 2018).
18. Sebek, P.; Bace, R.; Bartos, M.; Benes, J.; Chlumska, Z.; Dolezal, J.; Dvorsky, M.; Cizek, L. Does a minimal intervention approach threaten the biodiversity of protected areas? A multi-taxa short-term response to intervention in temperate oak-dominated forests. *For. Ecol. Manag.* **2015**, *358*, 80–89. [CrossRef]
19. Sutherland, W.J.; Pullin, A.S.; Dolman, P.M.; Knight, T.M. The need for evidence-based conservation. *Trends Ecol. Evol.* **2004**, *19*, 305–308. [CrossRef] [PubMed]
20. Convention on Biological Diversity (CBD). *CBD Sustainable Management of Non-Timber Forest Resources*; Technical Series No. 6; Convention on Biological Diversity: Montreal, QC, Canada, 2001.
21. Corona, P.; Chirici, G.; McRoberts, R.E.; Winter, S.; Barbati, A. Contribution of large-scale forest inventories to biodiversity assessment and monitoring. *For. Ecol. Manag.* **2011**, *262*, 2061–2069. [CrossRef]
22. Mattioli, W.; Mancini, L.D.; Portoghesi, L.; Corona, P. Biodiversity conservation and forest management: The case of the sweet chestnut coppice stands in Central Italy. *Plant Biosyst.* **2016**, *150*, 592–600. [CrossRef]

23. Cavalli, R.; Grigolato, S. Influence of characteristics and extension of a forest road network on the supply cost of forest woodchips. *J. For. Res.* **2010**, *15*, 202–209. [CrossRef]
24. Vusić, D.; Šušnjar, M.; Marchi, E.; Spina, R.; Zečić, T.; Picchio, R. Skidding operations in thinning and shelterwood cut of mixed stands—Work productivity, energy inputs and emissions. *Ecol. Eng.* **2013**, *61*, 216–223. [CrossRef]
25. Picchio, R.; Neri, F.; Maesano, M.; Savelli, S.; Sirna, A.; Blasi, S.; Baldini, S.; Marchi, E. Growth effects of thinning damage in a Corsican pine (*Pinus laricio* Poiret) stand in central Italy. *For. Ecol. Manag.* **2011**, *262*, 237–243. [CrossRef]
26. Picchio, R.; Spina, R.; Maesano, M.; Carbone, F.; Lo Monaco, A.; Marchi, E. Stumpage value in the short wood system for the conversion into high forest of an oak coppice. *For. Stud. China* **2011**, *13*, 252–262. [CrossRef]
27. Laschi, A.; Marchi, E.; González-García, S. Forest operations in coppice: Environmental assessment of two different logging methods. *Sci. Total Environ.* **2016**, *562*, 493–503. [CrossRef] [PubMed]
28. Edlund, J.; Keramati, E.; Servin, M. A long-tracked bogie design for forestry machines on soft and rough terrain. *J. Terramech.* **2013**, *50*, 73–83. [CrossRef]
29. Jamshidi, R.; Jaeger, D.; Raafatnia, N.; Tabari, M. Influence of two ground-based skidding systems on soil compaction under different slope and gradient conditions. *Int. J. Eng. Sci.* **2008**, *19*, 9–16.
30. Cambi, M.; Certini, G.; Neri, F.; Marchi, E. The impact of heavy traffic on forest soils: A review. *For. Ecol. Manag.* **2015**, *338*, 124–138. [CrossRef]
31. Cambi, M.; Certini, G.; Fabiano, F.; Foderi, C.; Laschi, A.; Picchio, R. Impact of wheeled and tracked tractors on soil physical properties in a mixed conifer stand. *iForest* **2016**, *9*, 89–94. [CrossRef]
32. Sánchez-Moreno, S.; Minoshima, H.; Ferris, H.; Jackson, L.E. Linking soil properties and nematode community composition: Effects of soil management on soil food webs. *Nematology* **2006**, *8*, 703–715. [CrossRef]
33. Baker, S.C.; Halpern, C.B.; Wardlaw, T.J.; Crawford, R.L.; Bigley, R.E.; Edgar, G.J.; Evans, S.A.; Franklin, J.F.; Jordan, G.J.; Karpievitch, Y.; et al. Short- and long-term benefits for forest biodiversity of retaining unlogged patches in harvested areas. *For. Ecol. Manag.* **2015**, *353*, 187–195. [CrossRef]
34. Arevalo, R.J.; Aboal, R.J. Species turnover during secondary succession in a laurel forest stand 50 years after clearcutting. *For. Syst.* **2015**, *24*, e007.
35. Powell, J.S.V.; Babbitt, K.J. Buffer-Mediated Effects of Clearcutting on In-Pool Amphibian Productivity: Can Aquatic Processes Compensate for Terrestrial Habitat Disturbance? *Forests* **2017**, *8*, 10. [CrossRef]
36. Guay-Picard, A.; Auty, D.; Munson, A.D.; Achim, A. Partial harvesting in boreal mixedwoods: A case for planned heterogeneity in industrial silvicultural prescriptions. *For. Ecol. Manag.* **2015**, *358*, 291–302. [CrossRef]
37. Drössler, L.; Fahlvik, N.; Wysocka, N.K.; Hjelm, K.; Kuehne, C. Natural Regeneration in a Multi-Layered *Pinus sylvestris-Picea abies* Forest after Target Diameter Harvest and Soil Scarification. *Forests* **2017**, *8*, 35. [CrossRef]
38. Puettmann, K.J.; Wilson, S.M.; Baker, S.C.; Donoso, P.J.; Drössler, L.; Amente, G.; Harvey, B.D.; Knoke, T.; Lu, Y.; Nocentini, S.; et al. Silvicultural alternatives to conventional even-aged forest management—What limits global adoption? *For. Ecosyst.* **2015**, *2*, 8. [CrossRef]
39. Maesano, M.; Picchio, R.; Lo Monaco, A.; Neri, F.; Lasserre, B.; Marchetti, M. Productivity and energy consumption in logging operation in a Cameroonian tropical forest. *Ecol. Eng.* **2013**, *57*, 149–153. [CrossRef]
40. Sist, P.; Nguyen-Thé, N. Logging damage and the subsequent dynamics of a dipterocarp forest in East Kalimantan (1990–1996). *For. Ecol. Manag.* **2002**, *165*, 85–103. [CrossRef]
41. Sist, P.; Sheil, D.; Kartawinata, K.; Priyadi, H. Reduced-impact logging in Indonesian Borneo: Some results confirming the need for new silvicultural prescriptions. *For. Ecol. Manag.* **2003**, *179*, 415–427. [CrossRef]
42. Tavankar, F.; Bonyad, A.E.; Majnounian, B. Affective factors on residual tree damage during selection cutting and cable-skidder logging in the Caspian forests, Northern Iran. *Ecol. Eng.* **2015**, *83*, 505–512. [CrossRef]
43. Parisi, V.; Menta, C.; Gardi, C.; Jacomini, C.; Mozzanica, E. Microarthropod communities as a tool to assess soil quality and biodiversity: A new approach in Italy. *Agric. Ecosyst. Environ.* **2005**, *105*, 323–333. [CrossRef]
44. Venanzi, R.; Picchio, R.; Piovesan, G. Silvicultural and logging impact on soil characteristics in Chestnut (*Castanea sativa* Mill.) Mediterranean coppice. *Ecol. Eng.* **2016**, *92*, 82–89. [CrossRef]
45. Rivas-Martínez, S. Global Bioclimatics. Clasificación Bioclimática de la Tierra. 2008. Available online: http://www.globalbioclimatics.org/book/bioc/global_bioclimatics_2.htm (accessed on 15 February 2018).

46. La Marca, O.; Mercurio, R.; Nocentini, L. Management and restoration on Italian afforestations under global change. *Reforesta* **2016**, *2*, 50–59. [CrossRef]
47. Hurlbert, S.H. Pseudorepliction and the design of ecological field experiments. *Ecol. Monogr.* **1984**, *54*, 187–211. [CrossRef]
48. Marchi, E.; Picchio, R.; Mederski, P.S.; Vusić, D.; Perugini, M.; Venanzi, R. Impact of silvicultural treatment and forest operation on soil and regeneration in Mediterranean Turkey oak (*Quercus cerris* L.) coppice with standards. *Ecol. Eng.* **2016**, *95*, 475–484. [CrossRef]
49. Picchio, R.; Maesano, M.; Savelli, S.; Marchi, E. Productivity and energy balance in conversion of a *Quercus cerris* L. coppice stand into high forest in Central Italy. *Croat. J. For. Eng.* **2009**, *30*, 15–26.
50. Braun-Blanquet, J. *Plant Sociology—The Study of Plant Communities*, 1st ed.; McGraw-Hill Book Company: London, UK, 1932; p. 472.
51. Pignatti, S. *"Flora d'Italia"*; Ed. Edagricole: Bologna, Italy, 1982.
52. Fiori, A. *Nuova Flora Analitica d'Italia*; Ed. ReInk Books: Firenze, Italy, 1923.
53. Castroviejo, S.; Aedo, C.; Cirujano, S.; Lainz, M.; Montserrat, P.; Morales, R.; Munozgarmendia, F.; Navarro, C.; Paiva, J.; Soriano, C. *Flora Iberica*; Real Jardìn Botanico, CSIC: Madrid, Spain, 2005.
54. Tutin, T.G.; Burges, N.A.; Charter, A.O.; Edmondson, J.R.; Heywood, V.M.; Moore, D.M.; Valentine, D.H.; Walters, S.M.; Webb, D.A. *Flora Europaea*, 2nd ed.; University Press: Cambridge, UK, 1993; Volume 1.
55. Tutin, T.G.; Heywood, V.M.; Burges, N.A.; Valentine, D.H.; Walters, S.M.; Webb, D.A. *Flora Europaea*; University Press: Cambridge, UK, 1969; Volume 2–5.
56. Tavankar, F.; Majnounian, B.; Bonyad, A.E. Felling and skidding damage to residual trees following selection cutting in Caspian forests of Iran. *J. For. Sci.* **2013**, *59*, 196–203. [CrossRef]
57. Pielou, E.C. The measurement of diversity in different types of biological collections. *J. Theor. Biol.* **1966**, *13*, 131–144. [CrossRef]
58. Burton, P.J.; Balisky, A.C.; Coward, L.P.; Cumming, S.G.; Kneeshaw, D.D. The value of managing for biodiversity. *For. Chron.* **1992**, *68*, 225–237. [CrossRef]
59. Begehold, H.; Rzanny, M.; Winter, S. Patch patterns of lowland beech forests in a gradient of management intensity. *For. Ecol. Manag.* **2016**, *360*, 69–79. [CrossRef]
60. Astolfi, S.; Zuchi, S.; De Cesare, F.; Badalucco, L.; Grego, S. Cadmium-induced changes in soil biochemical characteristics of oat (*Avena sativa* L.) rhizosphere during early growth stages. *Soil Res.* **2011**, *49*, 642–651. [CrossRef]
61. Moore, S.E.; Allen, H.L. *Plantation Forestry. Maintaining Biodiversity in Forest Ecosystems*; M.L. Hunter Cambridge University Press: New York, NY, USA, 1999; pp. 400–433.
62. Brosofske, K.D.; Chen, J.; Crow, T.R. Understory vegetation and site factors: Implications for a managed Wisconsin landscape. *For. Ecol. Manag.* **2001**, *146*, 75–87. [CrossRef]
63. Poorbabaei, H.; Poorrahmati, G. Plant species diversity in loblolly pine (*Pinus taeda* L.) and sugi (*Cryptomeria japonica* D. Don.) plantations in the western Guilan, Iran. *Int. J. Biodvers. Conserv.* **2009**, *1*, 38–44.
64. Carnevale, N.J.; Montagnini, F. Facilitating regeneration of secondary forests with the use of mixed and pure plantations of indigenous tree species. *For. Ecol. Manag.* **2002**, *163*, 217–227. [CrossRef]
65. Gratani, L.; Frattaroli, A.R.; Console, C. Regeneration of the undergrowth in reafforested areas with *Pinus nigra*, in the High Aterno Valley (Italy). *Belg. J. Bot.* **1994**, *127*, 61–66.
66. Lindgren, P.M.; Ransome, D.B.; Sullivan, D.S.; Sullivan, T.P. Plant community attributes 12 to 14 years following precommercial thinning in a young lodgepole pine forest. *Can. J. For. Res.* **2006**, *36*, 48–61. [CrossRef]
67. Cantiani, P.; De Meo, I.; Becagli, C.; Bianchetto, E.; Cazau, C.; Mocali, S.; Salerni, E. Effects of thinnings on plants and fungi biodiversity in a *Pinus nigra* plantation: A case study in central Italy. *For. Ideas* **2015**, *21*, 149–162.
68. Fredericksen, T.S.; Ross, B.D.; Hoffman, W.; Morrison, M.L.; Beyea, J.; Johnson, B.N.; Lester, M.B.; Ross, E. Short-term understory plant community response to timber harvesting intensity on non-industrial forestlands in Pennsylvania. *For. Ecol. Manag.* **1999**, *116*, 129–139. [CrossRef]
69. Thysell, D.R.; Carey, A.B. *Effects of Forest Management on Understory and Overstory Vegetation: A Retrospective Study*; USDA Forest Service General Technical Report PNW-GTR-488; U.S. Department of Agriculture: Portland, OR, USA, 2000.

70. Lust, N.; Muys, B.; Nachtergale, L. Increase of biodiversity in homogeneous Scots pine stands by an ecologically diversified management. *Biodivers. Conserv.* **1998**, *7*, 249–260. [CrossRef]
71. Götmark, F.; Paletto, H.; Nordén, B.; Götmark, E. Evaluating partial cutting in broadleaved temperate forest under strong experimental control: Short-term effects on herbaceous plants. *For. Ecol. Manag.* **2005**, *214*, 124–141. [CrossRef]
72. Cambi, M.; Hoshika, Y.; Mariotti, B.; Paoletti, E.; Picchio, R.; Venanzi, R.; Marchi, E. Compaction by a forest machine affects soil quality and *Quercus robur* L. seedling performance in an experimental field. *For. Ecol. Manag.* **2017**, *384*, 406–414. [CrossRef]
73. Fekete, I.; Varga, C.; Biró, B.; Tóth, J.A.; Várbíró, G.; Lajtha, K.; Szabó, G.; Kotroczó, Z. The effects of litter production and litter depth on soil microclimate in a central European deciduous forest. *Plant Soil* **2016**, *398*, 291–300. [CrossRef]
74. Blasi, S.; Menta, C.; Balducci, L.; Delia Conti, F.; Petrini, E.; Piovesan, G. Soil microarthropod communities from Mediterranean forest ecosystems in Central Italy under different disturbances. *Environ. Monit. Assess.* **2013**, *185*, 1637–1655. [CrossRef] [PubMed]
75. Marchi, E.; Picchio, R.; Spinelli, R.; Verani, S.; Venanzi, R.; Certini, G. Environmental impact assessment of different logging methods in pine forests thinning. *Ecol. Eng.* **2014**, *70*, 429–436. [CrossRef]

forests

MDPI

Article

A Crown Width-Diameter Model for Natural Even-Aged Black Pine Forest Management

Dimitrios Raptis [1], Vassiliki Kazana [1,*], Angelos Kazaklis [2] and Christos Stamatiou [1]

[1] Eastern Macedonia & Thrace Institute of Technology, Department of Forestry and Natural Environment Management, 1st km Drama-Mikrohori, 66100 Drama, Greece; dimmraptis@gmail.com (D.R.); stamatioy@gmail.com (C.S.)

[2] OLYMPOS, Centre for Integrated Environmental Management, 39 Androutsou Str., 55132 Kalamaria, Thessaloniki, Greece; akaz98@otenet.gr

* Correspondence: vkazana@teiemt.gr; Tel.: +30-25210-60422

Received: 26 July 2018; Accepted: 2 October 2018; Published: 3 October 2018

Abstract: Crown size estimations are of vital importance in forest management practice. This paper presents nonlinear models that were developed for crown width prediction of Black pine (*Pinus nigra* Arn.) natural, pure, even-aged stands in Olympus Mountain, central Greece. Using a number of measured characteristics at tree and plot level from 66 sample plots as independent variables, an attempt was made to predict crown width accurately, initially based on Least Square Analysis. At the second stage, nonlinear mixed effect modeling was performed in order to increase the fitting ability of the proposed models and to deal with the lack of between observations independence error assumption. Based on the same form, a generalized crown width model was developed by including six main regressors, such as the diameter at breast height, the total height, the canopy base height, the basal area, the relative spacing index and the diameter to quadratic mean diameter ratio, while at the final stage, the same model was expanded to mixed-effect. The proposed models were evaluated against independent crown width sample observations that were also obtained from the study area. The results showed that the two types of mixed-effect models performed equally well and, therefore, we propose those for use in forestry practice. Furthermore, the exact contribution of each inherent variable in crown width allometry was evaluated, thus providing a framework to facilitate field measurements for forest management predictions.

Keywords: crown width prediction; forest management; nonlinear forest models; mixed effects models; Black pine

---

1. Introduction

Crown size is defined as the horizontal space that is available for the display of leaves [1]. It forms the tree growing space, which is associated with the availability of all of the required supplies for the tree growth [2]. Therefore, crown size is a critical feature of any tree growing at stand or open conditions, particularly in mountain areas, where water limitations rarely occur. In forestry practice, crown size estimation is considered of vital importance because it is associated with many different factors of forest management concern, such as the photosynthetic capacity [3], stem volume [4], increment efficiency [5], tree competition [6,7], tree health and vigor [8], as well as carbon, water and energy exchange [9]. From an ecological point of view, crown size directly affects many of the understory flora and fauna components by regulating the amount of the penetrated sunlight and precipitation to the lowest layers, thus retaining the forest moisture [10]. In order to measure the dimensions of a crown, Gregoire and Valentine [11] suggested several approaches. In the current research, it was assumed that crown size at above-ground level depends mainly on the crown dimensions at different directions forming an approximately circular shape via vertical projection [12]. Hence, its mathematical expression involves crown width measurements, or the horizontal distance between the edges of each tree crown passing through the center of the trunk to at least two vertical directions.

Hemery et al. [3] have pointed out the importance of modeling crown width under a thinning regime for final crop spacing by linking the diameter at breast height and the crown width (crown diameter) with the available growing space. However, the increasing use of remote sensing techniques in forestry has generated a need for well-defined allometric relationships between the basic dendrometric parameters in order to perform accurate estimations of the standing wood volume. Popescu et al. [13] demonstrated that the crown width is a tree dimension that is possible to measure directly with a lidar photogrammetric technique, while Gering and May [14] inversed a simple linear model for crown width prediction in order to predict diameter at breast height from aerial photos or photogrammetric data. Therefore, the development of forest tree allometric statistical models would provide valuable tools in forest management decision making by allowing the prediction of the relevant stand attributes. Despite the importance of crown width modeling, a limited number of studies have been published to date in comparison to studies related to tree-height prediction. In crown width modeling, the typical variable that has been used as a regressor is the diameter at breast height (e.g., [3,12,15–17]), mainly due to the fact that this variable is included in almost all types of forest inventories and it is easy to measure it in the field. Incorporation of additional variables would improve the fitting ability of the candidate regression models; however, it would require additional sampling effort. From a practical point of view, therefore, it would be very useful to use any variable that could be easily measured in the field as an independent variable, such as the dominant height, instead of their mean values [18]. A possible disadvantage of using stand-level attributes as regressors is the assumption that stands with similar characteristics are described by similar linear or nonlinear models, which is not fully accurate [19].

In many cases, crown width modeling has been based on data that has been obtained from installed sample plots or even from repeated measurements along the same tree, which most often present a hierarchical nested structure, thus violating the assumption of between observations error independence [20]. Therefore, we applied non-linear mixed effect modeling in order to ensure the independency assumption by adding a plot-specific random-effect part along with a fixed part in each estimated parameter during the modeling process. Mixed-effect models of forest growth and yield have been recommended by several authors in the literature as an alternative solution to independency issues emanating from ordinary least squares application (e.g., [19,21–28]). As far as crown width modeling is concerned, application of the mixed-effect modeling approach has been more limited as yet (Sánchez-González et al. [23], Ducey [29], Fu et al. [30,31], Hao et al. [32], Yang and Huang [7] and Sharma et al. [33,34]). In most cases, the mixed-effect models were reported to have generated more accurate predictions than the fixed effect models, since the structural heterogeneity, at tree or at stand level, was reflected by the random-effect parameters. In any case, a subsample of observations would always be used for calibration, although this might be a disadvantage during the data analysis. However, in many cases, four to five observations would suffice for a reasonable accuracy of the mixed-modeling predictions [21].

Black pine (*Pinus nigra* Arn.) is an important commercial tree species in Greece that covers an area of 281,692 hectares. Its annual wood volume amounts to 13,892,819 m^3, corresponding to 10% of the country's total wood production [35]. Black pine geographically extends from the Iberian Peninsula to the Alps, Apennine Mountains, and the Balkan region, Asia Minor, Crimea and the USA and Canada. In Greece, Black pine is mainly located in the mountains of the continental country with some populations also extending to the islands of the North Aegean Sea. In Olympus Mountain, which has served as the study area for the research presented in the current article, the Black pine forest stands are, to a big extent, pure, even-aged and they are managed by the Hellenic Forest Service predominately for wood production. Black pine, in addition to its economic value, has the ability to grow in degraded ecosystems. This is the reason that Black pine has been widely used in afforestation of mountain areas [36]. The canopy of Black pine protects the vulnerable soils from erosion by creating a physical obstacle against erodibility factors, such as rain and wind and also by supplying material for the formation of a thick protective layer of pine litter [37]. Therefore, the prediction possibility of the required growing space of this species would support forest management decisions regarding afforestation. Furthermore, the development of crown width prediction models could facilitate standing volume estimations via remote sensing analysis and the establishment of the most appropriate silvicultural regimes in the framework of sustainable forest management.

In the context of our research, the main hypothesis was that the crown width-diameter allometry of Black pine (*Pinus nigra* Arn.) would be significantly influenced by a combination of basic tree- and stand-level parameters, such as diameter, basal area, stand density measures and tree characteristics. Furthermore, an additional hypothesis was that the generalized crown width-diameter models of Black pine would be more accurate than the single mixed ones. Therefore, the specific objectives of our research involved the following: (i) to develop a crown width prediction model of Black pine for use in Greece and other countries (ii) to evaluate the combinatorial effect of basic tree- and stand-level parameters along with random variables of mixed models on crown prediction and (iii) to compare different crown prediction strategies.

2. Materials and Methods

2.1. Data

The research reported in the current article was conducted over a two-year period in Olympus Mountain, Central Greece, where 66 non-permanent sample plots were installed, so as to measure a number of variables at tree and stand level (Figure 1). Measurements were carried out in a total of 3442 trees over an area of approximately 73,477 hectares that also included the Olympus National Park. The main forest species in the area is Black pine (*Pinus nigra* Arn.) and, to a lesser extent, the Macedonian fir (*Abies borisii-regis* Mattf.), intermingled with broadleaved species in some restricted locations. The climate is a typical sub-Mediterranean and the annual rainfall is about 715 mm. The geological substratum is mainly composed of limestones, granite outcrops and flysch. The elevation from the sea level in the established sample plots varied between 780 and 1520 m, with a mean value of 1148 m. Slope steepness ranged between 1.11% and 64.90%, and averaged at 36.51% from the horizontal level.

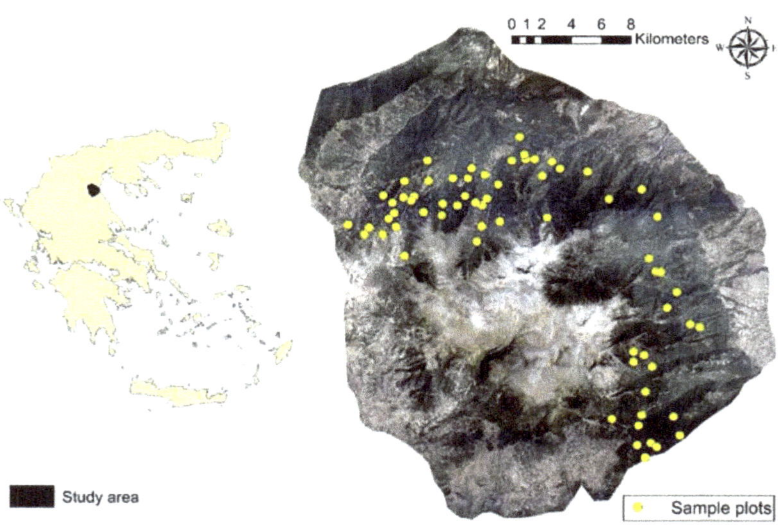

Figure 1. Study area location map and distribution of the sample plots.

In order to select forest stands with Black pine as dominant species, a representative vector-format file (shapefile) of vegetation type and cover was used in GIS environment (ArcGIS 10.2, ESRI, Redlands, CA, USA) and 70 potential plot locations were selected randomly, using the random point function. Due to the complex terrain of the study area, the random sampling process made it possible to cover all the potential site qualities of the area. The shape of the sample plots was circular with stable radius, covering an area of 500 m^2 in each case. In order to minimize the repeated measurements in each tree in an attempt to avoid the within-tree error correlation, we used the *"canopy spread"* module of LaserAce hypsometer (Trimble, Sunnyvale, CA, USA) to estimate the crown width (CW) in two vertical directions [38]. The diameter at breast height (DBH) was measured using a digital caliper to the nearest 0.1 m, and the total height (H) of each tree was measured by using the three-point measurement of LaserAce hypsometer. At stand level, the total basal area per hectare (BA in $m^2 \cdot ha^{-1}$), the number of stems per hectare ($N \cdot ha^{-1}$), the dominant height (mean height of the 100 largest trees per hectare or equally proportional—HDOM), the dominant diameter (mean diameter of the 100 largest trees per hectare or equally proportional—DDOM) and the quadratic mean diameter (QMD) were calculated. The HDOM and the DDOM were measured locally by selecting the five largest trees in each plot [39].

The Site Index in each plot was estimated using the Site Index Curves for Black Pine in Greece [40]. The total sample was composed of 3442 trees, the majority of which 90% was used for model development and about 10% for model validation [41]. The descriptive statistics of the total sample are shown in Table 1.

Table 1. Descriptive statistics of the total sample.

Variables	Fitting Data (n = 3080)				Validation Data (n = 362)			
	Mean	Min	Max	Standard Deviation	Mean	Min	Max	Standard Deviation
diameter (cm)	23.85	1.10	85.40	11.52	21.99	1.20	64.0	12.45
height (m)	14.75	1.40	30.40	5.41	13.26	2.30	23.40	4.48
canopy base height (m)	7.37	0.20	19.90	3.53	5.81	0.30	10.80	2.12
crown width (m)	4.04	0.20	13.70	1.92	4.58	0.60	11.60	2.18
height/diameter ratio	67.67	23.64	154.54	19.91	71.74	33.78	143.63	19.50
stems per hectare ($N \cdot ha^{-1}$)	1050	240	2100	396	1196	480	1500	386
dominant height (m)	20.17	13.58	27.92	3.04	18.00	1.06	20.80	2.20
dominant diameter (m)	38.19	25.34	72.90	8.89	34.91	26.32	45.24	6.17
site index (m)	17.15	8.80	22.20	2.99	17.72	12.10	19.90	2.39
basal area ($m^2 \cdot ha^{-1}$)	56.70	25.70	99.09	16.06	54.97	28.07	76.80	16.18
quadratic diameter (cm)	27.11	17.76	40.10	5.04	24.71	21.48	38.86	4.77
basal area of larger trees($m^2 \cdot ha^{-1}$)	41.63	1.68	99.09	20.76	36.27	1.68	76.80	18.59
diameter to dominant diameter	0.67	0.02	1.70	0.30	0.63	0.06	1.34	0.31
diameter to quadratic diameter	0.93	0.03	2.68	0.41	0.90	0.10	2.06	0.43
height to dominant height	0.73	0.06	1.15	0.24	0.68	0.12	1.10	0.24
reineke stand density index	1166.4	437.0	1960	312.8	1053.6	480.3	1513.3	293.4
relative spacing index	0.153	0.100	0.442	0.04	0.152	0.124	0.235	0.035

2.2. Model Development

Diameter at breast height (DBH) is the most commonly used independent variable in crown width modeling, since the relationship between DBH and crown width (CW) has been well established in the literature. The mathematical form that links DBH and CW could be either linear [12,14,16,42] or nonlinear [7,23,30,34]. For the needs of the current research, a number of candidate simple linear and nonlinear models were fitted and the best ones, in terms of their fitting ability, were selected for further analysis. The candidate models are shown in Table 2.

Table 2. Simple linear and nonlinear models for crown width modeling.

Model	Name	Mathematical Expression	Reference
M1	Linear	$CW = \beta_0 + \beta_1 DBH$	[12,14]
M2	Power	$CW = \beta_0 DBH^{\beta_1}$	[43]
M3	Monomolecular	$CW = \beta_0[1 - \exp(-\beta_1 DBH)]$	[23]
M4	Exponential	$CW = \beta_0 \exp(\beta_1 DBH)$	[7]
M5	Logistic	$CW = \beta_0/[1 + \beta_1 \exp(-\beta_2 DH)]$	[30]

CW = crown width, DBH = The diameter at breast height, M1, ... ,M5 = Candidate models.

2.3. Additional Variables for Prediction-A Generalized Crown Width Model

It has been well established that DBH and CW for open and stand grown trees are closely correlated [44]. The incorporation of DBH as the only predictive variable assumes that trees of the same stem diameter also have the same mean crown dimensions, regardless of the competition levels inside the stand; an assumption that does not hold. In order to relax this assumption, previous research efforts have instead used tree variables, such as the crown length or ratio, the tree height and the distance dependent or independent stand-level predictors, which reflect the competition status [7,15,23,34,45]. The inclusion of additional variables leads to the creation of generalized models, which usually improve the crown width predictability in large areas under different stand competition regimes. However, a potential disadvantage of the generalized crown width prediction models is the inherent assumption that stands with similar attributes also present similar trends in crown modeling.

The inclusion of a number of stand attributes in the base model, which affect crown size, constitutes a standard procedure for expanding a simple model to a generalized form. For the sake of model simplicity, a number of predictors at stand level were tested [31] and the final form of the model was created. In this context, tree attributes, such as the total height (H), the height/diameter ratio (HDR) and the canopy base height (CBH), were used as covariates for crown width prediction. Furthermore, one tree-centered competition variable that is the basal area of the largest trees (BAL) and three size ratios, the diameter/dominant-diameter (DBH/DDOM), the height/dominant height (H/HDOM) and the diameter/quadratic mean diameter (DQMD), were tested in order to explain a larger part of crown variation. The most common distance-independent predictors at stand level that have been incorporated in generalized crown width models were the dominant height or diameter, the number of stems per hectare, the basal area, the quadratic mean diameter and the Reineke's Stand Density Index (SDI), which was expressed as $SDI = (N\ per\ ha) \times (QMD/25.4)^{1.605}$ [46]. In the current research, in addition to the aforementioned stand level variables, the Relative Spacing Index (RSI), estimated as $RSI = \sqrt{\frac{1000}{N\ per\ ha}}\ /HDOM$ [34,47,48], was also evaluated during the development stage of a generalized nonlinear crown prediction model following a methodology which involved a direct addition of a variable into the model [33], thus approximating a forward selection method. The relationships of crown width with the basic variables at tree- and at stand- level are presented in the corresponding scatter plots (Figure 2). The candidate models were evaluated against each other using the ANOVA function for nested models in R statistical language (R Core Team) by using the Akaike information criterion (AIC), together with the log-likelihood (−2LL) ratio test and the statistical significance of each associated parameter.

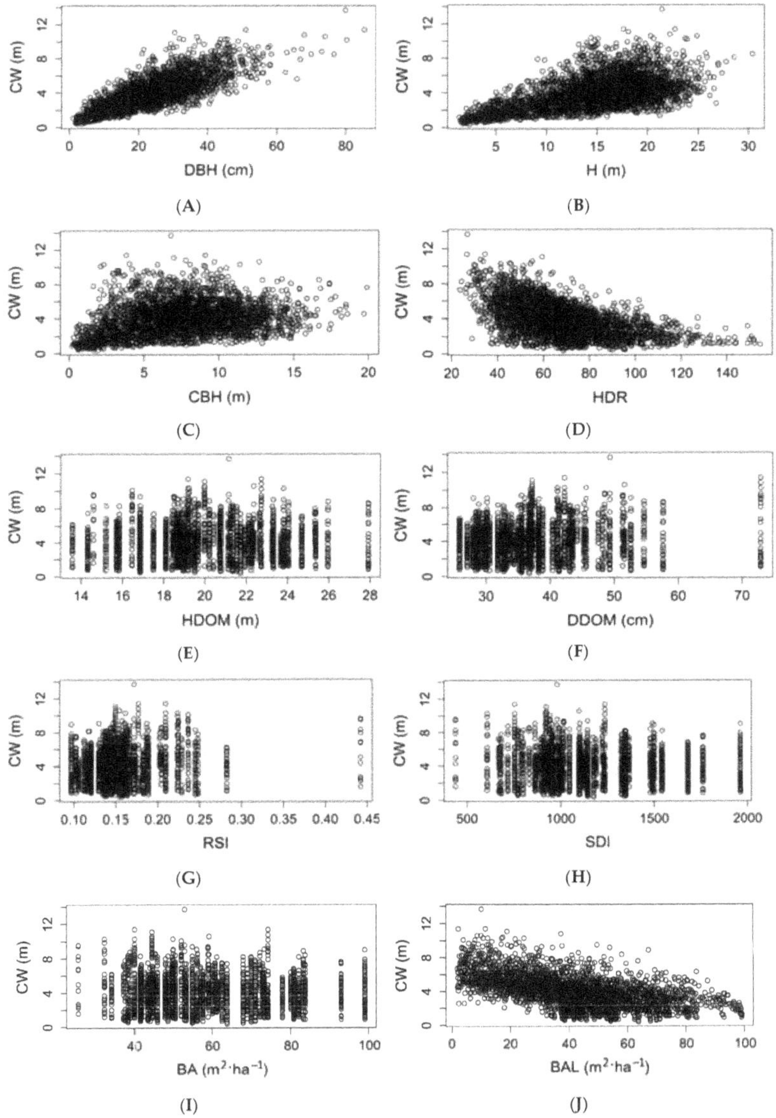

Figure 2. Scatter plots of crown width against basic parameters at tree and stand level of the fitting data. DBH = diameter at breast height, H = total height, CBH = canopy base height, HDR = height/diameter ratio, HDOM = height of the 100 largest trees per hectare or equally proportional, DDOM = diameter of the 100 largest trees per hectare or equally proportional, RSI = relative spacing index, SDI = Reineke's stand density index, BA = basal area, BAL = basal area of the largest trees. (**A**) diameter at breast height versus crown width (**B**) total height versus crown width (**C**) canopy base height versus crown width (**D**) height/diameter ratio versus crown width (**E**) dominant height versus crown width (**F**) dominant diameter versus crown width (**G**) relative spacing index versus crown width (**H**) Reineke's stand density index versus crown width (**I**) basal area versus crown width (**J**) basal area of large trees versus crown width.

2.4. Mixed Effect Models

Very often, the field measurements that are obtained from established sample plots present a hierarchical nested structure, which violates the basic least square assumption of between observations independence [32]. The mixed effect models provide an appropriate framework to overcome this limitation [49] by estimating plot average parameters along with a random part, which is related specifically to each sample plot and, consequently, to the stand average conditions. The general vector form of a mixed effect model, with respect to crown width modeling, was expressed as:

$$\boldsymbol{CW_i} = f(\boldsymbol{\Phi_i}, \boldsymbol{DBH_i}) + \boldsymbol{e_i} \tag{1}$$

where $\boldsymbol{CW_i}$ was a $(\boldsymbol{n_i} \times 1)$ vector of crown widths for the $\boldsymbol{i}$th plot, f was a nonlinear function, $\boldsymbol{\Phi_i}$ was a $(\boldsymbol{r_i} \times 1)$ parameter vector, with $\boldsymbol{r}$ to represent the total number of the fixed effects in the model, $\boldsymbol{DBH_i}$ was the $(\boldsymbol{n_i} \times 1)$ predictor vector for $\boldsymbol{n_i}$ observations and $\boldsymbol{e_i}$ was $(\boldsymbol{n_i} \times 1)$ the vector of the residuals. The $\boldsymbol{\Phi_i}$ vector consisted of one fixed effect part which was the same for the entire population and a random effect part which was unique for each plot [20]. The vector of random effects and the error vector were assumed to be uncorrelated and normally distributed with zero mean and variance-covariance matrices $\boldsymbol{D}$ and $\boldsymbol{R_i}$, respectively [7,27]. The $\boldsymbol{R_i}$ variance-covariance matrix was expressed as [50]:

$$\boldsymbol{R_i} = \sigma^2 \boldsymbol{G_i^{0.5}} \boldsymbol{\Gamma_i} \boldsymbol{G_i^{0.5}} \tag{2}$$

where σ^2 was a scaling factor for error dispersion [49], $\boldsymbol{G_i}$ was the diagonal matrix for within plot variance heteroscedasticity and $\boldsymbol{\Gamma_i}$ was an identity matrix that was related to the within plot autocorrelation structure of the residuals. When evidences of heteroscedasticity are present, a variance function may be applied, such as the power, the exponential and the constant/power functions [20,30,34,51], so as to link variance with a predictor. The diagonal elements of $\boldsymbol{G}$ matrix are provided by the selected variance function. Assuming that the $\boldsymbol{i}$ indicator represents the different plots and the $\boldsymbol{j}$ indicator represents each tree located within the plot, the variance functions were expressed through the following mathematical expressions [31]:

$$\boldsymbol{Var}(\varepsilon_{ij}) = \sigma^2 \boldsymbol{DBH_{ij}^{2\delta}} \text{ (power function)} \tag{3}$$

$$\boldsymbol{Var}(\varepsilon_{ij}) = \sigma^2 \mathbf{exp}(2\delta \boldsymbol{DBH_{ij}}) \text{ (exponential function)} \tag{4}$$

$$\boldsymbol{Var}(\varepsilon_{ij}) = \sigma^2 \left(\delta_1 + \boldsymbol{DBH_{ij}^{\delta_2}}\right)^2 \text{ (constant-power function)} \tag{5}$$

where δ, δ_1 and δ_2 are parameters to be estimated. The most suitable function was determined by visual inspection of the residuals and the Akaike Criterion (AIC).

2.5. Statistical Analysis

In order to evaluate the fitting ability of the ordinary least square and the mixed effect models, the following criteria were selected:

1. the root mean square error (RMSE)

$$RMSE = \sqrt{\frac{\sum_{j=1}^{n}\left(CW_j - \widehat{CW_j}\right)^2}{n}} \tag{6}$$

2. an efficiency index (EI) based on R-squared expression [28]:

$$EI = 1 - \frac{\sum_{j=1}^{n}\left(CW_j - \widehat{CW_j}\right)^2}{\sum_{j=1}^{n}\left(CW_j - \overline{CW_j}\right)^2} \tag{7}$$

3. the Akaike Information Criterion (AIC):

$$\mathbf{AIC = -2log(L) + 2p} \tag{8}$$

4. the mean prediction bias:

$$Bias = \frac{\sum_{j=1}^{n}\left(CW_j - \widehat{CW}_j\right)}{n} \tag{9}$$

where CW_j, $\widehat{CW}_j$ και $\overline{CW}_j$ represent the measured, estimated and mean values of crown width of the jth observation, n the total number of observations, p the number of the estimated model parameters and L the log-likelihood function of the fitted model. In addition, the model parameter estimates should be statistically different from zero. Visual inspection of the residuals was used, so as to detect heteroscedasticity trends. Following the suggestions of Temesgen et al. [52], the threshold value of bias was set to 10cm in order to preclude severely biased equations.

2.6. Calibrated Response

In general, the fitting procedure of crown width modeling requires a minimum number of observations to produce unbiased estimates at stand level. The corresponding required number of observations for height-diameter model fitting should be at least 20–25 per stand, according to van Laar and Akça [53]. An interesting property of mixed models is the ability to calibrate the random part of the parameters, even if a small number of observations would be available, and still obtain sufficiently accurate estimates. Usually four to five observations per stand would be sufficient enough to calibrate a mixed model and to obtain reliable parameter estimates [21,30]. In our case, the random effects vector prediction was based on the following Bayesian estimator [54]:

$$\hat{b}_i \approx \hat{D}\,\hat{Z}_i^T\left(\hat{R}_i + \hat{Z}_i\,\hat{D}\,\hat{Z}_i^T\right)^{-1}\hat{e}_i \tag{10}$$

where $\hat{D}$ was the estimated variance-covariance matrix for the random parameters, $\hat{R}_i$ the variance-covariance matrix for within plot variability, $\hat{Z}_i$ was the matrix of partial derivatives, estimated at $\hat{b}_i$, with respect to its fixed parameters [21]. Using the derivatives of Equation (11) for a total sample of j = 2 trees from ith plot, the $\hat{Z}_i$ matrix becomes $\left[\begin{array}{cc} DBH_{i1}^{b_0} & b_0 DBH_{i1}^{b_1}\,ln(DBH_{i1}) \\ DBH_{i2}^{b_0} & b_0 DBH_{i2}^{b_1}\,ln(DBH_{i2}) \end{array}\right]$. The algorithm was suitably implemented in *R* statistical language along with the *nlme* function.

The performance of each Black pine crown width model, generalized or mixed, was further evaluated against independent Black pine validation data, which were not included in model development. Two main sample scenarios were formulated in an effort to assess the predictive ability of each modeling technique separately:

1. No calibration, use of only fixed effect parameters.
2. Four observations of crown width-diameter [34].

The procedure was repeated 100 times for each plot, following different combinations of randomly selected trees [28]. The mean values of the statistical measures after 100 repetitions were reported for comparison.

The evaluation was based on the mean values of RMSE and bias, which were obtained through the fitting process of the proposed models in each of the six validation sample plots. The analysis was based on *R* statistical language, the *nls* function for nonlinear modeling and the *nlme* function for nonlinear mixed effect models.

3. Results

3.1. A Simple Mixed Effect Model

The fitting statistics of the simple least square nonlinear models are presented in Table 3.

Table 3. Fitting statistics of the simple models

Fitting Criteria	Simple Models				
	M1	M2	M3	M4	M5
EI	0.6445	0.6477	0.6420	0.5738	0.6394
RMSE	1.1202	1.1151	1.1241	1.2266	1.1282
Bias	0.0000	0.0031	0.0322	−0.0405	−0.0085

EI = efficiency index, RMSE = root mean square error, Bias = prediction bias.

All the parameter estimates were statistically significant at ($p < 0.01$) level. All models, except the M4 model, predicted crown width relatively well, although the linear M1 and the nonlinear M2 appeared more accurate during the calibration stage. Based on the fitting statistics of Table 3, the M2 model performed slightly better than the M1 and, therefore, it was selected as the basis for mixed effect development. At the first stage, it was assumed that both the M2 fixed parameters contained a random part [55], according to the following form (MM1):

$$CW = (\beta_0 + b_0)DBH^{(\beta_1 + b_1)} \tag{11}$$

where b_0 and b_1 were the random effects of the model. The model converged at the calibration stage, however we detected evidence of heteroscedasticity in the residual plots. In order to minimize the unequal variance trends, the variance was modeled with functions, which were compared in terms of their fitting performance.

According to the results presented in Table 4, the constant-power function provided significantly different results compared to the same model using equal residual variance. The parameter estimates and fit statistics of the final model are presented in Table 5. In order to examine the coverage extent of the mixed model on both the fitting and the validation data, the mixed model was localized for each plot by using four randomly selected trees (Figure 3). According to Figure 3, the simple mixed model covered the range of the observed crown width values of the total sample.

Table 4. Fitting statistics of the simple mixed model (MM1), using different types of variance functions.

Variance Function	Akaike Criterion (AIC)
σ^2	8534.211
constant–power	8169.929
exponential	8263.744
power	8208.646

Table 5. Estimated parameters and fitting statistics of the simple mixed model (MM2).

Fixed Parameters		Random Effects Covariance Elements				Parameters of Function Variance	
β_0	β_1	$Var\ (b_0)$	$Var\ (b_1)$	$Cov\ (b_0, b_1)$	σ^2	δ_1	δ_2
0.3459 (0.016)	0.7885 (0.014)	0.0095	0.0066	−0.0067	0.0503	0.0002	0.4515

Fit Statistics			
EI	RMSE	AIC	Bias
0.7613	0.9178	8169.9	−0.0034

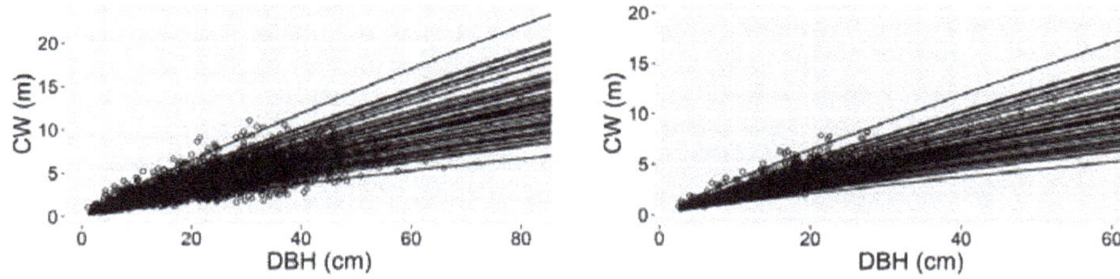

(**A**) (**B**)

Figure 3. Localized curves of simple mixed model (MM2) based on four randomly selected trees overlaid on the fitting data (**A**) and on validation data (**B**).

Hence, the final form of the simple mixed-effect model was

$$\textbf{MM1 + constant-power function MM2} \tag{12}$$

All the estimated parameters were significantly different than zero, according to the *t-statistic* test values.

3.2. Development of a Generalized Model

The following model form (13) presented the best fitting ability and it was selected as a generalized model for the prediction of black pine crown width. Using a total number of six predictors, it can be expressed as (GM1):

$$CW = \beta_0 DBH^{\beta_1 (BA^{\beta_2} + \beta_3 CBH + \beta_4 RSI + \beta_5 H + \beta_6 DQMD)} \tag{13}$$

The residual analysis of the GM1 model revealed trends of unequal variance (heteroscedasticity), since variance changed as diameter values were increasing. Following the methodology described by Ritz and Streibig [56], a weighted least square method was used with weights in form of a function. The power function yielded the best results, and the final form of the proposed ordinary least square model is presented in Table 6.

Table 6. Estimated parameters and fitting statistics of GM2.

Fixed Parameters							Parameters of Function Variance
β_0	β_1	β_2	β_3	β_4	β_5	β_6	δ
0.3601	1.3017 (0.061)	−0.1013 (0.008)	−0.0032 (0.001)	−0.2729 (0.0314)	−0.0019 (0.001)	0.0246 (0.006)	0.4809
Fit Statistics							
EI		RMSE		AIC			Bias
0.6917		1.0433		8570.3			0.0012

Hence, the final form of the generalized model was

$$\textbf{GM1 + power function GM2} \tag{14}$$

The fixed parameters of the GM2 model were statistically significant at the $p < 0.001$ level.

The contribution of each covariate of the GM1 model to the variations of crown width is presented in Figure 4. The curves were created by using the mean values of the variables in Table 1, the fixed parameters of Table 6 and equal intervals of the test variable, starting from the minimum to the maximum range [33,34].

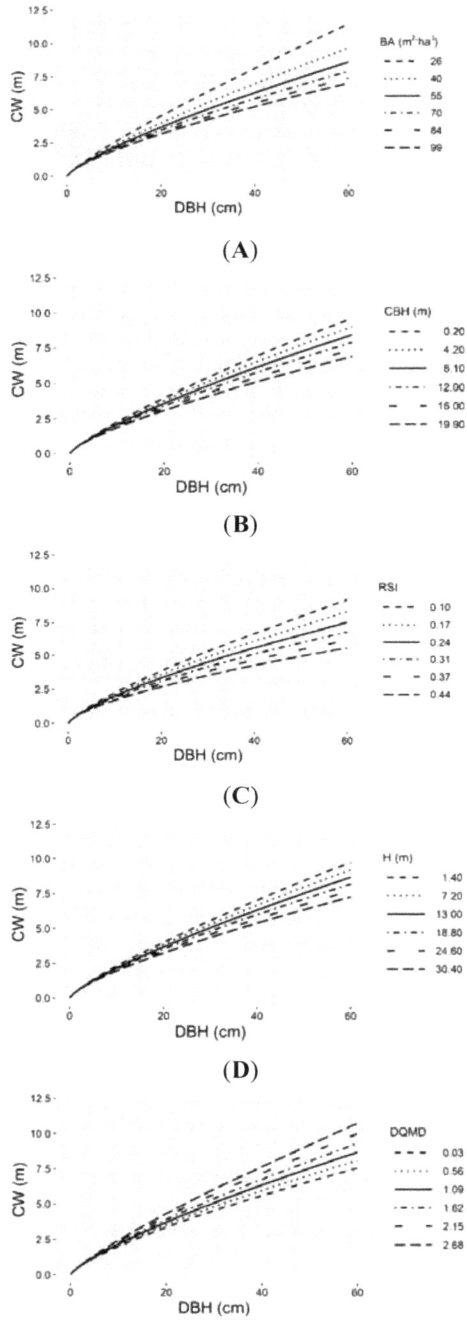

Figure 4. Effects of the GM2 model's covariates ((**A**) BA, (**B**) CBH, (**C**) RSI, (**D**) H, and (**E**) the diameter/quadratic mean diameter, DQMD) on the crown width of Black pine.

3.3. Development of a Generalized Mixed Effect Model

Starting from the mathematical form of the GM1 model, it was assumed that both fixed parameters, that is β_0 and β_1, contain a random part and that the model converged for the fitting data. However, the random part of the β_1 parameter revealed almost zero variation and the model was simplified by modifying the β_1 parameter to fixed. In addition, the parameter estimate (β_6) that was linked to DQMD was not significant according to the t-statistic criterion and the mixed model was precluded from further analysis. The GM1 model was further modified by excluding the DQMD variable and convergence was achieved with both two and one random parameters. The two nested models were compared between them with the ANOVA method [20] and no significant differences were detected between them following the results of the log-likelihood ratio test (−2LL) ($p > 0.05$). This resulted in the selection of the simplest form of the two models with one random variable, according to the following mathematical expression (GMM1):

$$CW = (\beta_0 + b_0)DBH^{\beta_1(BA^{\beta_2}+\beta_3 CBH+\beta_4 RSI+\beta_5 H)} \quad (15)$$

The residual analysis of the GMM1 model revealed treads of unequal variance and a constant-power function was used to model variance with diameter. The final form of the proposed generalized mixed model was (GMM2):

$$\textbf{GMM1 + constant-power function GMM2} \quad (16)$$

The parameter estimates and fit statistics of the proposed generalized mixed model are presented in Table 7. The fixed parameters of the GMM2 model were significant at the $p < 0.001$ level.

Table 7. Estimated parameters and fitting statistics of the GMM2.

Fixed Parameters							Parameters of Function Variance		
β_0	β_1	β_2	β_3	β_4	β_5	$Var(b_0)$	σ^2	δ_1	δ_2
0.2654 (0.012)	1.3794 (0.122)	−0.0773 (0.019)	−0.0044 (0.001)	−0.2852 (0.067)	−0.0015 (0.001)	0.0011	0.0290	0.0002	0.4515
Fit Statistics									
EI		**RMSE**			**AIC**			**Bias**	
0.7644		0.9119			8078.0			0.0018	

The extent of GMM2 coverage to crown width variation of the total sample was tested graphically by localizing the mixed-effect parameters to each sample plot using a random sample of four trees. According to Figure 5, the generalized mixed effect model covered the largest part of crown variation.

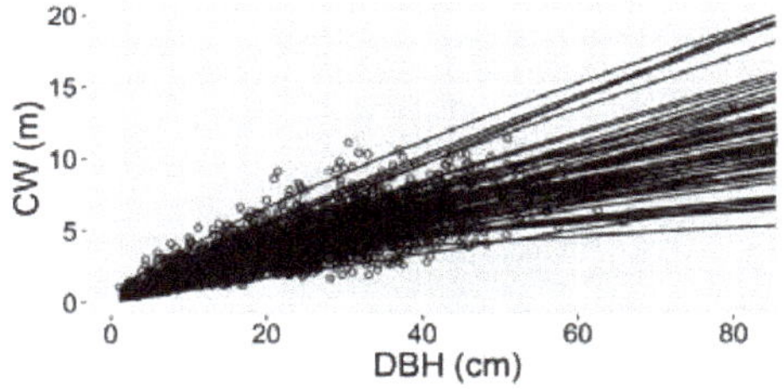

(A)

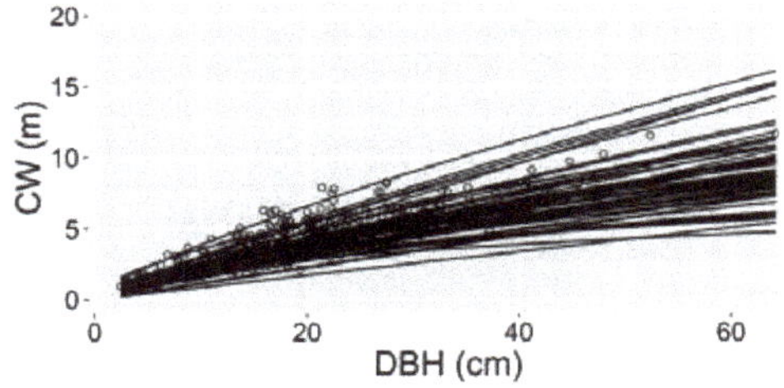

(B)

Figure 5. Localized curves of the generalized mixed model (GMM2) based on four randomly selected trees overlaid on the fitting data (**A**) and on validation data (**B**).

In an effort to compare the two modeling strategies (simple and generalized mixed modelling), the distribution of the standardized residuals was evaluated graphically (Figure 6).

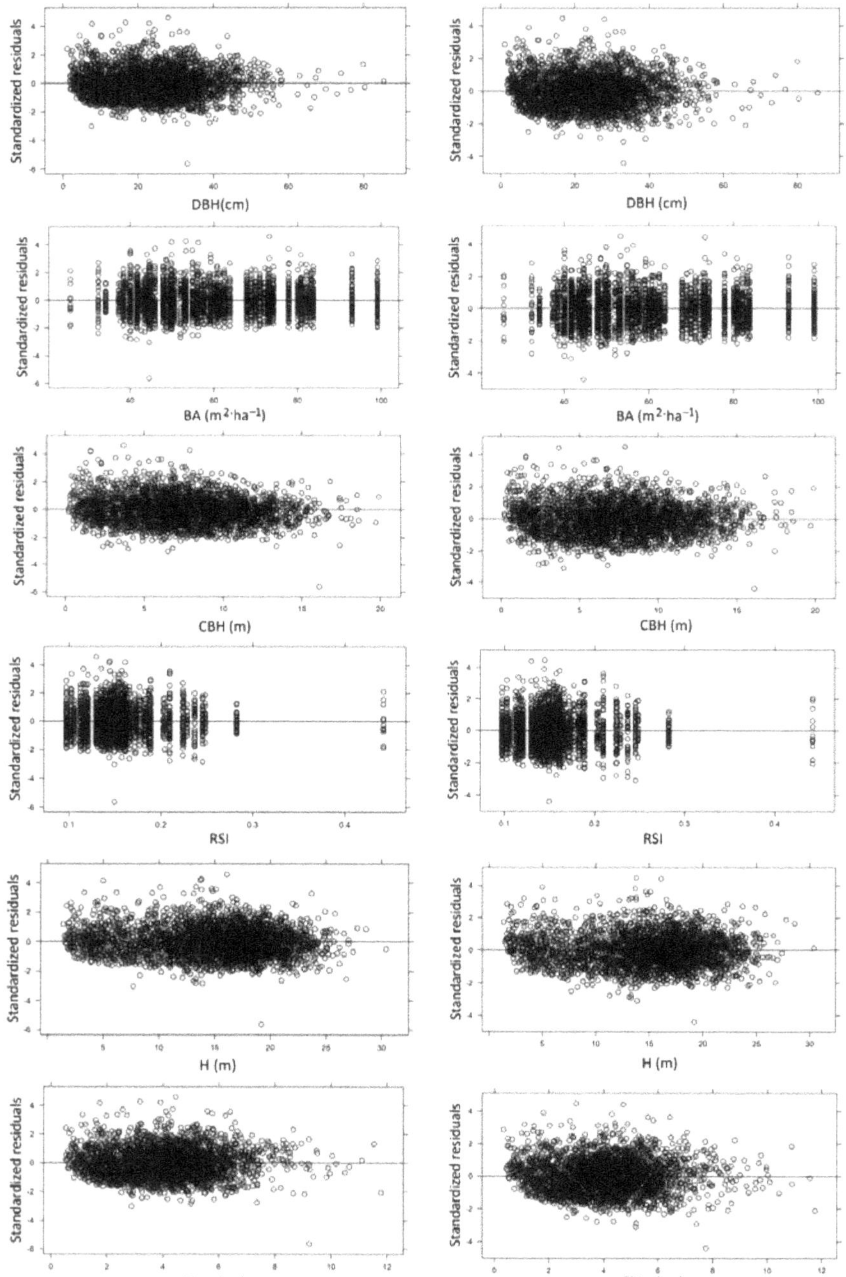

Figure 6. *Cont.*

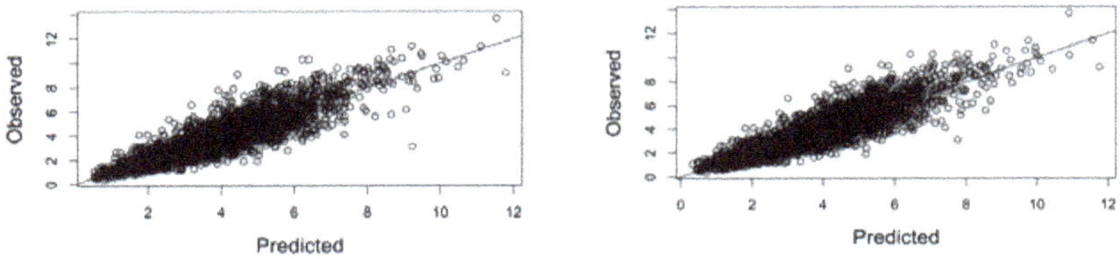

(A) Simple Mixed Model (MM2) (B) Generalized Mixed Model (GMM2)

Figure 6. Comparison of the two modelling strategies, simple mixed model (**A**) and generalized mixed model (**B**).

According to Figure 6, the two models performed equally well, with slightly better fitting of the GMM2 model. However, both models revealed no trends of unequal variance across the total range of the tested variables.

3.4. Evaluation

A total set of n = 362 trees from six randomly selected independent plots that were located at different stands within the study area was used in order to evaluate the predictive ability of the simple mixed (MM2) and the generalized mixed (GMM2) models. The fixed-effect (FE) corresponded to the fixed-effect part of the corresponding model.

The value attributes of the independent sample were among those that were used for model development, while the calibration of the mixed model random parameters was based on the random selection of four trees.

After a series of combinations of random tree selection, the mean RMSE values of all the modeling techniques were calculated, and they are presented in Figure 7. As it is shown in Figure 7, it can be concluded that both mixed models (MM2 and GMM2), for which a pre-sample for calibration of four random trees was used, could predict crown width with low RMSE, when compared to fixed models. The RMSE for all six sample plots was 0.966 m and the mean bias was 0.0021 m for the MM2 model, while the corresponding values for GMM2 were 0.933 m and 0.2557 m, respectively.

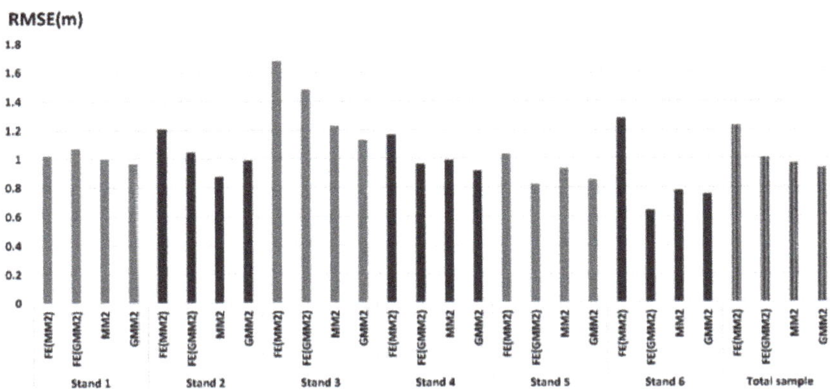

Figure 7. Root mean square error values per sample plot and total sample during model evaluation against independent data of crown width-diameter. FE = fixed-effect part of the corresponding model in the parenthesis.

The corresponding mean value as the minimum possible RMSE from the simple least square analysis for each separate plot was 0.782 m for the basic model and 0.994 m for GM2, whereas the fixed parts (FE) of the mixed models presented the highest RMSE, that is, 1.232 m for MM2 and 1005 m for GMM2. The mean bias of the generalized model (GM2) was about 0.445 m which is relevant to under prediction biases of crown width.

4. Discussion

The models presented in the current article explained the greatest part of the crown width variation, a finding that has also been pointed out by other relevant studies [33,34]. The crown width-diameter allometry tends to be explained by a linear model, despite the slightly better performance of the M2 model, which complies with the findings of Dawkins [42]. However, the power-type M2 model was considered as more flexible since it was easy to be linearized and then expanded to a mixed effect model. The same model for crown width modeling has also been used by Russell and Weiskittel [57] and Sharma et al. [33,34]. In its simple form, the stem diameter is the only predictor of crown horizontal dimension, which is not as accurate because such models seem to overestimate crown size for dense stands and to underestimate it for sparse stands [33,58]. Indeed, as it is shown in Figures 7 and 8, the simple model that only uses DBH as the independent variable, that is the FE (MM2) model, presented the largest error values along with the most biased estimates of all models that were tested. For this reason, a number of covariates at tree and at stand level were added to increase the percentage of crown variation that could be explained by the model.

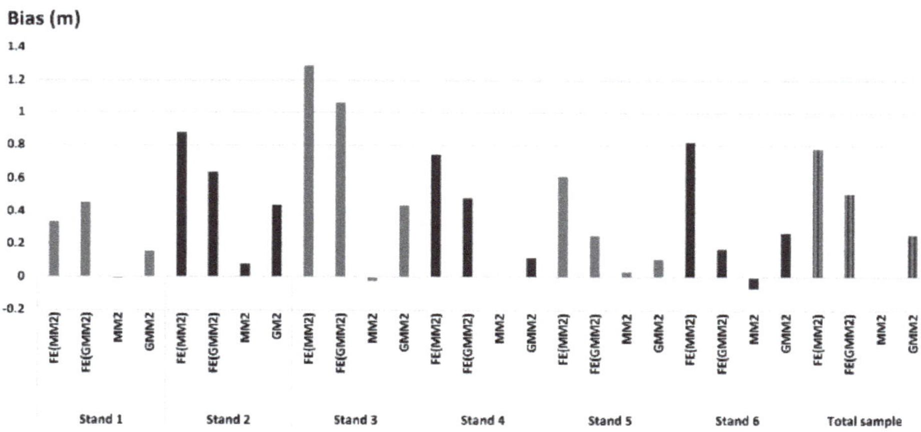

Figure 8. Mean prediction bias per sample plot and total sample during model evaluation against independent data of crown width-diameter.

Basal area, as a measure of stand density, is expected to affect crown width [59], while the canopy base height has been mentioned as a valid predictor of crown width by a number of relevant studies [31–34]. Tree height has also been used as a valid regressor of crown width allometry at tree level [31,32], while both the RSI and the fraction between diameter and quadratic mean diameter have been found to affect crown width variation according to Ducey [29]. Sharma et al. [33,34] have used the RSI in crown width modeling along with a number of other variables at tree or at stand level. In our current research, as it can be concluded from Figure 4, the contribution of RSI in crown width variation is increased, second only to basal area. However, the contribution of each parameter alone is quite different than altogether, as long as the crown width expression is concerned. In this sense, the differences that emerged were related to the overall contribution of each variable in crown width modeling. From Figure 2, it is quite clear that as height increases, crown width increases too, which is biologically valid. However, in conjunction with other variables (Equation (13) and Table 6), crown width decreases as height increases. This, however, was reported much earlier by Briegleb [60], who concluded that "for trees of given diameter, the shorter trees have larger crowns than the taller". As a result, the value of the GM2 model can be evaluated in a biological base, since it explains the crown width allometry by considering the combined effect of a number of parameters rather than the effect of each one independently. In this sense, dominant height is expected to affect crown width according to the findings of other relevant studies (e.g., [31,33,34]). In the current study, during the forward modeling process, HDOM appeared strongly related to crown width, however after the inclusion of BA and CBH, the associated parameter of HDOM turned to insignificant at the $p > 0.05$ level. In addition, despite the positive correlation between crown width and RSI, an inverse relationship was detected according to Equation (13). This phenomenon may be attributed to the inclusion of the basal area in the model, which seems to affect crown width variation more than the other inherent predictors—a hypothesis that is further supported by Figure 4. As such, RSI decreases as the number of trees and the dominant height increases, however the basal area and the tree diameter provide a biological limit which actually determines the size of the trees in the stand.

Another interesting outcome from the current study is the exclusion of DQMD variable from the GM2 model after its expansion to mixed-effect. The inclusion of a random part in the model replaced the contribution of the DQMD in the crown width variation, while the almost zero variance of β_1 parameter in Equation (15) can be attributed to the combined effect of the inherent variables.

The expansion of the base model M2 to mixed-effect significantly improved the predictive ability of the model. As can be observed in Tables 3 and 5, the efficiency index (similar to R^2) increased to 13% in comparison to that of the base model, indicating that the random part explained a great part of crown variation, which was not possible to explain before. This may be attributed to the increased flexibility of the MM2 model, since both parameters were assumed to contain a random part. However, during the evaluation stage, the sub-sample of four random trees at plot level was not sufficient enough to restrict the potential error to the minimum possible; instead, the unbiased estimations increased their overall efficiency. A possible disadvantage in using this technique is the typical lack of crown size measurements in standard forest inventories [61], which might limit the practical use of mixed effect modeling.

The inclusion of four predictors in the mixed model increased the model's precision by 0.41%, a relatively low rate, taking into consideration its increased complexity. During model evaluation, the GMM2 model presented reduced prediction error in comparison to the MM2 model, however its prediction bias increased, according to Figures 7 and 8, which was not observed during the model's calibration. Comparison between simple and generalized mixed models has been attempted earlier by several authors. Temesgen et al. [62] demonstrated that a generalized mixed-effect model which included both random and stand-level variables resulted in low RMSE reductions, compared to simple mixed models which included random effects and tree-level predictors. The results of the current study confirm this hypothesis, since the prediction improvement of the GMM2 model is relatively low compared to the MM2 model as far as crown width modeling is concerned. The random part of the simple MM2 model explained a large part of crown width variation, which the generalized GMM2 model managed to explain with the additional prediction variables.

Ease of use in forestry practice should be an important concern of the analysts towards the development of such models. DBH and basal area are the main variables at tree and stand level, respectively that are used extensively around the world in forest inventories and forest management planning. Often, crown width allometry is used inversed [13], aiming at stem diameter calculations via remote sensing techniques in order to estimate the wood standing volume within stands, a procedure that could be further facilitated by using simple models calibrated at stand level. In this case, the fixed part of the MM2 model is proposed for practical use in the field.

5. Conclusions

In conclusion, a crown width mixed effect model for crown size prediction of Black pine is proposed in the current article. The model, which uses the stem diameter as the basic regressor, improves the fitting ability of the simple fixed effect model. It is also more accurate than a model which includes a basic stand density covariate in its formulation, as it uses a number of random variables that explain a great part of crown variation. By defining crown width allometry, a very useful tool for wood volume estimation is provided, which can also be linked to remote sensing analysis in the frame of sustainable forest management.

Author Contributions: Conceptualization, D.R.; Methodology, D.R., V.K., A.K.; Formal Analysis, D.R., V.K., A.K.; Investigation, D.R., V.K., A.K., C.S.; Data Curation, D.R., C.S.; Writing—Original Draft Preparation, D.R.; Writing—Review & Editing, V.K., A.K.; Funding Acquisition, V.K., A.K.

Funding: This research was funded by the Green Fund of the Hellenic Ministry of Environment & Energy grant contract [Improving the sampling process for assessing wood volume as part of the policy planning review of forest management plan specifications].

Acknowledgments: We express our thanks to the editorial members and the three anonymous reviewers.

Conflicts of Interest: The authors declare no conflict of interest.

References

1. Bullock, S.H. Development patterns of tree dimensions in a Neotropical deciduous forest. *Biotropica* **2000**, *32*, 42–52. [CrossRef]
2. O'Hara, K.L. Stand structure and growing space efficiency following thinning in an even-aged Douglas-fir stand. *Can. J. For. Res.* **1988**, *18*, 859–866. [CrossRef]
3. Hemery, G.E.; Savill, P.S.; Pryor, S.N. Applications of the crown diameter–stem diameter relationship for different species of broadleaved trees. *For. Ecol. Manag.* **2005**, *215*, 285–294. [CrossRef]
4. Bonnor, G.M. Stem diameter estimates from crown width and tree height. *Commonw. For. Rev.* **1968**, *47*, 8–13.
5. Badoux, E. Relations entre le developpement de la crime et l'accroissement chez le pin silvestre. *Mitteilungen der Schweizerischen Anstalt fur das Forstliche Versuchswesen* **1946**, *24*, 405–516. (In French)
6. Krajicek, J.E.; Brinkman, K.A.; Gingrich, S.F. Crown competition—A measure of density. *For. Sci.* **1961**, *7*, 35–42.
7. Yang, Y.; Huang, S. Allometric modelling of crown width for white spruce by fixed- and mixed-effects models. *For. Chron.* **2017**, *93*, 138–147. [CrossRef]
8. Gottschalk, K.W.; MacFarlane, W.R. *Photographic Guide to Crown Condition of Oaks: Use for Gypsy Moth Silvicultural Treatments*; GTR- NE-168 Northeastern Forest Experimental Station; USDA Forest Service: Radnor, PA, USA, 1993.
9. Song, C. Estimating tree crown size with spatial information of high resolution optical remotely sensed imagery. *Int. J. Remote Sens.* **2007**, *28*, 3305–3322. [CrossRef]
10. Zarnoch, S.J.; Bechtold, W.A.; Stolte, K.W. Using crown condition variables as indicators of forest health. *Can. J. For. Res.* **2004**, *34*, 1057–1070. [CrossRef]
11. Gregoire, T.G.; Valentine, H.T. A sampling strategy to estimate the area and perimeter of irregularly shaped planar regions. *For. Sci.* **1995**, *41*, 470–476.
12. Foli, E.G.; Alder, D.; Miller, H.G.; Swaine, M.D. Modelling growing space requirements for some tropical forest tree species. *For. Ecol. Manag.* **2003**, *173*, 79–88. [CrossRef]
13. Popescu, S.C.; Wynne, R.H.; Nelson, R.H. Measuring individual tree crown diameter with LIDAR and assessing its influence on estimating forest volume and biomass. *Can. J. Remote Sens.* **2003**, *29*, 564–577. [CrossRef]
14. Gering, L.R.; May, D.M. The relationship of diameter at breast height and crown diameter for four species in Hardin County, Tennessee. *South. J. Appl. For.* **1995**, *19*, 177–181.
15. Bragg, D.C. A local basal area adjustment for crown with prediction. *North. J. Appl. For.* **2001**, *18*, 22–28.
16. Lockhart, B.R.; Robert, C.; Weih, J.R.; Keith, M.S. Crown radius and diameter at breast height relationships for six bottomland hardwood species. *J. Ark. Acad. Sci.* **2005**, *59*, 110–115.
17. Pretzsch, H.; Biber, P.; Uhl, E.; Dahlhausen, J.; Rötzer, T.; Caldentey, J.; Koike, T.; van Con, T.; Chavanne, A.; Seifert, T.; et al. Crown size and growing space requirement of common tree species in urban centres, parks, and forests. *Urban For. Urban Green.* **2015**, *14*, 466–479. [CrossRef]

18. López, C.A.; Gorgoso, J.J.; Castedo, F.; Rojo, A.; Rodríguez, R.; Álvarez, J.G.; Sánchez, F. A height-diameter model for Pinus radiata D. Don in Galicia (northwest Spain). *Ann. For. Sci.* **2003**, *60*, 237–245. [CrossRef]
19. Mehtätalo, L.; de-Miguel, S.; Gregoire, T.G. Modeling height diameter curves for prediction. *Can. J. For. Res.* **2015**, *45*, 826–837. [CrossRef]
20. Pinheiro, J.C.; Bates, D.M. *Mixed-Effects Models in S and S-PLUS*; Spring: New York, NY, USA, 2000.
21. Calama, R.; Montero, G. Interregional nonlinear height-diameter model with random coefficients for stone pine in Spain. *Can. J. For. Res.* **2004**, *34*, 150–163. [CrossRef]
22. Castedo-Dorado, F.; Diéguez-Aranda, U.; Barrio-Anda, M.; Sánchez, M.; von Gadow, K. A generalized height-diameter model including random components for radiata pine plantations in northeastern Spain. *For. Ecol. Manag.* **2006**, *229*, 202–213. [CrossRef]
23. Sánchez-González, M.; Cañellas, I.; Montero, G. Generalized height-diameter and crown diameter prediction models for cork oak forests in Spain. *For. Syst.* **2007**, *16*, 76–88. [CrossRef]
24. Sharma, M.; Patron, J. Height-diameter equations for boreal tree species in Ontario using a mixed-effects modeling approach. *For. Ecol. Manag.* **2007**, *249*, 187–198. [CrossRef]
25. Vargas-Larreta, B.; Castedo-Dorado, F.; Álvarez-González, J.G.; Barrio-Anta, M.; Cruz-Cobos, F. A generalized height-diameter model with random coefficients for uneven-aged stands in El Salto, Durango (Mexico). *Forestry* **2009**, *82*, 445–462. [CrossRef]
26. Crecente-Campo, F.; Tomé, M.; Soares, P.; Diéguez-Aranda, U. A generalized nonlinear mixed-effects height-diameter model for *Eucalyptus globulus* L. in northwestern Spain. *For. Ecol. Manag.* **2010**, *259*, 943–952. [CrossRef]
27. Corral-Rivas, S.; Alvarez-Gonzalez, J.G.; Crecente-Campo, F.; Corral-Rivas, J.J. Local and generalized height-diameter models with random parameters for mixed, uneven-aged forests in Northwestern Durango, Mexico. *For. Ecosyst.* **2014**, *6*, 1–9. [CrossRef]
28. Gómez-García, E.; Diéguez-Aranda, U.; Castedo-Dorado, F.; Crecente-Campo, F. A comparison of model forms for the development of height-diameter relationships in even-aged stands. *For. Sci.* **2014**, *60*, 560–568. [CrossRef]
29. Ducey, M.G. Predicting Crown Size and Shape from Simple Stand Variables. *J. Sustain. For.* **2009**, *28*, 5–21. [CrossRef]
30. Fu, L.; Sun, H.; Sharma, R.P.; Lei, Y.; Zhang, H.; Tang, S. Nonlinear mixed-effects crown width models for individual trees of Chinese fir (*Cunninghamia lanceolata*) in south-central China. *For. Ecol. Manag.* **2013**, *302*, 210–220. [CrossRef]
31. Fu, L.; Sharma, R.P.; Hao, K.; Tang, S. A generalized interregional nonlinear mixed effects crown width model for Prince Rupprecht larch in northern China. *For. Ecol. Manag.* **2017**, *389*, 364–373. [CrossRef]
32. Hao, X.; Sun, Y.J.; Wang, X.J.; Wang, J.; Yao, F. Linear mixed-effects models to describe individual tree crown width for China-fir in Fujian province, southeast China. *PLoS ONE* **2015**, *10*, 1–14. [CrossRef] [PubMed]
33. Sharma, R.P.; Vacek, Z.; Vacek, S. Individual tree crown width models for Norway spruce and European beech in Czech Republic. *For. Ecol. Manag.* **2016**, *366*, 208–220. [CrossRef]
34. Sharma, R.P.; Bílek, L.; Vacek, Z.; Vacek, S. Modelling crown width-diameter relationship for Scots pine in the central Europe. *Trees* **2017**, *31*, 1875–1889. [CrossRef]
35. Ministry of Agriculture. *Results of the First National Forest Inventory of Greece*; General Secretariat of Forests and Natural Environment: Athens, Greece, 1992; p. 134.
36. Rey, F.; Berger, F. Management of Austrian black pine on marly lands for sustainable protection against erosion (Southern Alps, France). *New For.* **2006**, *31*, 535–543. [CrossRef]
37. Raptis, D.I. Defining the Features of Natural Black Pine Stands in the Southeast Mt. Olympus under the Frame of Multi-Purpose Silviculture. Ph.D. Thesis, Aristotle University of Thessaloniki, Thessaloniki, Greece, 2011.
38. Ayhan, H.O. Crown diameter: D.b.h. relationships in Scots pine. *Arbor* **1973**, *5*, 15–25.
39. Gómez-García, E.; Fonseca, T.F.; Crecente-Campo, F.; Almeida, L.R.; Diéguez-Aranda, U.; Huang, S.; Marques, C.P. Height-diameter models for maritime pine in Portugal: A comparison of basic, generalized and mixed-effects models. *iForest* **2015**, *9*, 72–78. [CrossRef]
40. Apatsidis, L.D. Site quality and site indexes for Black Pine of Greece. *Das. Erevna* **1985**, *1*, 5–20. (In Greek)
41. Moore, J.A.; Zhang, L.; Stuck, D. Height–diameter equations for ten tree species in the Inland Northwest. *West. J. Appl. For.* **1996**, *11*, 132–137.

42. Dawkins, H.C. Crown diameters: Their relation to bole diameter in tropical forest trees. *Commonw. For. Rev.* **1963**, *42*, 318–333.
43. Huxley, J.S.; Teissier, G. Terminology of relative growth. *Nature* **1936**, *137*, 780–781. [CrossRef]
44. Hetherington, J.C. Crown diameter: Stem diameter relationship in managed stands of sitka spruce. *Commonw. For. Rev.* **1967**, *46*, 278–281.
45. Bechtold, W.A. Crown-diameter predictions models for 87 species of stand-grown trees in the Eastern United States. *South. J. Appl. For.* **2003**, *27*, 269–278.
46. VanderSchaaf, C.L. Reineke's stand density index: A quantitative and nonunitless measure of stand density. In Proceedings of the 15th Biennial Southern Silvicultural Research Conference, Asheville, NC, USA; 2013; pp. 577–579.
47. Zhao, D.; Kane, M.; Borders, B.E. Crown ratio and relative spacing relationships for loblolly pine plantations. *Open J. For.* **2012**, *2*, 101–115. [CrossRef]
48. Saud, P.; Lynch, T.B.; Anup, K.C.; Guldin, J.M. Using quadratic mean diameter and relative spacing index to enhance height-diameter and crown ratio models fitted to longitudinal data. *Forestry* **2016**, *89*, 215–229. [CrossRef]
49. Gregoire, T.G.; Schabenberger, O.; Barrett, J.P. Linear modeling of irregularly spaced, unbalanced, longitudinal data from permanent plot measurements. *Can. J. For. Res.* **1995**, *25*, 136–156. [CrossRef]
50. Davidian, M.; Giltinan, D.M. *Nonlinear Models for Repeated Measurement Data*; Chapman and Hall: New York, NY, USA, 1995.
51. Rupšys, P. Generalized fixed-effects and mixed-effects parameters height–diameter models with diffusion processes. *Int. J. Biomath.* **2015**, *8*, 1–23. [CrossRef]
52. Temesgen, H.; Hann, D.W.; Monleon, V.J. Regional height–diameter equations for major tree species of southwest Oregon. *West. J. Appl. For.* **2007**, *22*, 213–219.
53. Van Laar, A.; Akça, A. *Forest Mensuration*; Springer London: Dordrecht, The Netherlands, 2007; Volume 13.
54. Vonesh, E.F.; Chinchilli, V.M. *Linear and Nonlinear Models for the Analysis of Repeated Measurements*; Marcel Dekker Inc.: New York, NY, USA, 1997.
55. Pinheiro, J.C.; Bates, D.M. *Model Building for Nonlinear Mixed Effects Model*; Department of Statistics, University of Wisconsin: Madison, WI, USA, 1998.
56. Ritz, C.; Streibig, J.C. *Nonlinear Regression with R*; Springer: New York, NY, USA, 2008.
57. Russell, M.B.; Weiskittel, A.R. Maximum and largest crown width equations for 15 tree species in Maine. *North. J. Appl. For.* **2011**, *28*, 84–91.
58. Thorpe, H.C.; Astrup, R.; Trowbridge, A.; Coates, K.D. Competition and tree crowns: A neighborhood analysis of three boreal tree species. *For. Ecol. Manag.* **2010**, *259*, 1586–1596. [CrossRef]
59. Sun, Y.X.; Gao, H.L.; Li, F.R. Using Linear Mixed-Effects Models with Quantile Regression to Simulate the Crown Profile of Planted *Pinus sylvestris* var. *Mongolica Trees*. *Forests* **2017**, *8*, 446. [CrossRef]
60. Briegleb, P.A. An approach to density measurement in Douglas-fir. *J. For.* **1952**, *50*, 529–536.
61. Gill, S.J.; Biging, G.S.; Murphy, E.C. Modeling conifer tree crown radius and estimating canopy cover. *For. Ecol. Manag.* **2000**, *126*, 405–416. [CrossRef]
62. Temesgen, H.V.J.; Monleon, V.J.; Hann, D.W. Analysis and comparison of nonlinear tree height prediction strategies for Douglas-fir forests. *Can. J. For. Res.* **2008**, *38*, 553–565. [CrossRef]

forests

MDPI

Article

Palaeoecological Evidence for Survival of Scots Pine through the Late Holocene in Western Ireland: Implications for Ecological Management

Jenni R. Roche [1,2,*], Fraser J. G. Mitchell [2], Steve Waldren [2] and Bettina S. Stefanini [2]

[1] Dublin Bay Biosphere Partnership, Dublin City Council, Parks & Landscape Services Division, Civic Offices, Wood Quay, Dublin 8, Ireland

[2] Department of Botany, School of Natural Sciences, Trinity College Dublin, Dublin 2, Ireland; fraser.mitchell@tcd.ie (F.J.G.M.); stephen.waldren@tcd.ie (S.W.); stefanb@tcd.ie (B.S.S.)

* Correspondence: jennifer@tcd.ie; Tel.: +353-1-222-3370

Received: 29 April 2018; Accepted: 7 June 2018; Published: 13 June 2018

Abstract: The dynamics of Scots pine (*Pinus sylvestris* L.) in Europe during the Holocene have been spatially and temporally complex. The species underwent extirpation and reintroduction in several north-west European countries. This study investigated the late Holocene vegetation history of a present-day pinewood in western Ireland, to test the widely accepted hypothesis that *P. sylvestris* became extinct in Ireland *c.* AD 400. Palaeoecological, chronological and loss-on-ignition analyses were conducted on a sediment core extracted from an adjacent lake. The pollen profile showed no major *Pinus* decline and a *Pinus* macrofossil occurred *c.* AD 840, indicating localised survival of *P. sylvestris* from *c.* AD 350 to the present. The available archival maps and historical literature provide supporting evidence for continuity of forest cover. The hypothesis that *P. sylvestris* became extinct in Ireland is rejected. The implications for ecological management are significant. We argue that *P. sylvestris* should be considered native to Ireland, at least at this site. As Ireland's only putative native *P. sylvestris* population and the western limit of the species' native range, this site is of high conservation value and must be carefully managed and monitored. Seed-sourcing for ex-situ forest restoration must be compatible with the long-term viability of the population in-situ.

Keywords: conservation value; ecological management; forest ecology; native status; palaeoecology; *Pinus sylvestris*; pollen analysis; the Burren; woodland ecology

1. Introduction

Palaeoecological data provide a valuable long-term perspective on contemporary ecosystem dynamics but are under-utilised in conservation management [1]. Site-based palaeoecological studies can provide an evidence base for conservation management decisions, particularly in determining land use history, assessing naturalness and setting appropriate targets for restoration. However, published examples in which palaeoecology is used to inform practical management decisions are rare [2].

Palaeoecological studies have shown *P. sylvestris* to be one of Europe's most dynamic tree species. Its postglacial history in northern Europe is considered particularly well known due to the abundance of its pollen and macrofossils [3]. Its distribution in Europe over the last 13,000 years has been mapped [4] and, in Britain and Ireland, over the last 10,000 years [5]. These maps record large-scale range shifts but may not detect fine-scale, local distribution patterns [6].

P. sylvestris is a pioneer species with broad ecological tolerances. Its distribution is heavily influenced by competitive interactions but it forms stable vegetation communities on nutrient-poor soils [7]. It was most abundant in European forests during the early postglacial when, in response to climatic amelioration, it migrated rapidly across the northern European lowlands at up to 150 km per century. It formed pioneer *Pinus-Betula* forests, which were most extensive from *c.* 9500–7000 cal BP (calibrated radiocarbon years before AD 1950) [4,8]. *P. sylvestris* then declined in the south, while its northern range limits continued to expand, colonising north-west Scotland *c.* 9900 cal BP and northernmost Fennoscandia by *c.* 8500 cal BP [9,10]. *P. sylvestris* underwent marked range reductions in north-west Europe *c.* 4500 cal BP. This *Pinus* decline is an important pollen stratigraphic marker, observed in northern Scotland, England, Ireland and Finland, probably due to a large-scale climatic shift to wetter conditions and associated competitive exclusion [11–15]. It appears that *P. sylvestris* became extinct in several north-west European countries including Denmark, the Netherlands, Belgium, England, Wales and Ireland [5,16–19].

P. sylvestris colonised Ireland relatively early in the Holocene; one of its earliest records, evidenced by pollen and macrofossils, is from Gortlecka in the Burren *c.* 10,500 cal BP [20]. *Pinus* was the dominant arboreal pollen type in most western and upland sites for at least part of the early Holocene and an important component of raised bog, river valley and upland habitats. A major *Pinus* decline began *c.* 4500 cal BP, possibly due to climate change, competition with *Alnus glutinosa*, blanket bog expansion and human activity [5,6]. A late outpost occurred at Gortlecka; *Pinus* pollen and macrofossils were present *c.* 1050 ± 160 cal BP/AD 900 [20] but Watts, the author of that study, expressed concern that this date may be too young. The latest unambiguous record was a preserved stump from Clonsast Bog, a raised bog in County Offaly. This was directly dated to 1550 ± 140 cal BP/AD 400 [21], at which point *P. sylvestris* is widely believed to have become extinct in Ireland [6]. The species' supposed extirpation is of great interest as it is asynchronous between sites across Ireland and the apparent causal factors differ between sites [20].

P. sylvestris was reintroduced to Ireland in the mid-17th century and has been widely planted [22]. Due to its supposed extirpation and reintroduction, the native status of *P. sylvestris* is disputed, causing inconsistencies in conservation and forest management policy [23]. The Native Woodland Scheme provides funding to plant *P. sylvestris* when establishing or restoring native woodland [24] yet the Irish Peatland Conservation Council lists it as an invasive alien species [25]. Further palaeoecological research is urgently needed to determine if *P. sylvestris* became extinct in Ireland and to clarify its native status.

This study examines the late Holocene vegetation history of a pinewood in western Ireland, to test the widely accepted hypothesis that *P. sylvestris* became extinct in Ireland. Palaeoecological evidence is presented for localised survival of native *P. sylvestris* from 1600 cal BP/AD 350 to the present. The available historical sources and archival maps indicate a long history of woodland cover. Recommendations for the ecological management of the site are presented.

2. Materials and Methods

2.1. Site Description

Rockforest Lough (53.005, −8.958, Irish Grid Reference R 356 953) is situated 10 km north-east of Corofin, County Clare, western Ireland at 16 m above sea level (Figure 1). It lies on Carboniferous Limestone in the Burren, a karstic area covering over 300 km^2 and renowned for its rich archaeological heritage and plant diversity [26,27]. The climate is oceanic; mean rainfall is 1400–1600 mm and mean air temperature is 10 °C (1981–2010) annually. The prevailing winds at Shannon Airport, County Clare (1946–2010) are south-westerly [28].

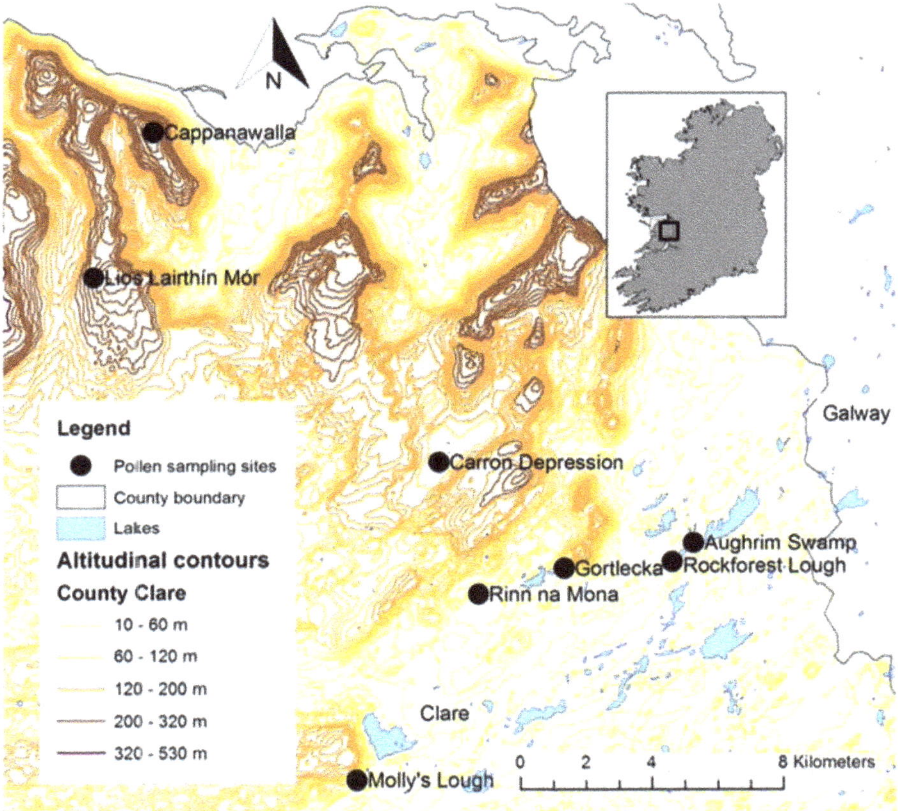

Figure 1. Location of Rockforest Lough and other pollen sampling sites mentioned in the text.

Satisfactory coring sites can be difficult to obtain in the karstic limestone of the Burren. Permanent lakes with brown algal mud are preferable to turloughs, swamps and lakes with carbonate mud, as pollen preservation is better and the hard water effect, which impairs radiocarbon dating, is less pronounced [20]. Rockforest Lough is a permanent lake *c.* 8 ha in area with a deep basin (7.48 m) and a shallower arm to the north-east. Its pollen source area is estimated to be 300–800 m (after [29]). The water level varies seasonally due to karstic hydrology [20] but the basin has not been known to dry out, at least in living memory (J. Cunningham, pers. comm.). It is bordered by reeds with pasture on glacial till to the south and limestone pavement to the north. Rockforest House is located on the south-western edge of Rockforest Lough (Figure 2). A pinewood occurs on limestone pavement *c.* 500 m to the south-west. Stunted, mature *P. sylvestris* is scattered through patchy *Corylus avellana* scrub. *P. sylvestris* regeneration is scarce. The species-rich vegetation was surveyed and classified as the *Corylus avellana–Brachypodium sylvaticum* pinewood type, which has affinities with Norwegian basiphilous pinewoods [23,30]. The woodland is located in the state-owned Burren National Park and the East Burren Special Area of Conservation, which is protected under the European Union (EU) Habitats Directive (92/43/EEC).

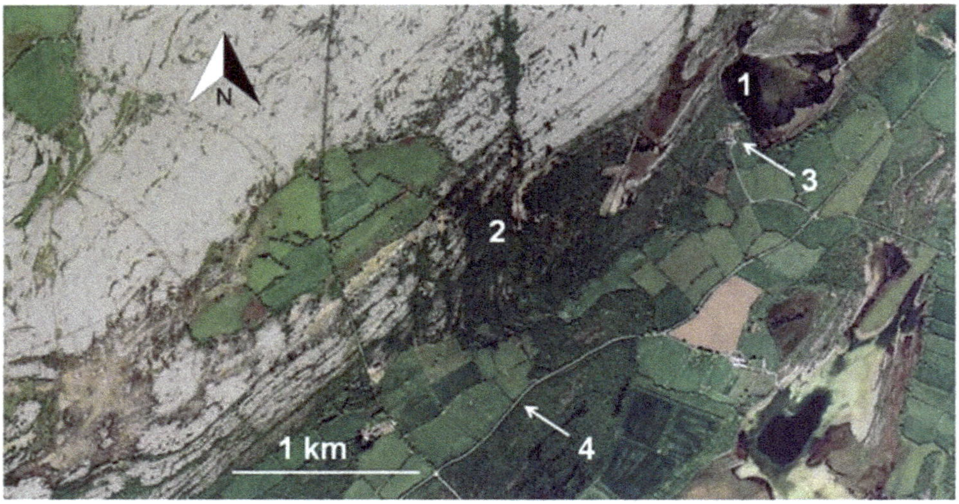

Figure 2. Aerial photograph (2000) showing the locations of (1) Rockforest Lough, (2) Rockforest Wood, (3) Rockforest House, (4) Corofin to Gort road/*Bealach an Fhiodhfail* [31].

2.2. Coring

Coring was conducted in the deepest area of Rockforest Lough on 18–19 June 2008. A short core (RFB, 81 cm) was extracted with a modified rod-operated plexiglass piston corer, preserving the sediment-water interface, and vertically extruded at 1 cm intervals on-site. The lower sediment (RFC) was sampled to a depth of 153 cm using a Livingstone corer [32]. The cores were wrapped on-site, stored at 4 °C and sliced at 1 cm intervals in the laboratory.

2.3. Loss-on-Ignition (LOI)

To ensure sufficient material, 2 cm thick samples were used for LOI analysis. Using a Thermolyne Type 6000 furnace, oven-dried, weighed samples were ignited for five hours at 550 °C to estimate organic content and three hours at 950 °C to estimate carbonate content [33]. The LOI profile was used to match the cores.

2.4. Chronology

Spheroidal Carbonaceous Particle (SCP) extraction followed Rose [34,35] with some modifications due to abundant fine material and low SCP numbers [14]. Sample dry weights were increased to 0.6–4.7 g; solvent volumes were increased accordingly. To remove fine material and facilitate counting, samples were filtered using a 10 μm sieve. To avoid the hard water effect, terrestrial plant macrofossils were sieved or hand-picked from the sediment. Eight macrofossil samples were radiocarbon dated using accelerator mass spectrometry (AMS). A chronology was generated using Bchron (version 3.1.4 with the IntCal13 calibration curve), a Bayesian modelling method which uses stochastic linear interpolation [36].

2.5. Pollen Analysis

Pollen analysis was conducted on 1 cm thick samples at 8 cm intervals, reducing to 4 cm in the upper core. Sediment subsamples of 0.5 cm^3 were treated by standard methods [14,37]. *Lycopodium* tablets were added to enable pollen concentration calculation [38]. Samples were mounted in silicone oil and counted with an Olympus BX40 microscope at 400× magnification and 1000× under oil immersion when required. Slides were systematically checked for *Pinus* stomata during routine pollen counting. Pollen and spores were identified following Moore et al. [37], the illustrations of Beug [39] and Reille [40] and a reference collection held by Trinity College. Nomenclature followed Moore et al. [37], excepting the aggregation of the Urticaceae, *Polypodium*, *Rumex* and Coryloid taxa. Coryloid pollen was assumed to be *Corylus avellana*, which is much more frequent than *Myrica gale* in the Burren [26]. A minimum of 400 identifiable terrestrial pollen and spores were counted from each sample. Indeterminate grains were also counted. A percentage pollen diagram was generated with TILIA version 2.0.19 [41]. The pollen sum was total terrestrial pollen and spores including indeterminate grains. The latter were included to reduce overrepresentation of *Pinus* as its distinctive pollen morphology makes it less likely to be classified as indeterminate. Concentrations were examined to ensure that no major changes arose that were not observed in the percentage pollen diagram.

3. Results

3.1. Loss-on-Ignition (LOI)

The sediment comprised homogeneous brown algal mud. Mineral input was relatively stable from 153–35 cm (Figure 3), with a mean of 32.0%. It increased abruptly at 33 cm, with a mean of 34.4% until the present. Peak values exceeded 38.6% at 7–11 cm.

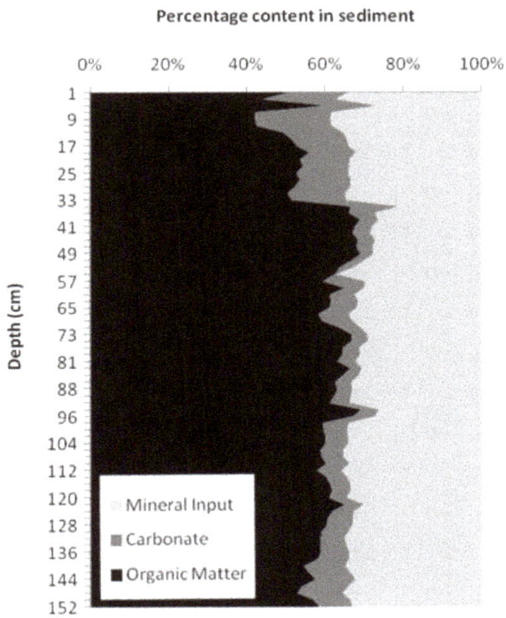

Figure 3. Loss-on-ignition (LOI) profile for Rockforest Lough.

3.2. Chronology

SCPs, produced from fossil fuel combustion and preserved in lake sediments, provide a historical record of atmospheric pollutant deposition. SCP concentration profiles are consistent and often regionally characteristic such that the main profile features can be used for sediment dating. The Rockforest SCP record began at 13 cm (Figure 4). SCP numbers were low initially, likely due to low levels of industrialisation in the region. A rapid increase and sub-surface peak appeared at 8 cm and 7 cm respectively. Based on mean dates from Irish SCP profiles [42], these features were assigned to the 1880s, 1960s and 1981 respectively (Table 1). A second rapid increase at 4 cm was assigned to 1985, when Moneypoint Power Station, *c.* 50 km to the south-west, was commissioned. The filtration of samples may have selectively removed SCPs generated by oil combustion, which are usually smaller than those from coal [34]. However, as the nearest power station is primarily coal-fired and the main profile features described by Rose et al. [42] were observed, selective removal does not appear to have been a significant issue.

The chronology is well-constrained in the upper metre but less so below as fewer dates were obtainable (Table 1). The model excluded an anomalously old date from bulked organic material at 72–74 cm (Figure 5). Sediment accumulation was relatively constant in the upper metre (0.7 mm year^{-1}) but considerably lower below (0.1 mm year^{-1}). Pollen analysis was confined to the upper 112 cm due to increased uncertainty in the chronology below the AMS date at 104–107 cm.

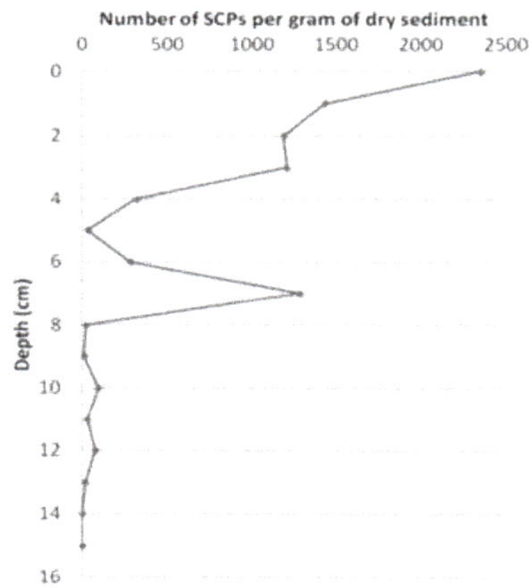

Figure 4. Spheroidal Carbonaceous Particle (SCP) concentration profile for Rockforest Lough.

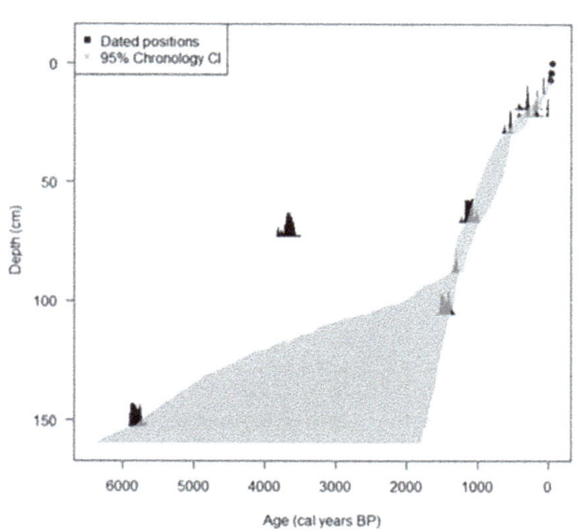

Figure 5. Bchron age-depth model for Rockforest Lough based on dates given in Table 1.

Table 1. Spheroidal Carbonaceous Particle (SCP) and Accelerator Mass Spectrometer (AMS) dates from Rockforest Lough.

Core	Depth (cm)	Laboratory Reference	Sample Description	^{14}C Year BP [1]	Calendric Age (AD)	IntCal 2.5% (cal Year BP)	IntCal 50% (cal Year BP)	IntCal 97.5% (cal Year BP)
			SCPs					
RFB	4	-	2nd rapid increase	-	1985	−9	−37	−35
RFB	7	-	Sub-surface peak	-	1981 ± 2	−36	−30	−24
RFB	8	-	Rapid increase	-	1965 ± 5	−2	0	2
RFB	13	-	Start of record	-	1885 ± 5	92	10	108
			Macrofossils					
RFB	19	Beta-247933	Wood	240 ± 40	-	2	284	423
RFB	22	Beta-247934	Plant material	230 ± 40	-	2	264	418
RFB	29	Beta-247935	Wood	520 ± 40	-	508	538	631
RFB	66–68	Beta-252787	Plant material	1180 ± 40	-	989	1107	1229
RFC	72–74	Beta-252788	Organic material	3410 ± 50	-	3511	3662	3819
RFC	88	Beta-253480	Plant material	1380 ± 40	-	1198	1300	1359
RFC	104–107	Beta-252790	Plant material	1540 ± 40	-	1347	1438	1522
RFC	153	Beta-252791	*Pinus* wood	5050 ± 40	-	5676	5814	5898

[1] AMS dates are quoted with a standard deviation of 2σ (95% confidence limit).

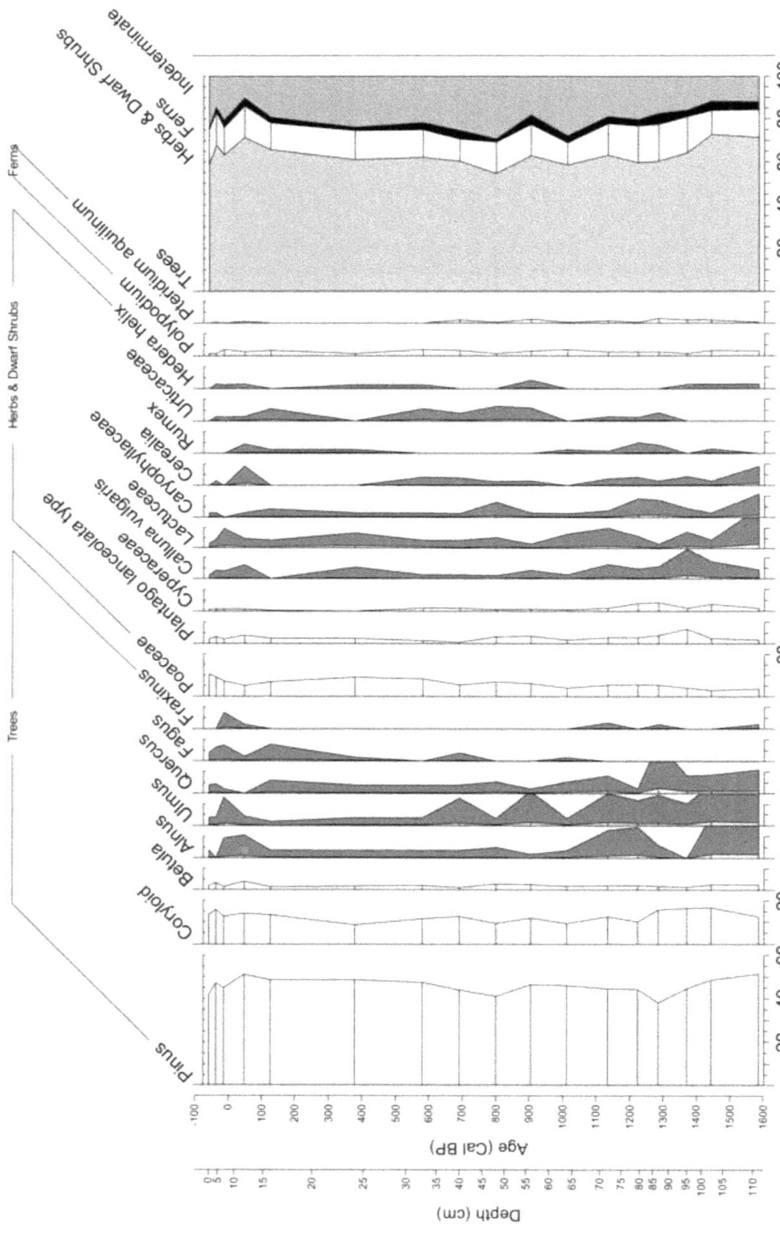

Figure 6. Percentage pollen diagram of selected taxa from Rockforest Lough (shading = Exaggeration × 10).

3.3. Pollen and Macrofossil Stratigraphy

Figure 6 spans from *c.* 1600 cal BP/AD 350 to the present; 68 terrestrial taxa were recorded. Pollen preservation was sub-optimal with frequencies of 10.3–29.1% indeterminate grains. These were included in the pollen sum to reduce overrepresentation of *Pinus*. Arboreal pollen (AP) dominates throughout, ranging from 54.6–73.0%. The *Pinus* signal is consistently high, peaking at the base of the profile (51.3%) and never falling below 38.2%. *Corylus* is frequent throughout. *Betula*, *Alnus*, *Quercus* and *Ulmus* are present at lower frequencies. *Fagus* appears in the upper half, becoming consistently present at 24 cm. Poaceae dominate the non-arboreal pollen, increasing gradually from 3.5–10.3%.

No *Pinus* stomata were found. Macrofossils were scarce but a *Pinus* needle and *Pinus* wood fragment were found at 66 and 153 cm respectively.

4. Discussion

The pollen profile is relatively static. It opens with relatively high AP frequencies (71%) from *c.* 1600–1450 cal BP/AD 350–500 (Figure 6). *Pinus* dominated, indicating an open woodland structure as *P. sylvestris* cannot tolerate heavy shade from other trees [7]. *Alnus* may represent regional pollen as *Alnus* macrofossils were not recorded at Gortlecka [20] and it is rare in the Burren today [26]. This period of high AP frequencies may represent the Late Iron Age Lull (LIAL), when declining farming activity resulted in widespread woodland regeneration. The LIAL appears in other pollen profiles from the area, at Lios Lairthín Mór and Molly's Lough (Figure 1), from *c.* 1950–1450 cal BP/AD 0–500, typically followed by declining AP indicating a resumption of human activity and woodland clearance [43,44]. Indeed, AP declined at Rockforest during the early medieval period *c.* 1450–1300 cal BP/AD 500–650. *Pinus* fell to its lowest frequency, 38.2%, *c.* 1300 cal BP/AD 650 but still dominated the AP.

For the remainder of the early medieval period, and until *c.* 800 cal BP/AD 1150, little change was observed overall. A *P. sylvestris* needle found in the sediment at 66 cm demonstrates that *Pinus* was locally present at 1110 cal BP/AD 840. From *c.* 800–50 cal BP/AD 1150–1900, AP and *Pinus* rose gradually. Mineral input increased abruptly at 600 cal BP/AD 1350, which may reflect intensified human activity in the lake catchment. However, this was not apparent in the pollen profile as AP and *Pinus* continued to rise. The non-native *Fagus* peaks *c.* 125 cal BP/AD 1825. Between *c.* 50 cal BP/AD 1900 and the present, mineral input and Poaceae reach maximum values, indicating intensive human activity and grassland expansion, while AP and *Pinus* decrease. The overall rise in Poaceae from 1600 cal BP/AD 350 to the present indicates the gradual opening up of the woodland and expansion of open grassland. The Cerealia signal, though discontinuous, indicates arable farming.

While palaeoecological methods are most effective in elucidating woodland history [45], archaeological and historical sources are essential in data interpretation [46]. The archaeology within a 1 km radius of Rockforest Lough includes a *fulacht fia* (putative cooking pit), earthwork, ringfort (residence or farmstead generally dating to AD 500–1000) and tower house (fortified house generally dating to AD 1400–1600) [47].

Rockforest is referenced in historical sources, which refer to both woodland cover and human activities. The ancient name of Rockforest Wood was *Coill Ó Flanchada*. The Wars of Thomond affected County Clare in the 12th–14th century. A strategically important pass, *Bealach an Fhiodhfail*, once the main route from Clare to Galway, now the Corofin to Gort road, went through the wood [48] (Figure 2). A battle was fought at the entrance to the wood to contest the pass in 1311 and a defeated force fled through it in 1314 [49]. The Annals of the Four Masters recorded that in 1599 Red Hugh O'Donnell's army massed at the "eastern extremity" of the wood and marched "through the centre of *Coill Ó Flanchada*, through *Bealach an Fhiodhfail*" [50]. In 1655, the Down Survey showed that the only timber woods in this barony (Inchiquin) lay in this parish (Kilkeedy), occurring in nearly every townland in the parish. Covering 850 ha, they likely formed one of the county's largest woods [51]. Henry Pelham's Grand Jury Map of County Clare (1787) depicts woodland at Rockforest but little in the surrounding landscape. Rockforest House was built in the late 18th century [52]; its estate included a grain silo, ice house and walled garden [53]. By 1808, over 30 ha of rocky soil of poor agricultural value had been planted with *Acer*, *Alnus*, *Betula*, *Fagus*, *Fraxinus*, *Larix*, *Picea*, *Pinus pinaster*, *Pinus sylvestris*, *Quercus*, *Ulmus* and other species [54]. A "finely planted demesne" extended almost a mile along the road [53]. The first (1840) and second (1899) editions of the Ordnance Survey six inch maps depict the area as wooded [14]. Selective felling and scrub clearance was undertaken in the 20th century (J. Cunningham, pers. comm.). Though fragmentary in nature, the available historical sources imply continuous presence of woodland cover at Rockforest despite ongoing human activity. However, historical sources are generally dated with a high level of certainty, while dating of palaeoecological data must be inferred from the age-depth model, making it prone to inherent errors; these data should be compared cautiously.

The continuously elevated AP signal and the available historical sources both suggest that Rockforest has a long history of woodland cover. This contrasts, however, with patterns seen in contemporaneous Irish pollen profiles. Prior to the Great Famine of 1845, Ireland's population reached almost nine million, causing severe land-use pressure. Poorer land was cleared for agriculture over much of Ireland [55]. In the Burren, AP frequencies were just 4% from 250–100 cal BP/AD 1700–1850 at Cappanawalla (Figure 1) [56]. Firewood was so scarce that *Pteridium aquilinum* and *Dryas octopetala* were used [57]. The dissimilarity may relate to the management of the Rockforest Estate. Nationally, the woodland resource steadily diminished until the 20th century, excepting woodland remnants within estates [58,59]. Rockforest appears to be one such estate.

Pinus pollen is usually abundantly produced and well-dispersed and so is generally overrepresented in the pollen rain [60]. A 'critical pollen percentage' of 20% was proposed [5] and later revised to 5% to indicate local presence of *P. sylvestris* [61]. However, analyses of fossil stomata provided unambiguous evidence of local presence of *P. sylvestris* when *Pinus* pollen frequencies were as low as 2.8% [62] and 0.4% [63]. From a study of modern pollen deposition in the Rockforest area, McGeever and Mitchell [64] concluded that a *Pinus* pollen value of 5% indicated local presence of *P. sylvestris* in this area. The *Pinus* curve from Rockforest Lough is consistently high, never less than 38.2%, and greatly exceeds the critical pollen percentages. This strongly suggests that *P. sylvestris* was locally present and was a significant component of the vegetation.

Furthermore, macrofossil evidence demonstrates local presence of *P. sylvestris* around Rockforest Lough during the Neolithic at 5810 cal BP/3860 BC and the early medieval period at 1110 cal BP/AD 840. The latter is significant as the species was presumed to have become extinct in Ireland during that period [6]. The latest unambiguous Irish specimen was a preserved stump from Clonsast Bog, which was directly dated to 1550 $\pm$ 140 cal BP/AD 400 [21]. The later Rockforest macrofossil (1110 $\pm$ 120 cal BP/AD 840) is contemporaneous with the last recorded *Pinus* macrofossils from Gortlecka (1050 $\pm$ 160 cal BP/AD 900) [20], which is located 3.3 km west of Rockforest Lough (Figure 1). Watts expressed concern that the date appeared too young but this new evidence strongly indicates that Watts' date was accurate and *Pinus* persisted in the Burren after its assumed disappearance from midland raised bogs.

The later Rockforest macrofossil coincides with a *Pinus* pollen frequency of 45%. Subsequently, the *Pinus* signal remains high, dominating the AP to the present. Though *Pinus* had been planted on the Rockforest Estate by 1808 [54], its pollen was previously strongly represented. Rockforest Wood, which is located 500 m upwind of the coring site and matches the vegetation type recorded in the pollen profile, is the likely source of this *Pinus* signal. The *P. sylvestris* population at Rockforest appears to have persisted through the late Holocene to the present.

These findings are also supported by those of McGeever and Mitchell's [64] analysis of a radiocarbon-dated terrestrial core from Aughrim Swamp, which is located 650 m north-east of the Rockforest Lough coring site (Figure 1). Pollen preservation was good throughout the pollen profile, which extends from the present to 1600 cal BP/AD 350. A continuous *Pinus* signal was observed. Although a decline began *c.* 1550 cal BP/AD 400, reaching minimum values of *c.* 8% of total terrestrial pollen (TTP) *c.* 1350 cal BP/AD 600, the *Pinus* signal recovered quickly, reaching sustained levels of *c.* 15–25% of TTP, before declining to *c.* 5% at the top of the core. It is likely that Rockforest

Lough (8 ha) mainly samples extralocal pollen from a pollen source area of 300–800 m [29], while the small wetland of Aughrim Swamp (0.4 ha) mainly samples local pollen, making it more sensitive to localised vegetation change [65]. The upper halves of undated cores from Rinn na Mona [20] and the Carron Depression [66] (Figure 1) also show continuous *Pinus* signals during the period of supposed extirpation, with frequencies of *c.* 3% and up to *c.* 7% of TTP respectively. While this may indicate that *P. sylvestris* at Rockforest dispersed pollen to these sites or that other localised stands were present, the frequencies in question are rather low in relation to the critical pollen percentage. Macrofossil data were not presented for these sites.

Floristically similar pinewoods existed on limestone pavement elsewhere in Ireland at various times during the Holocene but are no longer extant. Pollen and macrofossil evidence suggest that open woodland composed of *Pinus*, *Ulmus*, *Corylus* and *Betula* existed at Gortlecka (Figure 1) in the early Holocene [20]. Fine spatial resolution pollen analysis indicates that woodland rich in *Pinus*, *Corylus*, *Quercus* and *Ulmus* at Reenadinna, County Kerry developed into *Taxus* woodland *c.* 5730 cal BP/3780 BC [67]. Fine spatial resolution pollen analysis also suggests that open, species-rich woodland dominated by *Pinus* and *Corylus*, with *Calluna* and *Succisa*, existed at Cappanawalla (Figure 1) in the north-west Burren uplands from *c.* 3450–2450 cal BP/1500–500 BC. *Pinus* appeared to have been extirpated there *c.* 500 BC due to clearance by humans, followed by an expansion in grassland [56]. Pinewoods on limestone are present today at Rockvale in the Burren, Castletaylor and Coole Park in County Galway and Keel Bridge and Ballykine in County Mayo but most appear to have originated from introduced *P. sylvestris* [30]. Pinewoods on limestone no longer occur in Britain but pollen and macrofossil evidence indicate that they once existed at Malham, northern England [68]. They are extant and widespread in Fennoscandia [23,69].

These findings provoke the question of why *P. sylvestris* could have survived at Rockforest, in contrast to the decline and extirpation observed elsewhere. Climate is the main determinant of large-scale forest composition but edaphic factors, succession and human disturbance become increasingly important at the local scale [8]. The Burren's patchy, shallow soils suffer periodic drought and low phosphorus and nitrogen availability [70]. *P. sylvestris* tolerates these stresses, giving it a competitive advantage [5]. During its decline in Ireland, *Pinus* was replaced by blanket peat in the uplands and *Alnus* in the lowlands, probably due to a climatic shift to wetter conditions [6]. In the karstic Burren lowlands, blanket peat did not develop and *Alnus* is not thought to have been a significant component of the Holocene vegetation [20,26]. Fine spatial resolution pollen and charcoal analyses have shown that late outposts of *Pinus* in Counties Sligo and Kerry died out *c.* 1800 cal BP/AD 150 and 1700 cal BP/AD 250 respectively, likely due to human activity including woodland clearance using fire [45,71–74]. The aforementioned dated pine stump (1550 ± 140 cal BP/AD 400) from Clonsast Bog came from a layer of stumps found on a recurrence surface [21] i.e., an abrupt stratigraphic transition from highly humified peat to unhumified peat, indicating increased surface wetness caused by a climatic shift to wetter conditions [75]. Waterlogging is unlikely to have been significant at Rockforest Wood due to the karstic hydrology of limestone pavement. While substantial deforestation occurred in the Burren and *Pinus* declines there have been attributed to human activity [56,76], the pollen data and historical sources suggest that Rockforest was an exception to this pattern. The area was subject to continued human activity but the level of disturbance does not appear to have been sufficient to eradicate *P. sylvestris*.

5. Conclusions

The absence of a *Pinus* decline strongly indicates that a relict population of *P. sylvestris* persisted at Rockforest from at least 1600 cal BP/AD 350 to the present (Figure 6). This is supported by the presence of a *P. sylvestris* macrofossil dated to 1110 cal BP/AD 840, which demonstrates that *Pinus* was locally present. The widely accepted hypothesis that *P. sylvestris* became extinct in Ireland is therefore rejected.

Existing research on the postglacial dynamics of *Pinus* in Ireland should be re-evaluated in light of these findings. When *Pinus* pollen was recorded during the period of presumed extinction in Ireland, authors questioned the validity of dating analyses [20] or invoked redeposition or long distance transport [66]. Localised survival of *Pinus* should be considered as a potential source of such a signal, at least in the Burren. A review of the postglacial dynamics of *Pinus* in Ireland is recommended, incorporating relevant studies completed since those of Bradshaw and Browne [6] and Bennett [5] and utilising Geographic Information Systems.

Further site-specific research and conservation measures are needed. Research on the genetics of *P. sylvestris* at Rockforest is ongoing (C. Kelleher, unpublished). Data on the number of individuals, age structure and spatial extent of this population are urgently required to determine its conservation status. *P. sylvestris* has been placed on the waiting list of the Irish Red Data List, pending further research to enable assessment [77]. Based on our current understanding of the distribution of native *P. sylvestris* in Ireland as being limited to a single known location, it could be assessed as critically endangered i.e., facing an extremely high risk of extinction in the wild. The insect fauna should be studied to determine whether any pine-dependent species, many of which are considered extinct in Ireland [78–80], occur there.

On the basis of the evidence presented, we argue that Rockforest Wood is Ireland's only known native *P. sylvestris* population. At a longitude of 8°57′ W, Rockforest Wood appears to be the western limit of the global native range of *P. sylvestris*, previously thought to be the north-west Iberian Peninsula at *c.* 8° W [3,16]. This population is of high conservation value but its rarity increases its extinction risk. Furthermore, reintroduced *P. sylvestris* in the vicinity may threaten the genetic integrity of the genepool. Rockforest Wood is located within a protected area but, given the scarcity of *P. sylvestris* regeneration, should be carefully managed and monitored. While ex-situ conservation is recommended, any seed-sourcing for native woodland restoration must be compatible with the long-term viability of the population in-situ. Coordinated action between conservation and forestry agencies will be required to ensure the continued survival of native *P. sylvestris* at Rockforest and to develop opportunities for the restoration of native pinewoods in Ireland.

Author Contributions: F.J.G.M. and S.W. conceived the research idea. J.R.R., F.J.G.M. and B.S.S. conducted coring. J.R.R. and B.S.S. collected the data. J.R.R. and F.J.G.M. analysed the data. J.R.R. wrote the paper. All authors contributed to revisions.

Funding: This work was funded by an Irish Research Council Government of Ireland Postgraduate Scholarship (RS/2005/180). The Society of Irish Foresters funded the article processing charge.

Acknowledgments: We thank Emma Glanville (National Parks and Wildlife Service) and Joseph Cunningham for site access and information and Richard Bradshaw, Daniel Kelly and the anonymous reviewers for their comments on the manuscript.

Conflicts of Interest: The authors declare no conflict of interest. The funding sponsors had no role in the study design; collection, analysis, or interpretation of data; writing of the paper, or the decision to publish the results.

References

1. Willis, K.J.; Birks, H.J.B. What is natural? The need for a long-term perspective in biodiversity conservation. *Science* **2006**, *314*, 1261–1265. [CrossRef] [PubMed]
2. Davies, A.L.; Columbo, S.; Hanley, N. Improving the application of long-term ecology in conservation and land management. *J. Appl. Ecol.* **2013**, *51*, 63–70. [CrossRef]
3. Willis, K.J.; Bennett, K.D.; Birks, H.J.B. The late Quaternary dynamics of pines in Europe. In *Ecology and Biogeography of Pinus*; Richardson, D.M., Ed.; Cambridge University Press: Cambridge, UK, 1998; pp. 107–121, ISBN 0521551765.
4. Huntley, B.; Birks, H.J.B. *An Atlas of Past and Present Pollen Maps for Europe 0-13,000 Years Ago*; Cambridge University Press: Cambridge, UK, 1983; ISBN 0521237351.
5. Bennett, K.D. The post-glacial history of *Pinus sylvestris* in the British Isles. *Quat. Sci. Rev.* **1984**, *3*, 133–155. [CrossRef]

6. Bradshaw, R.H.W.; Browne, P. Changing patterns in the postglacial distribution of *Pinus sylvestris* in Ireland. *J. Biogeogr.* **1987**, *14*, 237–248. [CrossRef]
7. Carlisle, A.; Brown, A.H.F. Biological flora of the British Isles: *Pinus sylvestris* L. *J. Ecol.* **1968**, *56*, 269–307. [CrossRef]
8. Huntley, B. European post-glacial forests: Compositional changes in response to climatic change. *J. Veg. Sci.* **1990**, *1*, 507–518. [CrossRef]
9. Froyd, C.A.; Bennett, K.D. Long-term ecology of native pinewood communities in East Glen Affric, Scotland. *Forestry* **2006**, *79*, 279–291. [CrossRef]
10. Birks, H.J.B. Holocene isochrone maps and patterns of tree-spreading in the British Isles. *J. Biogeogr.* **1989**, *16*, 503–540. [CrossRef]
11. Lageard, J.G.A.; Chambers, F.M.; Thomas, P.A. Climatic significance of the marginalization of Scots pine (*Pinus sylvestris* L.) *c.* 2508 BC at White Moss, south Cheshire, UK. *Holocene* **1999**, *9*, 321–331. [CrossRef]
12. Smith, A.G.; Pilcher, J.R. Radiocarbon dates and vegetational history of the British Isles. *New Phytol.* **1973**, *72*, 903–914. [CrossRef]
13. Eronen, M.; Huttunen, P. Radiocarbon-dated sub-fossil pines from Finnish Lapland. *Geogr. Ann. A* **1987**, *69*, 297–304. [CrossRef]
14. Roche, J.R. The Vegetation Ecology and Native Status of Scots Pine (Pinus sylvestris L.) in Ireland. Ph.D. Thesis, Trinity College Dublin, Dublin, Ireland, 2010.
15. Gear, A.J.; Huntley, B. Rapid changes in the range limits of Scots pine 4000 years ago. *Science* **1991**, *251*, 544–547. [CrossRef] [PubMed]
16. Jalas, J.; Suominen, J. *Atlas Florae Europaeae, 2. Gymnospermae*; Cambridge University Press: Cambridge, UK, 1973; ISBN 9519108017.
17. Le Maitre, D.C. Pines in cultivation: A global view. In *Ecology and Biogeography of Pinus*; Richardson, D.M., Ed.; Cambridge University Press: Cambridge, UK, 1998; pp. 407–431, ISBN 0521551765.
18. Lust, N.; Geudens, G.; Olsthoorn, A.F.M. Scots pine in Belgium and the Netherlands. *Investig. Agrar. Sist. Recur. For.* **2000**, *9*, 213–231. [CrossRef]
19. Mirov, N.T. *The Genus Pinus*; Ronald Press: New York, NY, USA, 1967.
20. Watts, W.A. The Holocene vegetation of the Burren, western Ireland. In *Lake Sediments and Environmental History*; Haworth, E.Y., Lund, J.W.G., Eds.; Leicester University Press: Leicester, UK, 1984; pp. 359–376, ISBN 9780816613649.
21. McAulay, I.R.; Watts, W.A. Dublin radiocarbon dates. *Radiocarbon* **1961**, *3*, 26–38. [CrossRef]
22. McCracken, E. *Irish Woods since Tudor Times*; David & Charles: Newton Abbot, UK, 1971; ISBN 9780715350089.
23. Roche, J.R.; Mitchell, F.J.G.; Waldren, S.; Bjørndalen, J.E. Are Ireland's reintroduced *Pinus sylvestris* forests floristically analogous to their native counterparts in north-west Europe? *Biol. Environ.* **2015**, *115*, 97–114. [CrossRef]
24. Cross, J.R.; Collins, K.D. *Management Guidelines for Ireland's Native Woodlands*; National Parks & Wildlife Service, Forest Service: Dublin, Ireland, 2017; ISBN 9781902696782.
25. Malone, S.; O'Connell, C. *Ireland's Peatland Conservation Action Plan 2020-Halting the Loss of Peatland Biodiversity*; Irish Peatland Conservation Council: Rathangan, Ireland, 2009; ISBN 1874189277.
26. Webb, D.A.; Scannell, M.J.P. *Flora of Connemara and the Burren*; Royal Dublin Society, Cambridge University Press: Cambridge, UK, 1983; ISBN 052123395X.
27. O'Connell, J.W.; Korff, A. (Eds.) *The Book of the Burren*; Tír Eolas: Kinvara, Ireland, 1991; ISBN 187382100X.
28. Long-Term Climate Averages for Ireland 1981–2010. Available online: http://edepositireland.ie/handle/2262/74915 (accessed on 18 April 2018).
29. Sugita, S. Pollen representation of vegetation in Quaternary sediments-theory and method in patchy vegetation. *J. Ecol.* **1994**, *82*, 881–897. [CrossRef]
30. Roche, J.R.; Mitchell, F.J.G.; Waldren, S. Plant community ecology of *Pinus sylvestris*, an extirpated species reintroduced to Ireland. *Biodivers. Conserv.* **2009**, *18*, 2185–2203. [CrossRef]
31. National Inventory of Architectural Heritage. Available online: www.buildingsofireland.ie/cgi-bin/displayimage.cgi?ic=3536&size=f&type=a1 (accessed on 1 April 2018).
32. Livingstone, D.A. A lightweight piston sampler for lake deposits. *Ecology* **1955**, *36*, 137–139. [CrossRef]
33. Grimshaw, H.M. Analysis of soils. In *Chemical Analysis of Ecological Materials*; Allen, S.E., Ed.; Blackwell Scientific: Oxford, UK, 1989; pp. 7–45, ISBN 0632003219.

34. Rose, N.L. A method for the extraction of carbonaceous particles from lake sediment. *J. Paleolimnol.* **1990**, *3*, 45–53. [CrossRef]
35. Rose, N.L. A note on further refinements to a procedure for the extraction of carbonaceous fly-ash particles from sediments. *J. Paleolimnol.* **1994**, *11*, 201–204. [CrossRef]
36. Haslett, J.; Parnell, A. A simple monotone process with application to radiocarbon-dated depth chronologies. *J. R. Stat. Soc.* **2008**, *57C*, 399–418. [CrossRef]
37. Moore, P.D.; Webb, J.A.; Collinson, M.E. *Pollen Analysis*; Blackwell Scientific: Oxford, UK, 1991; ISBN 0865428956.
38. Stockmarr, J. Tablets with spores used in absolute pollen analysis. *Pollen et Spores* **1971**, *13*, 615–621.
39. Beug, H.J. *Leitfaden der Pollenbestimmung Für Mitteleuropa und Angrenzende Gebiete*; Pfeil: München, Germany, 2004; ISBN 3899370430.
40. Reille, M. *Pollen et Spores d'Europe et d'Afrique du Nord*; Laboratoire de Botanique Historique et Palynologie: Marseille, France, 1992; ISBN 2950717535.
41. Grimm, E.C. *TILIA and TILIAGRAPH Software Package*; Illinois State Museum: Springfield, IL, USA, 1991.
42. Rose, N.L.; Harlock, S.; Appleby, P.G.; Battarbee, R.W. Dating of recent lake sediments in the United Kingdom and Ireland using spheroidal carbonaceous particle (SCP) concentration profiles. *Holocene* **1995**, *5*, 328–335. [CrossRef]
43. Jeličić, L.; O'Connell, M. History of vegetation and land use from 3200 B.P. to the present in the north-west Burren, a karstic region of western Ireland. *Veg. Hist. Archaeobot.* **1992**, *1*, 119–140. [CrossRef]
44. Lamb, H.; Thompson, A. Unusual mid-Holocene abundance of *Ulmus* in western Ireland-human impact in the absence of a pathogen? *Holocene* **2005**, *15*, 447–452. [CrossRef]
45. Mitchell, F.J.G. The vegetational history of the Killarney oakwoods, SW Ireland: Evidence from fine spatial-resolution pollen analysis. *J. Ecol.* **1988**, *76*, 415–436. [CrossRef]
46. Van Geel, B.; Middeldorp, A.A. Vegetational history of Carbury Bog (Co. Kildare, Ireland) during the last 850 years and a test of the temperature indicator value of 2H/1H measurements of peat samples in relation to historical sources and meteorological data. *New Phytol.* **1988**, *109*, 377–392. [CrossRef]
47. National Monuments Service Sites and Monuments Record. Available online: http://webgis.archaeology.ie/historicenvironment (accessed on 1 April 2018).
48. Frost, J. *The History and Topography of the County of Clare, from the Earliest Times to the Beginning of the 18th Century*; Sealey, Byers and Walter: Dublin, Ireland, 1893.
49. MacCraith, S.M.R. *Caithréim Thoirdhealbhaigh*; Irish Texts Society: London, UK, 1929.
50. O'Donovan, J. *Annals of the Kingdom of Ireland by the Four Masters, from the Earliest Period to the Year 1616*; Hodges and Smith: Dublin, Ireland, 1851.
51. Westropp, T.J. The forests of the counties of the Lower Shannon Valley. *Proc. R. Ir. Acad.* **1909**, *27C*, 270–300.
52. Weir, H. *Historical Genealogical Architectural Notes on Some Houses of Clare*; Ballinakella Press: Whitegate, UK, 1999; ISBN 094653828X.
53. Lewis, S. *County Clare: A History and Topography*, 2nd ed.; Clasp Press: Ennis, Ireland, 1837.
54. Dutton, H. *Statistical Survey of the County of Clare*; Graisberry and Campbell: Dublin, Ireland, 1808.
55. Cole, E.E.; Mitchell, F.J.G. Human impact on the Irish landscape during the late Holocene inferred from palynological studies at three peatland sites. *Holocene* **2003**, *13*, 507–515. [CrossRef]
56. Feeser, I.; O'Connell, M. Fresh insights into long-term changes in flora, vegetation, land use and soil erosion in the karstic environment of the Burren, western Ireland. *J. Ecol.* **2009**, *97*, 1083–1100. [CrossRef]
57. Foot, F.J. On the distribution of plants in the Burren, County of Clare. *Proc. R. Ir. Acad.* **1864**, *24*, 143–160.
58. Hickie, D. *Native Trees & Forests of Ireland*; Gill & Macmillan: Dublin, Ireland, 2002; ISBN 0717134113.
59. Everett, N. *The Woods of Ireland. A History, 700–1800*; Four Courts Press: Dublin, Ireland, 2014; ISBN 9781846825910.
60. Broström, A.; Nielsen, A.B.; Gaillard, M.J.; Hjelle, K.; Mazier, F.; Binney, H.; Bunting, J.; Fyfe, R.; Meltsov, V.; Poska, A.; et al. Pollen productivity estimates of key European plant taxa for quantitative reconstruction of past vegetation: A review. *Veg. Hist. Archaeobot.* **2008**, *17*, 461–478. [CrossRef]
61. Bennett, K.D. Post-glacial dynamics of pine (*Pinus sylvestris* L.) and pinewoods in Scotland. In *Our Pinewood Heritage*; Aldhous, J.R., Ed.; Forestry Commission, Royal Society for the Protection of Birds, Scottish Natural Heritage: Inverness, UK, 1995; pp. 23–39, ISBN 0855383259.

62. Fossitt, J.A. Late-glacial and Holocene vegetation history of western Donegal, Ireland. *Biol. Environ.* **1994**, *94*, 1–31.
63. Froyd, C.A. Fossil stomata reveal early pine presence in Scotland: Implications for postglacial colonization analyses. *Ecology* **2005**, *86*, 579–586. [CrossRef]
64. McGeever, A.H.; Mitchell, F.J.G. Re-defining the natural range of Scots Pine (*Pinus sylvestris* L.): A newly discovered microrefugium in western Ireland. *J. Biogeog.* **2016**, *43*, 2199–2208. [CrossRef]
65. Jacobson, G.L.; Bradshaw, R.H.W. The selection of sites for paleovegetational studies. *Q. Res.* **1981**, *16*, 80–96. [CrossRef]
66. Crabtree, K. Evidence for the Burren's forest cover. In *Archaeological Aspects of Woodland Ecology. Symposia of the Association for Environmental Archaeology No. 2*; Bell, M., Limbrey, S., Eds.; British Archaeological Reports; BAR International Series 146: Oxford, UK, 1982; pp. 105–113.
67. Mitchell, F.J.G. The history and vegetation dynamics of a yew wood (*Taxus baccata* L.) in S.W. Ireland. *New Phytol.* **1990**, *115*, 573–577. [CrossRef]
68. Peterken, G.F. *Woodland Conservation and Management*; Chapman & Hall: London, UK, 1981; ISBN 0412128209.
69. Bjørndalen, J.E. Some synchorological aspects of basiphilous pine forests in Fennoscandia. *Vegetatio* **1985**, *59*, 211–224. [CrossRef]
70. Jeffrey, D.W. Grasslands and heath: A review and hypothesis to explain the distribution of Burren plant communities. *Biol. Environ.* **2003**, *103*, 111–123. [CrossRef]
71. Dodson, J.R.; Bradshaw, R.H.W. A history of vegetation and fire, 6600 BP to present, County Sligo, Western Ireland. *Boreas* **1987**, *16*, 113–123. [CrossRef]
72. Little, D.J.; Mitchell, F.J.G.; von Engelbrechten, S.; Farrell, E.P. Assessment of the impact of past disturbance and prehistoric *Pinus sylvestris* on vegetation dynamics and soil development in Uragh Wood, SW Ireland. *Holocene* **1996**, *6*, 90–99. [CrossRef]
73. Ghilardi, B.; O'Connell, M. Fine-resolution pollen-analytical study of Holocene woodland dynamics and land use in north Sligo, Ireland. *Boreas* **2013**, *42*, 623–649. [CrossRef]
74. Watts, W.A. Contemporary accounts of the Killarney woods 1580–1870. *Ir. Geogr.* **1984**, *17*, 1–13. [CrossRef]
75. Allaby, M. *A Dictionary of Ecology*, 4th ed.; Oxford University Press: Oxford, UK, 2010.
76. Feeser, I.; O'Connell, M. Late Holocene land-use and vegetation dynamics in an upland karst based on pollen and coprophilous fungal spore analyses: An example from the Burren, western Ireland. *Veg. Hist. Archaeobot.* **2010**, *19*, 409–426. [CrossRef]
77. Wyse Jackson, M.; FitzPatrick, Ú.; Cole, E.; Jebb, M.; McFerran, D.; Sheehy Skeffington, M.; Wright, M. *Ireland Red List No. 10: Vascular Plants*; National Parks & Wildlife Service: Dublin, Ireland, 2016; ISSN 2009-2016.
78. Speight, M.C.D. The extinction of indigenous *Pinus sylvestris* in Ireland: Relevant faunal data. *Ir. Nat. J.* **1985**, *21*, 449–453.
79. Reilly, E. An ever closing gap? Modern ecological and palaeoecological contributions towards understanding the Irish postglacial insect fauna. In *Mind the Gap. Postglacial Colonization of Ireland*; Davenport, J.L., Sleeman, D.P., Woodman, P.C., Eds.; Irish Naturalists' Journal: Belfast, UK, 2008; pp. 63–71.
80. Whitehouse, N. The Holocene British and Irish ancient forest fossil beetle fauna: Implications for forest history, biodiversity and faunal colonization. *Q. Sci. Rev.* **2006**, *25*, 1755–1789. [CrossRef]

forests

MDPI

Article

Sustainability Assessment of Alternative Thinning Operations in Mediterranean Softwood Plantations

Janine Schweier [1,*], Raffaele Spinelli [2], Natascia Magagnotti [2], Bernhard Wolfslehner [3] and Manfred J. Lexer [4]

1 Albert-Ludwigs-University of Freiburg, Werthmannstraße 6, 79085 Freiburg, Germany
2 CNR–IVALSA, Via Madonna del Piano 10, I-50019 Sesto Fiorentino (FI), Italy; spinelli@ivalsa.cnr.it (R.S.); magagnotti@ivalsa.cnr.it (N.M.)
3 Central-East and South-East European Regional Office of the European Forest Institute, Department of Economics and Social Sciences, University of Natural Resources and Life Sciences, Vienna (BOKU), Feistmantelstr. 4, A-1180 Vienna, Austria; bernhard.wolfslehner@boku.ac.at
4 Department of Forest and Soil Sciences, University of Natural Resources and Life Sciences, Vienna (BOKU), Peter-Jordan-Strasse 82, A-1190 Vienna, Austria; mj.lexer@boku.ac.at
* Correspondence: Janine.Schweier@foresteng.uni-freiburg.de; Tel.: +49-761-203-97616

Received: 20 April 2018; Accepted: 15 June 2018; Published: 21 June 2018

Abstract: Mediterranean pines account for 14,000 ha in Tuscany alone, where they form large and homogeneous stands and represent an important resource for the forest economy. Among the harvesting systems applied to thinning operations, the most popular are whole-tree and cut-to-length harvestings. Both systems can be deployed with different levels of mechanization. The decision about which system might be the best option in a specific case possibly leads to conflicts due to different management goals, for instance when the shift from cut-to-length to whole-tree harvesting systems results in a decrease of costs and an increase of environmental burdens. Thus, an accurate determination of economic, environmental and social indicators is crucial to make balanced decisions. With that in mind, we carried out a sustainability impact assessment of typical forest-wood chain alternatives applied to young Mediterranean pine plantations and made a comparative evaluation by means of multi-criteria analyses. Trials were carried out in umbrella pine (*Pinus pinea* L.) plantations in Tuscany. The analyzed cases considered four thinning operations and included the processes of harvesting, extraction and chipping. In the analysis, 12 indicators were considered (e.g., global warming potential, fatal accidents). Results of the investigation allow quantifying possible sustainability impacts and, thus, supporting management decisions.

Keywords: motor-manual harvesting; mechanized harvesting; WT harvesting; CTL harvesting; energy chips; life cycle assessment; social indicators; sustainable forest management; decision support; multi-criteria analysis

1. Introduction

Mediterranean pines represent a group of highly adaptable species that include *Pinus brutia*, *Pinus halepensis*, *Pinus nigra*, *Pinus pinaster* and others such as *Pinus pinea* L. as the main representatives. These are specifically adapted to the fire regime that characterizes the area and play a central role in the vegetation dynamics of the Mediterranean regions [1]. Their aggressive colonizing strategy explains the rapid occupation of abandoned fields, as well as a relatively simple silviculture [2]. The total surface occupied by these pines is estimated at 13 million ha, or 25% of the total forest area of the Mediterranean basin. In regions like North Africa and Anatolia, Mediterranean pines may account for up to 75% of the total forest area [3]. The adaptability, rapid early growth and general colonizing capacity explain why Mediterranean pines have been traditionally used for reforestation projects and often form extensive plantations.

Umbrella pine (*Pinus pinea* L.) likely occupies the smallest area among Mediterranean pines, and yet, it accounts for almost 14,000 ha in Tuscany alone, where it forms large and homogeneous stands along the coastal region and represents an important resource for the forest economy [4].

The mainstream silvicultural prescription for umbrella pine stands is two to four thinning operations, followed by clear-cutting and replanting or re-naturalization, if the quality of the hardwood understory is good [5]. Among the many harvesting systems applied to thinning operations, the most popular are whole-tree (WT) and cut-to-length (CTL) harvesting. The former consists of felling trees and extracting them whole (stem, top and branches) to the landing, which offers the advantages of simplified in-forest handling and increased volume recovery. The latter describes a logging method where the felled trees are processed into commercial assortments already in the forest [6]: that requires laborious in-stand tree processing, but results in lower organic matter removals, which is especially desirable for poorer sites, where organic fertility may be a serious issue [7,8].

Both WT and CTL harvesting can be deployed with different levels of mechanization. In the motor-manual (MM) version, trees are felled or felled and processed with chainsaws, and the product is extracted with skidders, forwarders or forestry-fitted farm tractors [9]. Motor-manual harvesting is especially suited to self-employed forest owners and small contractors, with limited investment capacity [10]. At the other end of the scale, fully-mechanized (FM) harvesting is implemented: fully-mechanized whole-tree (FM-WT) harvesting is applied by feller-bunchers and grapple skidders. Self-propelled feller-bunchers are designed to fell standing trees and arrange them in bunches along skid roads, ready for pick-up by grapple skidders. Fully-mechanized cut-to-length (FM-CTL) harvesting is applied by harvesters, which are designed to combine felling and processing. Harvesters are also capable of stacking cut logs into neat piles by the strip road, where forwarders pick them up and carry them to the landing.

The intersection of the two harvesting systems with the two mechanization levels generates four possible options, each with its own specific pros and cons, especially as concerns financial performance, environmental impact and social consequences. These should be discussed carefully before deciding for or against any specific option.

Different management goals will favor different aspects of system performance, and yet, most managers will aim at financial sustainability and minimum adverse impacts on the natural environment and the community. For this reason, it is important to determine with some accuracy the financial, environmental and social performance of each individual option. With that in mind, we carried out a sustainability impact assessment (SIA) for the product obtained from the thinning of umbrella pine plantations, under the four technical choices derived from the intersection of the two harvesting systems and the two mechanizations described above and customarily applied to these stands. SIA methods have been specifically developed for this task, namely to evaluate how management choices influence the economic, environmental and social dimensions of system sustainability [11]. In addition, a multi-criteria analysis (MCA) was applied to compare alternative management scenarios on a cumulative preference scale. The results of our investigation will support management decisions by offering accurate information about sustainability indicators. Results for different indicators are often conflicting, and an accurate determination of these indicators is crucial to make balanced decisions that may reflect specific management goals. In particular, the specific objectives of this study are to apply an SIA to the typical forest-wood chain alternatives that can be applied to the thinning of umbrella pine plantations and to make comprehensive evaluations of the alternative forest-wood chains. Strictly speaking, the external validity of the study results is limited to umbrella pine plantations under similar conditions to those encountered in the study. However, the general trends found with this experiment could be extended to young pine plantations, although with much caution.

2. Materials and Methods

2.1. Approach

The SIA was carried out to determine the impacts with regard to sustainability of four alternative forest-wood chains, applied to umbrella pine plantations in Central Italy. This method was proposed by [12,13]. According to them, a proper SIA should be comprised of the following characteristics: (i) supply chains are described as a set of processes; (ii) each process is characterized by a set of sustainability indicators; (iii) the total amount of material flowing through the processes is the basis for assessing the overall sustainability impact; and (iv) an analysis of the trade-offs between the characteristics is carried out to assess holistically the impact of changes between the proposed alternatives.

2.2. Supply Chain Modelling

The so-called "cradle-to-gate" approach was applied, meaning that the analysis was restricted to a selected life cycle stage [14]. In our case, the study concentrates on the harvesting, extraction and chipping of trees. Four alternative thinning operations were modelled as forest-wood chains, as follows: Chain 1 = 1-MM-WT (motor-manual whole-tree harvesting); Chain 2 = 2-MM-CTL (motor-manual cut-to-length harvesting); Chain 3 = 3-FM-WT (fully-mechanized whole-tree harvesting) and Chain 4 = 4-FM-CTL (fully-mechanized cut-to-length harvesting). The chipping operations did not differ between the cases. According to the modelling rules [13] in each process, the wood material changes its appearance and/or moves to another location. Thus, the SIA builds on the conceptual representation of forest-wood chains as chains of value-adding production processes [15].

Modelling was carried out with the dedicated software Umberto (v 5.6), developed by IFU Hamburg GmbH [16]. With Umberto, material flow networks are created allowing one to model material and energy flows occurring in the system.

2.3. System Description and Boundaries

In all cases, the system boundaries covered the felling, extraction and chipping processes used in four alternative second thinning operations. In 1-MM-WT and 2-MM-CTL, trees were felled with chainsaws (Husqvarna 357XP, 3.2 kW) by professional operators and skidded to the roadside landing with a forestry-fitted farm tractor (Valtra 6400, 75 kW), equipped with a forestry winch. In 2-MM-CTL, trees were also delimbed and crosscut into random lengths (4–7 m) before skidding. In 3-FM-WT, trees were felled and bunched with a 27-t tracked feller buncher (JD759 J, 164 kW) and skidded to the roadside with a rubber-tired grapple skidder (JD460 G, 127 kW). In 4-FM-CTL, trees were felled, delimbed and crosscut to random lengths (4–7 m) with a 14-t four-wheel harvester (JD870 B, 114 kW), while logs were extracted to the roadside with a 10-t capacity forwarder (JD1110 B, 121 kW). In all cases, chipping was performed about four months after harvesting and extraction by a forwarder-mounted drum chipper (Erjo 12/90, 370/129 kW).

System boundaries were designated to be from where machines are brought to the working sites to where the wood chips are discharged into chip vans. Therefore, transport of personnel and equipment to the work site, as well as transport of the chips to the final destination were not considered. The building of roads and road maintenance were not included. In contrast, the production of materials and machines, so-called indirect or upstream processes, was considered within the boundaries. The CO_2 uptake due to tree growth and its release to the environment after biomass oxidation at the end of the life cycle was not considered, as well as changes in the soil organic matter stocks, the latter due to rare long-term data.

2.4. Selection of Sustainability Indicators

The indicators selected for the calculations were relevant and balanced with regard to economic, environmental and social sustainability, as well as feasibility in terms of data availability and quality [17]. In the present study, a set of 12 indicators was chosen (Table 1) to be analyzed based on existing indicator sets, e.g., [18,19]. The most relevant economic indicators were production costs, productivity and working delays. Production costs include all fixed and variable machine costs, as well as personnel costs. Productivity describes the machine performance per productive machine hour, and delays express nonproductive working times caused by mechanical, personnel or operational issues.

Table 1. Applied sustainability indicators.

No.	Indicator	Description	Unit
1	Productivity	Rate of product output per unit of time for a production system. A productivity ratio may also be calculated for resources other than time.	PMH gt^{-1}
2	Costs	Sum of production costs (fixed costs accruing regardless of the rate of activity inclusive of personnel costs, as well as variable costs that vary with the quantity of production).	€ gt^{-1}
3	Delays	Interruptions of the work process that can be related back to the organization of the work; commonly subdivided into the categories mechanical (e.g., repair), personnel (e.g., rest breaks) and operational delays (e.g., waiting times).	% of total working time
4	GWP	The potential of global warming is mainly caused by the release of greenhouse gas emissions due to anthropogenic activities such as fossil fuel combustion, chemical processing and transportation.	kg CO_2-eq. gt^{-1}
5	CED of fossil energy	The cumulative energy demand of fossil energy investigates the energy use throughout the overall life cycle, including the use of direct and indirect consumption of energy due to the use of materials.	MJ gt^{-1}
6	EP	Potential eutrophication due to some substances, calculated through the conversion factor of phosphorous and nitrogen compounds into phosphorous equivalents.	kg PO_4-eq. gt^{-1}
7	POPC (low NOx)	The potential of photochemical ozone creation, also known as summer smog, contains nitrogen oxides and volatile organic compounds.	kg ethylene gt^{-1}
8	AP	Potential acidification due to atmospheric deposition of sulfur and nitrogen.	kg SO_2-eq. gt^{-1}
9	Employment	Rate of full-time employments related to forest operations.	FTE 1000 gt^{-1}
10	Fatal accident	Fatal accidents related to forest operations.	Fatalities 10^{-6} gt^{-1}
11	Tree damage	Percentage of stand damage after forest operations (excluding wounds with an exposed surface <10 cm^2).	% of total trees
12	Soil compaction	The increase of bulk density was used to show possible soil compaction.	Difference in % (untrafficked/trafficked)

Note: PMH = productive machine hours, exclusive of delays; gt = green tonnes (fresh weight); min = minutes; CED = cumulated energy demand; MJ = megajoules; EE = energy efficiency; GWP = global warming potential; CO_2 = carbon dioxide; EP = eutrophication potential; PO_4 = phosphate; POPC = Photochemical ozone creation potential; AP = acidification potential; SO_2 = sulfur dioxide; FTE = full-time equivalent.

Concerning site impact, field measurements covered tree damage (frequency and severity), soil compaction and biomass retention, i.e., the amount of tops and branches left on site after harvesting. Biomass retention may offer a first indication about the potential for nutrient depletion, even when the actual amount of nutrients left on site is not determined. Biomass retention is relatively easy to gauge and is used here for comparing the risk for soil nutrient impact under the two treatments considered with this study. The basic concept is that a larger biomass retention for one of the treatments will point at a proportionally lower risk for soil nutrient depletion compared with the other treatment. This indicator does not define risk in absolute terms (i.e., quantity of nutrients released per ha), but only in relative terms, which is consistent with the comparative character of the study.

Besides the well-known environmental impact category global warming potential (GWP) and the cumulated energy demand (CED) of fossil energy, results were presented in the following environmental impact categories: eutrophication potential (EP), acidification potential (AP) and photochemical ozone creation potential (POPC). All of them are important categories for biomass cultivation and distribution and are highly influenced by nitrous and carbon oxides, which are of special interest to coastal pine plantations along the Tyrrhenian coast and generally in Central Italy, where most of such stands are included within natural reserves, under special conservation rules (e.g., Parco Regionale di San Rossore, Tenuta di Castel Porziano, Parco Nazionale del Circeo).

When it comes to societal aspects, attention was focused on fatal accident and employment rate, both of which are strongly impacted by the treatment of young stands.

2.5. Indicator Calculation

The system modelling outcome served as the basis for the comparative environmental and social-economic analysis. The analysis of environmental impacts (GWP, EP, AP, POPC) focused on technical aspects of the alternative operations and followed the ISO 14040 guidelines [20], which prescribe the inclusion of indirect impacts (e.g., machinery used, material transportation; cf. [11]). The database Ecoinvent (vs. 2.3) developed by the Swiss Centre for Life Cycle Inventories [21] was used to determine the impacts of the different thinning operations. Thereby, specific characteristics of the analyzed machines (weight, lifetime, duration of use) were considered. The Ecoinvent database is implemented in the modelling software Umberto and provides emission data for numerous materials (e.g., diesel). These emission data are then linked to the environmental categories to which they contribute (e.g., CO_2 to GWP).

In order to report the energy efficiency (EE), the CED of fossil energy was calculated, also by use of Umberto and Ecoinvent. The EE shows how much fossil energy is required to produce one unit of renewable energy.

Damage of trees by the harvest operations was determined by inspecting all residual trees according to the method described by [22]. Results refer to the total amount of trees damaged and, thus, do not relate to single operations. Wounds with an exposed surface <10 cm^2 were not recorded. Soil compaction was determined by collecting undisturbed core samples before and after harvests, as described in detail in [23]. Samples do not refer to single operations, but to the total area.

Statistical data are neither available for the accidents occurring during the thinning of Italian coastal pine plantations, nor for working accidents in Italian forestry in general, since the Italian work accident statistics lump forestry and agriculture together. Therefore, it is not possible to determine with any accuracy the number and severity of work accidents per unit product, nor to discriminate between different forest-wood chains and mechanization levels. However, a good proxy was found in the statistical data provided in the international scientific literature. In particular, Ref. [24] provide the fatal accidents among professional forest workers reported in the national statistics of Austria, Finland, Slovenia, Sweden and Switzerland. While these data do not discriminate between different forest-wood chains (i.e., CTL- vs. WT harvesting), they can be used to discriminate between mechanized and motor-manual operations, on the assumption that over 90% of the operations in Finland and Sweden are fully mechanized, while operations in the other countries are motor-manual to a large extent [25]. Based on this reasoning, the number of fatalities per m^3 recorded in the two Nordic countries was converted into a per tonne figure and the result, 0.03 fatalities per million tonnes, was assumed as the reference for mechanized operations. The fatality rate for motor-manual operations was estimated from the mean data for Austria, Slovenia and Switzerland and amounted to 0.22 fatalities per million green tonnes (10^{-6} gt).

The effect of the alternative operations on employment rate was calculated from the productivity data actually recorded in the study, after inflating these data for crew size and assuming 1510 h per year as full employment of one worker unit (FTE). Machine costs referred to Euros (€) per productive machine hour (PMH) and were calculated with the harmonized method developed within the scope of European COST Action FP0902 [26]. Delay time was reported separately in order to calculate delay factors [27,28], i.e., the ratio of delay time to productive working time. Data related to time input, fuel consumption and machine productivity were determined with a time study as reported in detail in [23] (see the Data Collection Section).

2.6. Biomass Removal and Retention

Table 2 indicates how much biomass was removed per alternative; cf. [23]. The concentration of nutrients within different biomass components was not determined. Conversely, biomass retention (tops and branches) was determined on twenty 1 × 1-m sample plots per treatment, using an improved version of the protocol developed by the Australian Forest Operations Research Alliance at the University of the Sunshine Coast [29]. Before locating the plots, the sampled area was divided in two strata according to the residual biomass load, in order to increase the accuracy of sampling and reduce the number of required sample plots. After that, 50 sample plots per corridor were located systematically on the terrain, and each of them was attributed to one of the strata. From the original 50 plots, 20 plots were selected randomly for measurement, reflecting the proportion between the strata. Then, all tops and branches on each plot were manually collected and weighed, using a portable electronic scale, accurate to ±10 g. Since stumps were not harvested under either treatment and the assessment was a comparative one, stump biomass was not determined.

Table 2. Biomass removal, in green tonnes (fresh weight).

Forest-Wood Chain	Biomass Volume (gt ha^{-1})
1-MM-WT	51.0
2-MM-CTL	29.6
3-FM-WT	53.9
4-FM-CTL	33.5

Note: MM-WT = motor-manual whole-tree harvesting; MM-CTL = motor-manual cut-to-length harvesting; FM-WT = fully-mechanized whole-tree harvesting; FM-CTL = fully-mechanized cut-to-length harvesting.

2.7. Multi-Criteria Analysis

MCA is a methodology to structure a decision problem (e.g., by means of indicators) and provide a formal model to compare a finite number of alternatives on a one-dimensional preference scale [30]. MCA has been richly taken up as a decision-support tool in forest management planning and practice as documented in scientific literature; cf. [31,32]. As in [33], the Promethee method [34] was employed in this application to compare four thinning operations via 12 environmental, economic and social impact indicators. In this method, preference information is required at two levels: (i) weights of importance for indicators; and (ii) preferences regarding when one alternative dominates another with regard to one of the indicators (i.e., the selection and definition of preference functions). The pair-wise comparisons of alternatives with regard to the evaluation criteria (i.e., indicators) result in a summed-up and weighed degree of dominance of one alternative over another and a cardinal ranking of all examined alternatives.

These two sources of preference information were modelled in an experimental way. First, we assumed equal weights for clusters of economic, environmental and social indicators (1/3 each). This means that in clusters with more indicators (4 economic, 6 environmental, 2 social indicators), these have a lower relative weight because of the distributive effect, which is a feasible way to smoothen an imbalance of representation in a decision-making problem [35]. A second arrangement, in absence of empirical preference information, is to set the preference threshold, i.e., where strict preference of one alternative over another will be applied, for each indicator to 10% of the highest value. For all indicators, a V-shape preference function has been applied; cf. [34].

2.8. Stand

Comparative trials were carried out in pine plantations located near Pisa, Italy, inside the Regional Park of San Rossore, which encloses a surface of about 3000 ha and is covered in a large part by umbrella pine plantations. This area is a typical example for the wider cohort of pine plantations growing along the coastlines of Southern Europe, where they have been established and managed for many centuries [3].

When the first thinning was conducted in 2002, all alternate rows had been removed, so that spacing at the time of the second thinning was 5 × 2.5 m. In the second thinning in 2014, trials were conducted of a 15-hectare tract, consisting of a 21-year old umbrella pine plantation. Selection of trees to remove was left to the logger, whose main task was to create enough space around candidate trees. Detailed information regarding site and thinning characteristics, as well as machineries is reported in [23,36].

2.9. Data Collection and Assumptions

Fuel input was determined by refilling all machine tanks at the end of each working day and recording the amount of fuel used during that day. This figure was divided by stopwatch hours and prorated to each plot based on the hours actually needed to harvest it. Wood moisture content was determined according to the European standard CEN/TS 14774-2 on the determination of moisture content, on one 500-g chip samples per load. These samples were obtained from the reduction of larger 3 L samples collected at different places within the same load. Data about utilization and maintenance of machines and value recovery were obtained directly from the machine owners and are described in [23].

3. Results

3.1. Material Flow and System Productivity

Table 2 shows the amount of harvested biomass for each alternative. The volume harvested was much lower when CTL harvesting was applied, compared with WT harvesting. Average productivity was also lower with CTL, for the same mechanization level (0.0237–0.0165 PMH gt^{-1} in 2-MM-CTL and 4-FM-CTL compared to 0.0196–0.0092 PMH gt^{-1} in 1-MM-WT and 3-FM-WT). Figure 1 shows the specific productivity by treatment and work phase. As reported by [23], felling and extraction productivity were significantly different between treatments, whereas chipping productivity was not. Mechanization allowed a dramatic increase of labor productivity. With mechanization, felling-processing productivity increased eight times and felling-bunching productivity 20 times. Extraction productivity increased from 6–8 times, as a result of mechanization. Shifting from CTL to WT harvesting resulted in productivity increases between 40% and 270%, depending on the work step and mechanization level.

The highest share of delays occurred in the chains with manual harvesting operations. In both cases, delay time was 29.7% of the total working time compared to 11.7–14.4% in the mechanized cases (4-FM-CTL, 3-FM-WT) (Table 3). When looking at the harvesting process only, the share of delays was 54.0% in 1-MM-WT and 2-MM-CTL, while it was 10.0–19.0% in 4-FM-CTL and 3-FM-WT. The resulting delay factors were 42 in 1-MM-WT and 2-MM-CTL, 17 in 3-FM-WT and 13 in 4-FM-CTL.

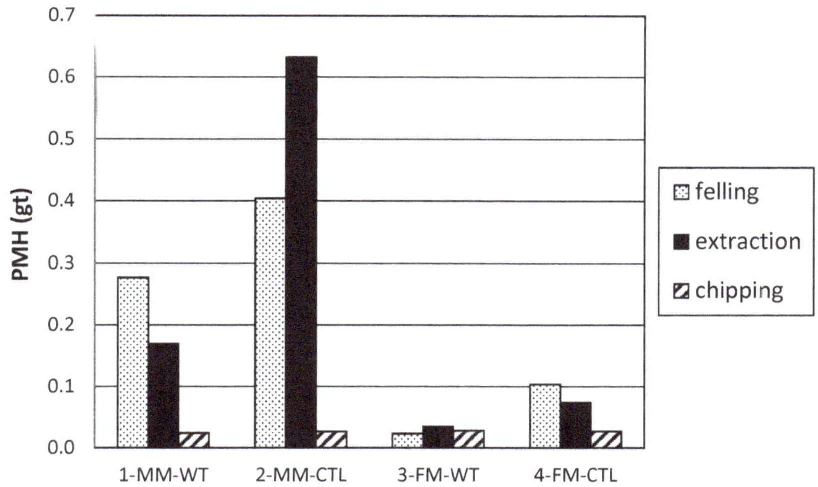

Figure 1. Specific productivity by treatment and work phase. Note: MM-WT = motor-manual whole-tree harvesting; MM-CTL = motor-manual cut-to-length harvesting; FM-WT = fully-mechanized whole-tree harvesting; FM-CTL = fully-mechanized cut-to-length harvesting.

3.2. Indicator Results

Relative indicator results per gt of wood chips are presented in Table 3. The sum is shown in bold letters, and the process causing highest impacts in each chain is highlighted in italic letters.

The economic indicator production costs followed the same pattern as the system productivity: forest-wood chains with motor-manual harvesting operations were significantly more expensive (40.36–77.19 € gt^{-1} in 1-MM-WT and 2-MM-CTL) than those applying fully-mechanized harvesting systems (9.99–19.28 € gt^{-1} in 3-FM-WT and 4-FM-CTL, Table 3): in 1-MM-WT, the manual felling was responsible for 69% and in 2-MM-CTL for 52% of the production costs. In the latter case, extraction was very time consuming and expensive (42% of costs).

When considering social indicators, the amounts of fatal accidents and FTEs were significantly higher in manual than in mechanized systems. The rate of accidents varied between 0.2108 and 0.2168 fatalities 10^{-6} gt^{-1} (2-MM-CTL, 1-MM-WT) compared to 0.0283–0.0259 fatalities 10^{-6} gt^{-1} (4-FM-CTL, 3-FM-WT) and, thus, was 7.5–8.4 times higher in manual cases (Table 3).

The rate of FTE 1000 gt^{-1} was 6.4–14.7-times higher in manual cases, as well, compared to mechanized cases: indicator results varied between 1.7659 and 0.9729 FTE 1000 gt^{-1} (2-MM-CTL, 1-MM-WT) vs. 0.1520 and 0.0664 FTE 1000 gt^{-1} (4-FM-CTL, 3-FM-WT) (Table 3).

It turned out that the percentage of tree damage also increased when shifting from FM to MM and from CTL to WT harvesting. Between 1.5% and 15% of the residual trees presented wounds larger than 10 cm^2 (Table 3). However, the 15% figure (2-MM-CTL) could be excluded from the comparison (cf. [23]), which results in a wounding frequency range between 1.5% and 6%.

In contrast, the other environmental indicators significantly depended on the choice of the applied system (CTL/WT). The GWP is highly influenced by the fuel consumption. As shown previously [23], mechanization entailed a reduction of fuel consumption between 10% and 40%. Shifting from mechanized CTL to manual WT harvesting allowed reducing fuel consumption by 19%. Differences were statistically significant between motor-manual CTL harvesting and all other treatments, as well as of harvesting system and mechanization level. The calculated GWP was 2.37–2.49 kg CO_2-eq. gt^{-1} (3-FM-WT, 1-MM-WT) when WT harvesting was applied and 3.85–5.55 kg CO_2-eq. gt^{-1} (4-FM-CTL, 2-MM-CTL) when CTL harvesting was applied (Table 3). Potential impacts caused by the extraction process in 2-MM-CTL were seven-times higher compared to the best case (3-FM-WT). In 2-MM-CTL, the use of fuel and lubricants contributed 43% to this result and the machinery 57%, while it was almost the opposite (58% and 42%) in the case of 3-FM-WT, which shows that the extraction process was sub-optimal in 2-MM-CTL.

The CED of fossil energy (Table 3) was much lower, as well, when WT harvesting was applied. It varied between 137.75 and 156.14 MJ gt^{-1} (3-FM-WT, 1-MM-WT) compared to 227.92 and 314.27 MJ gt^{-1} (4-FM-CTL, 2-MM-CTL). Assuming an average energy content of about 9200 MJ per tonne (moisture content = 45%), the resulting energy ratios are respectively 66.7, 59.0, 40.4 and 20.3. In the first named, 3-FM-WT, this means that 66.7 units of fossil energy are necessary to generate one unit of renewable energy from biomass. The analysis of the other environmental categories EP, POPC and AP confirmed that the CTL harvesting system led to significantly higher impacts than WT harvesting systems (Table 3), in the manual case (2-MM-CTL) due to the low productivity in the extraction system and in the mechanized case (4-FM-CTL) due to the high fuel consumption during the harvesting operation. Furthermore, results showed that only CTL harvesting caused soil compaction. However, the increase was low, ranging from 8.8–11.9% of the original value (Table 3). In 1-MM-WT, soil compaction was indicated to be negative (−1.2%) because the machines loosened up the soil and, thus, caused disturbance.

When looking at the total indicator results per hectare (Table 4), one needs to have in mind that a comparable amount of trees was harvested in all cases, but a greater amount of biomass was mobilized in 1-MM-WT and 3-FM-WT (Table 2). This may lead to changing results, e.g., in the category EP, the lowest impacts were expected in 3-FM-WT when looking at results per gt, while the value was equal in 3-FM-WT and 4-FM-CTL when looking at the results per ha (cf. GWP and accidents) (Table 4).

Table 3. Relative indicator results.

Indicator	Unit	Process	1-MM-WT	2-MM-CTL	3-FM-WT	4-FM-CTL
Costs	€ gt^{-1}	felling	*27.70*	*40.40*	2.76	*8.99*
		extraction	8.72	32.50	2.69	5.90
		chipping	3.94	4.29	*4.55*	4.39
		total value	**40.36**	**77.19**	**9.99**	**19.28**
Delay	%	felling	*54.00*	*54.00*	*19.00*	10.00
		extraction	28.00	28.00	10.00	10.00
		chipping	17.00	17.00	17.00	*17.00*
		total value	**29.70**	**29.70**	**14.40**	**11.70**
GWP	kg CO_2-eq. gt^{-1}	felling	0.43	0.58	0.68	*1.73*
		extraction	*1.16*	*4.01*	0.58	1.01
		chipping	0.89	0.96	*1.12*	1.11
		total value	**2.49**	**5.55**	**2.37**	**3.85**
CED of fossil energy	MJ gt^{-1}	felling	29.89	39.98	34.93	*100.50*
		extraction	*65.04*	*209.23*	33.02	57.37
		chipping	61.21	65.06	*69.80*	70.05
		total value	**156.14**	**314.27**	**137.75**	**227,92**
EP	kg PO_4-eq gt^{-1}	felling	0.0006	0.0021	0.0012	*0.0031*
		extraction	*0.0023*	*0.0120*	0.0010	0.0017
		chipping	0.0017	0.0034	*0.0018*	0.0018
		total value	**0.0046**	**0.0175**	**0.0040**	**0.0066**
POPC (low NO_x)	kg ethylene gt^{-1}	felling	0.0001	0.0001	0.0002	*0.0005*
		extraction	*0.0004*	*0.0010*	0.0002	0.0003
		chipping	0.0003	0.0007	*0.0004*	0.0004
		total value	**0.0008**	**0.0019**	**0.0008**	**0.0012**
AP	kg SO_2-eq gt^{-1}	felling	0.0038	0.0051	0.0042	0.0120
		extraction	*0.0091*	*0.0301*	0.0036	0.0065
		chipping	0.0079	0.0084	*0.0077*	*0.0075*
		total value	**0.0208**	**0.0437**	**0.0156**	**0.0260**

Table 3. *Cont.*

Indicator	Unit	Process	1-MM-WT	2-MM-CTL	3-FM-WT	4-FM-CTL
Employment	FTE 1000 gt^{-1}	felling	*0.7976*	*1.1634*	0.0188	*0.0764*
		extraction	0.1561	0.5817	*0.0254*	0.0542
		chipping	0.0192	0.0209	0.0221	0.0214
		total value	**0.9729**	**1.7659**	**0.0664**	**0.1520**
Fatal accident	fatalities 10^{-6} gt^{-1}	felling	n.a.	n.a.	n.a.	n.a.
		extraction	n.a.	n.a.	n.a.	n.a.
		chipping	n.a.	n.a.	n.a.	n.a.
		total value	***0.2168***	**0.2108**	**0.0259**	**0.0283**
Tree damage	% trees	felling	n.a.	n.a.	n.a.	n.a.
		extraction	n.a.	n.a.	n.a.	n.a.
		chipping	n.a.	n.a.	n.a.	n.a.
		total value	**6.0**	***15.7***	**2.5**	**1.5**
Soil compaction	% difference (untrafficked/ trafficked)	felling	n.a.	n.a.	n.a.	n.a.
		extraction	n.a.	n.a.	n.a.	n.a.
		chipping	n.a.	n.a.	n.a.	n.a.
		total value	**−1.2**	**8.8**	**2.5**	***11.9***

Note: The sum is shown in bold. The highest value in each alternative is shown in italics. The incidence of delay time over total harvesting time is the average delay incidence recorded for each single work step, weighted for the contribution of each work step to total harvesting time. MM-WT = motor-manual whole-tree harvesting; MM-CTL = motor-manual cut-to-length harvesting; FM-WT = fully-mechanized whole-tree harvesting; FM-CTL = fully-mechanized cut-to-length harvesting; PMH = productive machine hours; CED = cumulated energy demand; GWP = global warming potential; CO_2 = carbon dioxide; EP = eutrophication potential; PO_4 = phosphate; POPC = photochemical ozone creation potential; AP = acidification potential; SO_2 = sulfur dioxide; FTE = full-time equivalent.

Table 4. Total indicator results per hectare.

Indicator	Unit	1-MM-WT	2-MM-CTL	3-FM-WT	4-FM-CTL
Productivity	PMH ha^{-1}	24.01	31.45	4.60	6.84
Costs	€ ha^{-1}	2058.52	2284.92	538.55	645.89
Delays	%	29.70	29.70	14.40	11.70
GWP	kg CO_2-eq. ha^{-1}	126.91	164.21	127.68	128.83
CED of fossil energy	GJ-eq. ha^{-1}	7.96	16.03	7.03	11.62
EP	kg PO_4-eq. ha^{-1}	0.23	0.52	0.22	0.22
POPC (low NO_x)	kg ethylene ha^{-1}	0.04	0.06	0.04	0.04
AP	kg SO_2-eq. ha^{-1}	1.06	1.29	0.84	0.87
Employment	FTE 1000 ha^{-1}	49.62	52.27	3.58	5.09
Fatal accident	fatalities 10^{-6} ha^{-1}	11.06	6.24	1.40	0.95
Tree damage	% trees	6.00	15.70	2.50	1.50
Soil compaction	% difference	−1.20	8.80	2.50	11.90

Note: MM-WT = motor-manual whole-tree harvesting; MM-CTL = motor-manual cut-to-length harvesting; FM-WT = fully-mechanized whole-tree harvesting; FM-CTL = fully-mechanized cut-to-length harvesting. PMH = productive machine hours; CED = cumulated energy demand; GWP = global warming potential; CO_2 = carbon dioxide; EP = eutrophication potential; PO_4 = phosphate; POPC = photochemical ozone creation potential; AP = acidification potential; SO_2 = sulfur dioxide; FTE = full-time equivalent.

3.3. Multi-Criteria Analysis of Harvesting Operations

The MCA of the four harvesting alternatives (Figures 2 and 3) was based on per hectare results for the 12 indicators (Table 1).

The overall ranking results of the four alternative thinning operations showed a robust preference for both FM alternatives (Figure 2). The difference in the overall preference value φ between FM-WT (0.3612) and FM-CLT (0.2887) was relatively small, while MM alternatives showed both negative overall preference values with MM-CLT (−0.5156) having a much larger preference than MM-WT (−0.1343) (Figure 2).

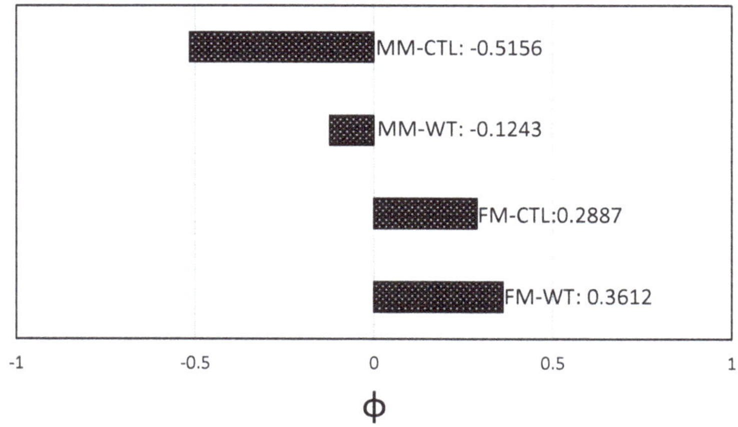

Figure 2. Ranking results of the multi-criteria analysis. Overall ranking of four alternative thinning operations. Phi (φ) indicates the degree of dominance of one alternative over another. Equal weights for the economic, environmental and social indicator cluster. Note: MM-WT = motor-manual whole-tree harvesting; MM-CTL = motor-manual cut-to-length harvesting; FM-WT = fully-mechanized whole-tree harvesting; FM-CTL = fully-mechanized cut-to-length harvesting.

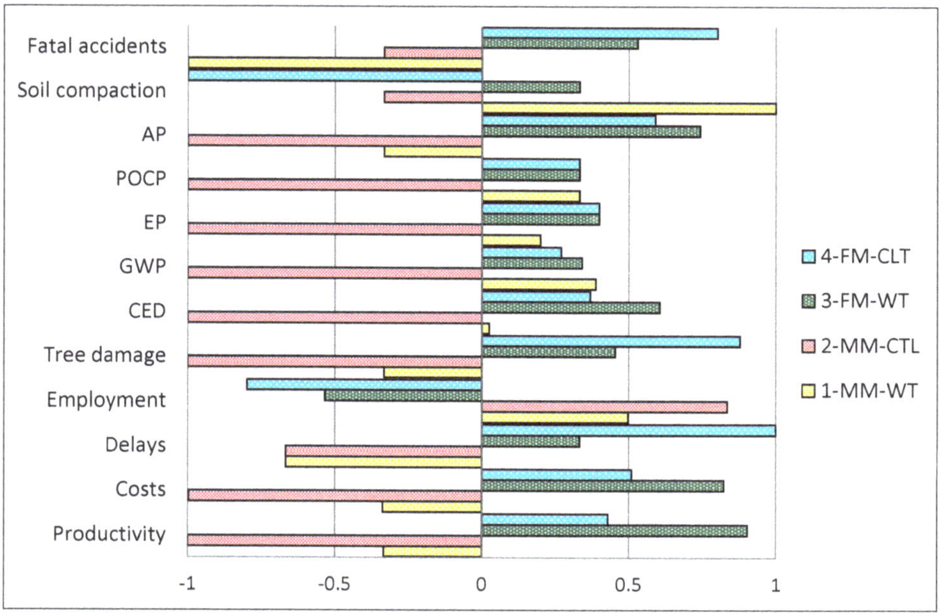

Figure 3. Preference results of the multi-criteria analysis. Preferences (Phi-values) for the four alternative thinning operations over all impact indicators. Note: MM-WT = motor-manual whole-tree harvesting; MM-CTL = motor-manual cut-to-length harvesting; FM-WT = fully-mechanized whole-tree harvesting; FM-CTL = fully-mechanized cut-to-length harvesting.

A deeper look into the action profiles of the alternatives, i.e., their performance in each indicator, reveals the reasons for the overall ranking (Figure 3). It shows that FM alternatives consequently have positive net flows apart from employment, while MM-CTL is not competitive in eight out of 12 indicators.

Pre-empting a full sensitivity analysis, the examination of stability intervals unravels the sensitivity of the ranking of alternatives to changing indicator weights. This analysis shows only increased weights for indicators 'tree damage' (weight > 21.69%), 'GWP' (weight > 79.52%), 'soil compaction' (weight > 22.02%), 'fatal accidents' (weight > 34.24%) and 'employment' (weight > 37.24%), while maintaining the initial weight relations in other indicators could change the overall ranking results.

4. Discussion

A major challenge of forest management is to consider the consequences of different management strategies and to estimate the financial, environmental and social performance of each individual option before an action is carried out. It is important to consider different pillars of sustainability and to link environmental impacts to socio-economic activities in order to guide decision-makers in their actions and to ensure that the impacts of their decisions are measured. With that in mind, we carried out an SIA for felling, processing and extraction of four alternative thinning operations in Mediterranean softwood plantations. The system boundaries included all steps necessary for turning standing trees in the forest into whole-tree chips loaded on trucks and ready for delivery to the mill.

Twelve indicators were considered to be important and feasible with regard to data collection. Input data were gathered mainly from field studies and in the case of fatal accidents from statistics. Respective indicator values were calculated by the use of different tools, e.g., potential environmental impacts of exhaust gases under the use of the Ecoinvent database and Umberto, a tool for life cycle assessments. Previous studies have shown that the use of several indicators and the combination of different methods to calculate indicator values lead to a strong analytical power for embracing financial, technological, environmental and other aspects of a production system, e.g., [37]. In particular, the feature of Umberto to include 'own' indicators and data, as well as data from the database Ecoinvent turned out to be efficient and productive. However, although a significant effort was made to collect realistic input data, there were some limitations, e.g., in the accuracy of the indicators, as results reflect average cost and emission values only or in the generalization of the results, as soil compaction for example is highly complex. Furthermore, system boundaries need to be considered when interpreting results. For instance, the transport of personnel and equipment to the work site was not included in our study, which would be a weak point if this were of major importance. There are numerous variables influencing the overall sustainability of a product, a process or a system, and the challenge is to find the most important ones in a specific case.

Furthermore, it is well known that WT harvesting leads to a higher removal of biomass (cf. Table 2) and, thus, of nutrients [38], because branches and foliage account for a significant proportion of the total nutrients bound in trees. However, it is still uncertain whether the lower biomass retention after WT harvesting actually has any visible effects on forest growth, stability or diversity. Kaarakk et al. [39] assessed the effects of WT harvesting and conventional stem-only harvesting and found out that 10 years after the final harvest, repeated WT harvesting had a decreasing effect on total carbon and nitrogen pools in the combined organic and mineral soil layer. In another study [40], the effects of WT harvesting (needles left on site) and conventional stem-only harvesting were compared 30 years after clear-cutting of Norway spruces, and any significant treatment effects with regard to the amount of organic matter, the amounts of nutrients in the forest floor or the concentrations of foliar nutrients were found. In [41], the impacts of WT harvesting were distinguished according to the harvesting intensity (thinning, clear-cutting). The authors figured out that negative effects of WT harvesting were lower in thinning than in clear-cutting operations [41]. In [42], soil carbon and nitrogen contents were determined over a period of 15–16 years after clear-cutting of conifers in Sweden, and any conclusive evidence was obtained about a general effect of harvest intensity on soil carbon or nitrogen pools, but only site-specific impacts. To conclude, the long-term impacts of intensified forest harvesting on nutrient losses are not clear (cf. [43]), and it is most likely that regional site interactions influence the overall effect of increased removal [42,43]. As we did not determine the concentration of nutrients within the different biomass components, we can state from our results only that seven-times more residue biomass (34 vs. 4.3 t ha^{-1}) was retained with CTL harvesting compared to WT harvesting.

The results have shown that highest productivities were reached in FM harvesting operations. Shifting from CTL to WT harvesting systems resulted in further productivity increases (Figure 1). The resulting monetary cost of felling, processing and extraction varied between 10 and 77 € gt^{-1} (Table 3) (≙ 19–147 € per oven dry tonne). Forest-wood chains with motor-manual harvesting operations were significantly more expensive (40–77 € gt^{-1} in 1-MM-WT and 2-MM-CTL, Table 3) than those applying fully-mechanized harvesting systems (10–19 € gt^{-1} in 3-FM-WT and 4-FM-CTL, Table 3). In addition, the highest share of delays occurred in the manual cases, in particular in the felling processes (Tables 3 and 4). However, not only the level of mechanization, but also the applied harvesting system had a significant effect on the costs, as well as on the productivity: it decreased with increasing level of mechanization and when shifting from CTL to WT harvesting. Other studies comparing WT and CTL harvesting confirm the better performance of forest-wood chains applying WT harvesting, e.g., [44,45].

In contrast, environmental indicator values (CED, GWP, EP, POPC, AP) depended on the applied harvesting system and were much better when WT harvesting was applied compared to CTL harvesting. This can be explained due mainly to the productivity: in 4-FM-CTL, the felling process was four-times less productive than in 3-FM-WT, and in 2-MM-CTL, the extraction process was 3.7-times less productive and, thus, more fuel consuming than in 1-MM-WT (Table 3). As stated above, the use of a forestry trailer would have been more appropriate. As potential environmental impacts strongly depend on exhaust gases and the use of machineries, an increase in productivity would result in lower impacts at the same time.

With regard to soil compaction, it should be noted that only the CTL treatments caused a compaction. However, such an increase was very low, ranging from 8.8–11.9% of the original value. The density increase recorded for the mechanized CTL treatment was significantly higher than for the manual CTL treatment. As soil compaction is highly complex and a very recent research topic [46–50], further analysis should be conducted if this indicator is of crucial importance.

The rate of FTE 1000 gt^{-1} was 9.8–14.6-times higher in the manual cases (Table 4). However, one should have in mind that it is increasingly difficult to find skilled labor in many regions. A very important argument to further support the process of mechanization in felling operations is that many fatal accidents occur during manual felling operations [51], as also shown in this study. Therefore, replacing manual with mechanized felling is likely to result in a reduction of accident frequency and severity.

Based on the indicator results, an MCA was carried out. In general, an MCA provides a proven suite of methods for sustainability evaluation in the context of forest management [31,52–55]. In particular, its use in SIA of forest-wood chains has been demonstrated previously [33,56,57]. MCA strives at contributing a cumulative assessment of alternative options that combines all indicator performances into one common preference ranking. Having in mind that the total surface occupied by homogeneous pine stands is estimated at about 25% of the total forest area of the Mediterranean basin (=ca. 13 million ha), indicator results per hectare (Table 4) were quite impressive, such that the MCA is a very useful method to support the comparison of given alternatives, because associated impacts were huge and conflicting in some cases (e.g., 3-FM-WT and 2-MM-CTL). To give one example: managing "just" the 13,800-ha umbrella pine plantations in Italy with the thinning system presented in 3-FM-WT would result in overall production costs of 7.4 million €. In contrast, managing them with 2-MM-CTL would cost 31.5 million €.

The MCA results in the current application showed very robust preferences for the FM harvesting operations, while the difference between FM-WT and FM-CLT was rather small (<0.1 of total preference value φ). This indicates that a final selection of the most appropriate option cannot be generalized, but has to be scrutinized in more detail in a specific decision-making context. It is remarkable that MM-CLT is by far the least preferred option, since it has the highest negative preference values in eight out of 12 indicators. MM-WT might be a compromise solution, if entrepreneurial limitations (e.g., investment capital for mechanization, skills and workforce) are a major obstacle to pursue FM options. MCA has been used is an exploratory way, i.e., no real indicator weights by decision-makers or stakeholders were used. While this renders the MCA application incomplete to a certain extent, the sensitivity analysis is to give insights into the impact of changing weights. In this respect, the analysis of stability intervals shows very robust rankings, unless indicator weights for five of the indicators are increased beyond 100% of what has been assumed for the preference modelling. Even if a decision maker would state such preferences in a specific decision-making situation, such an unbalanced weighting scheme might compromise the underlying assumptions of a well-balanced concept of all sustainability aspects within the current understanding of sustainable forest management.

5. Conclusions

An SIA was conducted for typical forest-wood chain alternatives that can be applied to the thinning of pine plantations, which are widespread in the Mediterranean region. Four technical choices derived from the intersection of the two harvesting systems and two levels of them were analyzed with regard to 12 indicators considering economic, environmental and social aspects.

The results offered quantitative support to the assessments made by field experts, namely that mechanized harvesting operations should be preferred unless there are entrepreneurial limitations to pursue mechanized options and that motor-manual harvesting with a cut-to-length system is by far the least preferred option, due mainly to the time-extraction process by the forestry-fitted farm tractor.

Author Contributions: R.S. and N.M. performed the field experiments and calculated the economic indicators. J.S. set up the outline of the manuscript and performed the LCA. B.W. and M.J.L. ran the MCA. J.S., R.S. and N.M. analyzed the data. J.S. wrote the paper with contributions from all authors.

Acknowledgments: The corresponding author is supported by the European Social Fund and by the Ministry of Science, Research and Arts Baden-Württemberg. This research received no external funding.

Conflicts of Interest: The authors declare no conflict of interest. The founding sponsors had no role in the design of the study; the collection, analyses or interpretation of the data; the writing of the manuscript; nor in the decision to publish the results.

Abbreviations

MM	motor-manual
FM	fully-mechanized
WT	whole-tree
CTL	cut-to-length
SIA	sustainability impact assessment
MCA	multi-criteria analysis
PMH	productive machine hours, exclusive of delays
gt	green tonne (fresh weight)
CED	cumulated energy demand
MJ	megajoules
EE	energy efficiency
GWP	global warming potential
CO_2	carbon dioxide
EP	eutrophication potential
PO_4	phosphate
POPC	photochemical ozone creation potential
AP	acidification potential
SO_2	sulfur dioxide
FTE	full-time equivalent

References

1. Tapias, R.; Climent, J.; Pardos, J.; Gil, L. Life histories of Mediterranean pines. *Plant Ecol.* **2004**, *171*, 53–68. [CrossRef]
2. Mercurio, R.; Spinelli, R. Exploring the silvicultural and economic viability of gap cutting in Mediterranean softwood plantations. *For. Stud. China* **2012**, *14*, 63–69. [CrossRef]
3. Barbero, M.; Loiser, R.; Quezel, P.; Richardson, D.; Romane, F. Pines of the Mediterranean basin. In *Ecology and Biogeography of Pinus*; Richardson, D., Ed.; Cambridge University Press: Cambridge, UK, 1998; pp. 153–170. ISBN 052-155-176-5.
4. Mondino, G.; Bernetti, G. *I tipi Forestali (Forest types)*; Regione Toscana, Giunta Regionale, Dipartimento dello Sviluppo Economico: Firenze, Italy, 1998; Available online: http://www.regione.toscana.it/-/i-tipi-forestali (accessed on 15 October 2017).
5. Zerbe, S. Restoration of natural broad-leaved woodland in Central Europe on sites with coniferous forest plantations. *For. Ecol. Manag.* **2002**, *167*, 27–42. [CrossRef]
6. Klvač, R. Forest Biomass Glossary; Cost Action FP0902; 2013. Available online: http://www.forestenergy.org/pages/cost-action-fp0902/glossary/ (accessed on 16 May 2018).
7. Jacobson, S.; Kukkola, M.; Mälkönen, E.; Tveite, B. Impact of whole-tree harvesting and compensatory fertilization on growth of coniferous thinning stands. *For. Ecol. Manag.* **2000**, *129*, 41–51. [CrossRef]
8. Smolander, A.; Kitunen, V.; Tamminen, P.; Kukkola, M. Removal of logging residue in Norway spruce thinning stands: Long-term changes in organic layer properties. *Soil Biol. Biochem.* **2010**, *42*, 1222–1228. [CrossRef]
9. Kellogg, L.D.; Bettinger, P.; Studier, D. *Terminology of Ground-Based Mechanized Logging in the Pacific Northwest*; Research Contribution 1; Forest Research Laboratory, Oregon State University: Corvallis, OR, USA, 1993.
10. Lindroos, O.; Lidestav, G.; Nordfjell, T. Swedish non-industrial private forest owners: A survey of self-employment and equipment investments. *Small-Scale For. Econ. Manag. Policy* **2005**, *4*, 409–425. [CrossRef]
11. Tuomasjukka, D.; Athanassiadis, D.; Vis, M. Threefold sustainability impact assessment method comparison for renewable energy value chains. *Int. J. For. Eng.* **2017**, *28*, 116–122. [CrossRef]
12. Lindner, M.; Suominen, T.; Palosuo, T.; Garcia-Gonzalo, J.; Verweij, P.; Zudin, S.; Päivinen, R. ToSIA. A tool for sustainability impact assessment of forest-wood-chains. *Ecol. Model.* **2010**, *221*, 2197–2205. [CrossRef]
13. Päivinen, R.; Lindner, M.; Rosén, K.; Lexer, M.J. A concept for assessing sustainability impacts of forestry-wood chains. *Eur. J. For. Res.* **2012**, *131*, 7–19. [CrossRef]
14. Institute for Environment, and Sustainability of European Commission. *ILCD Handbook: Specific Guide for Life Cycle Inventory Data Sets*; Publication Office of the European Union: Luxembourg, 2010; ISBN 978-92-79-19093-3.
15. Päivinen, R.; Lindner, M. Assessment of sustainability of forestwood chains. In *The Multifunctional Role of Forests-Policies, Methods and Case Studies*; EFI Proceedings; Cesaro, L., Gatto, P., Pettenella, D., Eds.; European Forest Institute: Joensuu, Finland, 2008; pp. 153–160.
16. Institut für Umweltinformatik (IFU). *Umberto 5.6*; IFU: Hamburg, Germany, 2011.
17. Pülzl, H.; Prokofieva, I.; Berg, S.; Rametsteiner, E.; Aggestam, F.; Wolfslehner, B. Indicator development in sustainability impact assessment: Balancing theory and practice. *Eur. J. For. Res.* **2012**, *131*, 35–46. [CrossRef]
18. OECD. *Environmental Indicators. Towards Sustainable Development*; OECD Publications: Paris, France, 2001.
19. MCPFE (Ministerial Conference on the Protection of Forests in Europe). *Improved Pan-European Indicators for Sustainable Forest Management*; MCPFE Liaison Unit: Vienna, Austria, 2003; Available online: http://www.foresteurope.org/documentos/improved_indicators.pdf (accessed on 18 October 2017).
20. International Organization of Standardization. *ISO 14040: 2006. Environmental Management—Life Cycle Assessment—Principles and Framework*; ISO: Geneva, Switzerland, 2006.
21. Ecoinvent. *Swiss Centre for Life Cycle Inventories*; Ecoinvent Centre, Empa: St. Gallen, Switzerland, 2010; Available online: http://www.ecoinvent.org/ (accessed on 1 October 2014).
22. Meng, W. Baumverletzungen durch Transportvorgänge bei der Holzernte—Ausmaß und Verteilung, Folgeschäden am Holz und Versuch ihrer Bewertung. Doctoral thesis, Selbstverl. d. Landesforstverwaltung Baden-Württemberg, Stuttgart, Germany, 1978; p. 159.

23. Spinelli, R.; Lombardini, C.; Magagnotti, N. Effect of mechanization level and harvesting system on the thinning cost of Mediterranean softwood plantations. *Silva Fenn.* **2014**, *48*. [CrossRef]
24. Klun, J.; Medved, M. Fatal accidents in forestry in some European countries. *Croat. J. For. Eng.* **2007**, *28*, 55–62.
25. Blombäck, P.; Poschen, P.; Lövgren, M. *Employment Trends and Prospects in the European Forest Sector*; Geneva Timber and Forest Discussion Papers, European Forest Sector Outlook Study (EFSOS); United Nations: Geneva, Switzerland, 2003.
26. Ackerman, P.; Belbo, H.; Eliasson, L.; de Jong, A.; Lazdins, A.; Lyons, J. The COST model for calculation of forest operations costs. *Int. J. For. Eng.* **2014**, *25*, 75–81. [CrossRef]
27. Björheden, R.; Apel, K.; Shiba, M.; Thompson, M.A. *IUFRO Forest Work Study Nomenclature*; Swedish University of Agricultural Science, Department of Operational Efficiency: Garpenberg, Sweden, 1995.
28. Spinelli, R.; Visser, R. Analyzing and estimating delays in harvester operations. *Int. J. For. Eng.* **2008**, *19*, 35–40.
29. Ghaffariyan, M.R.; Andonovski, V.; Brown, M. Application of slash-bundler for collecting harvest restudies in Eucalyptus plantation. *Silva Balcan* **2011**, *1–2*, 83–89.
30. Triantaphyllou, E.; Shu, B.; Nieto Sanchez, S.; Ray, T. Multi-Criteria Decision Making: An Operations Research Approach. In *Encyclopedia of Electrical and Electronics Engineering*; John Wiley & Sons: New York, NY, USA, 2015; Volume 15, pp. 175–186.
31. Diaz-Balteiro, L.; Romero, C. Making forestry decisions with multiple criteria: A review and an assessment. *For. Ecol. Manag.* **2008**, *255*, 3222–3241. [CrossRef]
32. Acosta, M.; Corral, S. Multicriteria decision analysis and participatory decision support systems in forest management. *Forests* **2017**, *8*, 116. [CrossRef]
33. Wolfslehner, B.; Brüchert, F.; Fischbach, J.; Rammer, W.; Becker, G.; Lindner, M.; Lexer, M.J. Exploratory multi-criteria analysis in sustainability impact assessment of forest-wood chains: The example of a regional case study in Baden-Württemberg. *Eur. J. For. Res.* **2012**, *131*, 47–56. [CrossRef]
34. Brans, J.P.; Vincke, P.; Mareschal, B. How to select and how to rank projects: The PROMETHEE method. *Eur. J. Oper. Res.* **1986**, *24*, 228–238. [CrossRef]
35. Wolfslehner, B.; Vacik, H.; Lexer, M.J. Application of the Analytic Network Process in multi-criteria analysis of sustainable forest management. *For. Ecol. Manag.* **2005**, *207*, 157–170. [CrossRef]
36. Spinelli, R.; Magagnotti, N.; Picchi, G. Complete tree harvesting as an alternative to mulching in early thinnings. *For. Prod. J.* **2009**, *59*, 79–84.
37. Schweier, J.; Molina-Herrera, S.; Ghirardo, A.; Grote, R.; Díaz-Pinés, E.; Kreuzwieser, J.; Haas, E.; Butterbach-Bahl, K.; Rennenberg, H.; Schnitzler, J.-P.; et al. Environmental impacts of bioenergy wood production from poplar short rotation coppice grown at a marginal agricultural site in Germany. *GCB Bioenergy* **2016**, *9*, 1207–1221. [CrossRef]
38. Achat, D.L.; Deleuze, C.; Landmann, G.; Pousse, N.; Ranger, J.; Augusto, L. Quantifying consequences of removing harvesting residues on forest soils and tree growth—A meta-analysis. *For. Ecol. Manag.* **2015**, *348*, 124–141. [CrossRef]
39. Kaarakka, L.; Tamminen, P.; Saarsalmi, A.; Kukkola, M.; Helmisaari, H.-S.; Burton, A.J. Effects of repeated whole-tree harvesting on soil properties and tree growth in a Norway spruce (*Picea abies* (L.) Karst.) stand. *For. Ecol. Manag.* **2014**, *313*, 180–187. [CrossRef]
40. Wall, A.; Hytönen, J. The long-term effects of logging residue removal on forest floor nutrient capital, foliar chemistry and growth of a Norway spruce stand. *Biomass Bioenergy* **2011**, *35*, 3328–3334. [CrossRef]
41. Wall, A. Risk analysis of effects of whole-tree harvesting on site productivity. *For. Ecol. Manag.* **2012**, *282*, 175–184. [CrossRef]
42. Olsson, B.A.; Staaf, H.; Lundkvist, H.; Bengtsson, J.; Rosen, K. Carbon and nitrogen in coniferous forest soils after clear-felling and harvests of different intensity. *For. Ecol. Manag.* **1996**, *82*, 19–32. [CrossRef]
43. Clarke, N.; Gundersen, P.; Jönsson-Belyazid, U.; Kjønaas, O.J.; Persson, T.; Sigurdsson, B.D.; Stupak, I.; Vesterdal, L. Influence of different tree-harvesting intensities on forest soil carbon stocks in boreal and northern temperate forest ecosystems. *For. Ecol. Manag.* **2015**, *351*, 9–19. [CrossRef]
44. Adebayo, A.; Han, H.; Johnson, L. Productivity and cost of cut-to-length and whole-tree harvesting in a mixed-conifer stand. *For. Prod. J.* **2007**, *57*, 59–69.

45. Coates, E.; Cronin, B.; Kent, T. A comparison of biomass production and machine system productivity using three harvesting methods in a conifer first thinning. *Irish For.* **2016**, *73*, 122–140.
46. Labelle, E.R.; Jaeger, D. Soil compaction caused by cut-to-length forest operations and possible short-term natural rehabilitation of soil density. *Soil Sci. Soc. Am. J.* **2011**, *75*, 2314–2329. [CrossRef]
47. Kleibl, M.; Klvač, R.; Lombardini, C.; Porhaly, J.; Spinelli, R. Soil Compaction and Recovery after Mechanized Final Felling of Italian Coastal Pine Plantations. *Croat. J. For. Eng.* **2014**, *35*, 63–71.
48. Allman, M.; Jankovský, M.; Messingerová, V.; Allmanová, Z.; Ferenčík, M. Soil compaction of various Central European forest soils caused by traffic of forestry machines with various chassis. *For. Syst.* **2015**, *24*, e038. [CrossRef]
49. Jourgholami, M.; Fathi, K.; Labelle, E.R. Effects of foliage and traffic intensity on runoff and sediment in skid trails after trafficking in a deciduous forest. *Eur. J. For. Res.* **2018**. [CrossRef]
50. Rodrigues, C.K.; da Silva Lopes, E.; Polizeli, K.M.V.C.; Müller, M.M.L. Soil compaction due to wood harvesting traffic at different extraction distances. *Floresta e Ambiente* **2018**, *25*, e20160045. [CrossRef]
51. Albizu-Urionabarrenetxea, P.M.; Tolosana-Esteban, E.; Roman-Jordan, E. Safety and health in forest harvesting operations. Diagnosis and preventive actions. A review. *For. Syst.* **2013**, *22*, 392–400. [CrossRef]
52. Huth, A.; Drechsler, M.; Köhler, P. Using multicriteria decision analysis and a forest growth model to assess impacts of tree harvesting in Dipterocarp lowland rain forests. *For. Ecol. Manag.* **2005**, *207*, 215–232. [CrossRef]
53. Kiker, G.A.; Bridges, T.S.; Varghese, A.; Seager, T.P.; Linkovjj, I. Application of Multicriteria Decision Analysis in Environmental Decision Making. *Integr. Environ. Assess. Manag.* **2005**, *1*, 95–108. [CrossRef] [PubMed]
54. Nordström, E.-M.; Eriksson, L.O.; Öhman, K. Integrating multiple criteria decision analysis in participatory forest planning: Experience from a case study in northern Sweden. *For. Policy Econ.* **2010**, *12*, 562–574. [CrossRef]
55. Eggers, J.; Holmgren, S.; Nordström, E.-M.; Lämås, T.; Lind, T.; Öhman, K. Balancing different forest values: Evaluation of forest management scenarios in a multi-criteria decision analysis framework. *For. Policy Econ.* **2017**. [CrossRef]
56. Wolfslehner, B.; Huber, P.; Lexer, M.J. Smart use of small-diameter hardwood—A forestry-wood chain sustainability impact assessment in Austria. *Scand. J. For. Res.* **2013**, *28*, 184–192. [CrossRef]
57. Ghaffariyan, M.R.; Brown, M. Selecting the efficient harvesting method using multiple-criteria analysis: A case study in south-west Western Australia. *J. For. Sci.* **2013**, *12*, 479–486. [CrossRef]

forests

MDPI

Article

Radial Growth Behavior of Pines on Romanian Degraded Lands

Ciprian Valentin Silvestru-Grigore [1], Florin Dinulică [2,*], Gheorghe Spârchez [2], Aureliu Florin Hălălişan [2], Lucian Constantin Dincă [3], Raluca Elena Enescu [3] and Vlad Emil Crişan [3]

1 Pîrscov Forest District, Buzău Forest Directorate, National Forest Administration, Pîrscov, 127450 Buzău, Romania; ciprian_silvestru@yahoo.com

2 Faculty of Silviculture and Forest Engineering, Transilvania University of Braşov, 500123 Braşov, Romania; sparchez@unitbv.ro (G.S.); aureliu.halalisan@unitbv.ro (A.F.H.)

3 Romanian National Institute of Research and Development in Forestry "Marin Drăcea", Braşov Branch, 500040 Braşov, Romania; dinka.lucian@gmail.com (L.C.D.); raluk.enescu@yahoo.com (R.E.E.); vlad_crsn@yahoo.com (V.E.C.)

* Correspondence: dinulica@unitbv.ro; Tel.: +40-751-137-007

Received: 20 February 2018; Accepted: 9 April 2018; Published: 17 April 2018

Abstract: More than a third of Romania's surface has low-productive soils, at the same time exposed to risks of climatic phenomena and generating high economic loss. Afforestation with pine has been the most common solution for the recovery of sheet erosion. Many of the pines grown on such land have run down. This paper presents the results of the first dendroecological investigation of degraded lands in Romania, 80 years after the first ecological reconstruction. In this way, the effects of reconstruction were assessed, supporting the adoption of future solutions for the improvement and efficiency of recovered ecosystems. Reconstructed radial growth was set against rainfall, air temperature, and management history. A total of 330 black pine and Scots pine trees (*Pinus sylvestris* L. and *Pinus nigra* Arn.) of different ages and social positions from 11 stands of different densities were cored for retrospective tree-ring analysis. Scots pine has made better use of these sites, with a better growth rate than black pine especially in plantations with lower survival and on dominant trees. The dynamics of radial growth distinguish the two pine species, with Scots pine showing an accentuated juvenile growth spurt and bigger growth range. The growth decline is predominantly a maturation effect that begins when the tree is around 40 years old and seems to be irreversible. After this age, weak or moderated removal is not enough to revive growth. The contribution of climate (air temperature and rainfall) to the last radial increments in decline is 3–57% and is higher than in the previous decades. On moderately degraded land by farming and grazing, the mixture of Scots pine and black pine, rather than monocultures, proved to be a sustainable solution. Dendrochronological surveying of restored ecosystems allows development of management strategies, which becomes critically important in the circumstances of climate warming.

Keywords: afforestation; bad lands; black pine; climate change; dendroecology; growth decline; plantations; Scots pine

1. Introduction

The large ecological amplitude of pine species [1–4], supported by their geographical distribution [5] and their ability to capitalize lands unsuited to forest vegetation through superior production [6–8], have been long been acknowledged [9]. These characteristics have favored the use of pines as species for afforestation or reforestation of naturally or anthropically degraded fields [10–15].

The diversity of climate, geological substratum, relief, and vegetal cover and their interactions, as well as the effects of the anthropogenic factor, make the natural environment in Romania vulnerable to geomorphologic, hydrologic, and climatic risks [16,17]. The recent national inventory indicates 8.34 million ha of degraded or unproductive land, affected especially by pluvial erosion [18], which amounts to 35% of the country's surface and nearly half of the agricultural land. By 2008, 0.3 million ha of degraded land was reforested, mostly with pine species [18]. Between 4000 and 10,000 seedlings per hectare were used for reforestation, depending on the erosion harshness, densities considerably higher than usual [15]. The postwar ecological reconstruction fulfilled its purpose, managing to eradicate the erosion areas 5–15 years after the afforestation [19].

Using pine in the recovery of degraded land is limited to sheet erosion [10]. The black pine proved to have sustained high growth even on highly eroded lands [20]. Pines can offer good protection of the land and good yields: 3–8 m^3 yearly per hectare in forest steppe, 5–10 m^3 yearly per hectare in hill mixed hardwood area, 5–8.5 m^3 yearly per hectare in oaks and beech area, and 7–12 m^3 yearly per hectare in Norway spruce area, at the age of 30 years and on moderately pluvial eroded land [20]. On rock outcrops, the yields are inconspicuous. On heavy soils, the volume of black pine growth is up to 15% less than on coarse-textured soils [10]. Mixes of pines with broad-leaved trees (wild cherry, ash, maple) are the most productive. The yields of Scots pine are superior to those of black pine on degraded fields in the same site [10]. Twenty-five years marks the maximum growth in height of Scots pines on degraded fields that were afforested during the last century [21]. In Romanian natural stands of Scots pine, 50 years is the age of maximum growth of the basal area [22], while black pine crops from the USA accelerate their growth just after this age [23]. The soil's pluvial erosion intensity does not influence the growth of Scots pines in the first 5–10 years [10].

Pines are particularly receptive to annual climate fluctuations [24–27]. In natural stands, for example, the sensitivity of Scots pine to rainfall is above that of Norway spruce and firs [22]. The dry sites accentuate the sensitivity of the trees' growth structure to fluctuations of the environment [28] and their tolerance to drought [29]. However, the pines' tolerance to the stress caused by drought is limited [26]. The limits have become more visible in the last decades through massive and repeated death occasioned by severe droughts [30,31], requiring care in choosing pines [32] and avoiding planting them in forest steppe [20].

However, diagnosing the trees' growth decay is difficult. Dryness, soil scarcity, nutrient deficiency in soil, excessive plantation density, and the biotic stressors incite a decrease in growth [33–37]. Growth depression may also be the symptom of tree maturation [38–40], as a growth increase can be a consequence of not only favorable climatic circumstances, but also a decrease in competition through logging, for example [41–44]. Pollution and fires can have a long-lasting and stimulating effect on the growth of Scots pines [45] respectively of black pine [46]. The contribution of genotype to the radial growth of the Scots pine is low, at least in the first 30 years, allowing for a stronger influence of ontogenetic factors (related to age or resource availability) [47].

Decreasing wood bioaccumulation is also a symptom of tree vitality decline [48]. Consequently, on degraded lands, the survival rate of pines will be lower [11]. Their fragility [49] and low biodiversity [50], as well as the subsequent natural colonization by rustic deciduous trees [51], make pine stands rather temporary solutions [52,53]. After 15–20 years, Scots pine plantations also show reduced density as a result of wind and snow breaking or tearing down the trees [51]. Almost a third of the pine forests on Romanian degraded lands require urgent improvement, promoting deciduous trees as the main species [54].

No dendroecology research has been carried out so far on reforested degraded lands in Romania. Established plantations have been followed for their survival, yield, competition, and health [55]. There has been no species-specific inquiry on radial growth dynamics, especially under different conditions of soil, climate, and vegetation imposed on pine plantations [20]. Increasingly, climatic extremes, deepening of land torrentiality [56], aridity of lands around the steppe, and dieback of pine plantations are a continuing challenge for forest management in choosing sustainable solutions for afforestation [57]. The degradation of plantations through the decline of the tree layer can reactivate erosion processes [58]. Surveillance of plantations with dendroecological methods allows us to predict and check the effects of various solutions for the ecological reconstruction of degraded lands.

This study was conducted in one of the most pluvial erosive areas of Romania [59]. We sought to delineate the dynamic behavior of pines (Scots pine and black pine) by means of growth ring time series, queried in relation to rainfall, air temperature, stand density, tree age and social position, and logging. The following hypotheses will be verified: (a) growth behavior is different between the two pines due to higher tolerance of black pine to drought and heavy soil, especially on degraded land [10,60–63]; (b) growth decline is climatically triggered by drought and warming [64]; and (c) human disturbance may have a highly xylogenetic contribution similar to that of the climate, by growth revival due to reduced competition following logging [65].

2. Materials and Methods

2.1. Erosion Background

The research was done in the Curvature Carpathians (Figure 1), the second greatest erosion area in Romania [16].

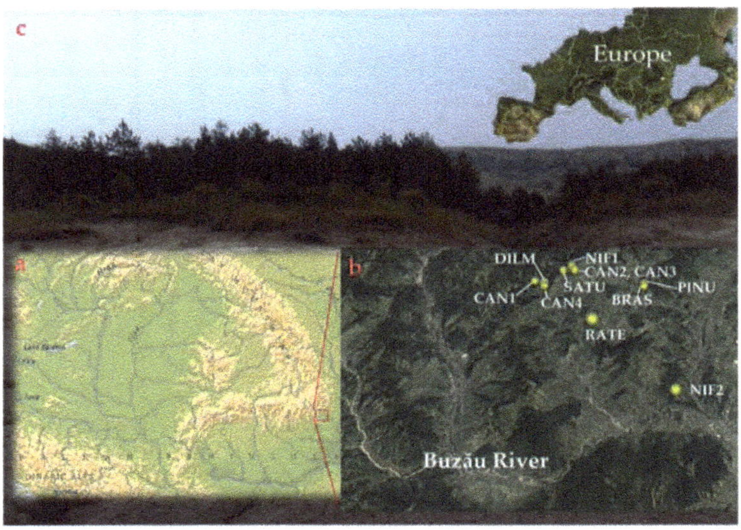

Figure 1. (**a**) Geographical location of the study area in the Carpathians area (South-eastern Europe); (**b**) Map of the sample plots area (the yellow dots mark the sampled stands); (**c**) A pine plantation on a Buzau' badland.

The dynamics of the erosion process are so active that it has led to elimination of large areas from agriculture, called Buzau badlands [59]. The erosivity of the pelitic formation in the clay-marl substratum explains the pluviodenudation in the hollows [66].

The sites have a mostly moderate temperate continental climate. The amount of precipitation (665 mm·year^{-1} at 420 m altitude) is exceeded by evapotranspiration (695 mm·year^{-1}) [67], so there is a water deficit over the entire vegetation season, totaling 173–195 days. These data confirm the relative aridity of the eroded slopes. Heavy rains are very aggressive: high intensity (1.35 mm·min^{-1} at 420 m altitude) and runoff [68].

Before afforestation, the sites were intensively exploited for agriculture and grazing, which led to the loss of productive capacity and protective functions. The soil-limiting factors for tree growth are shallow depth, weak trophicity, and damage to bioactive horizons.

2.2. Study Stands

The plantations discussed were established between 1935 and 1992 using native seedlings (*Pinus sylvestris* var. *romanica* Svoboda and *Pinus nigra* var. *austriaca* (Höss) Novak), with barren roots, 3 years old, planted at 2 × 1 m spacing, in 30 × 30 × 30 cm holes. All plantations started with 5000 trees per hectare, to achieve the stand closure by the age of 10 [20]. Slopes over 40° were terraced and reinforced with vegetation (especially sea-buckthorn stems, branches, and suckers). Along monocultures, mixtures of hardwoods were also planted. The plantations were followed yearly until they reached canopy closure, which occurred at 8–15 years. Where necessary, gap-filling from the same species was done. In the first decade, there were annual cleanings. After 25 years, some plantations were thinned only once, which led to 2–21% removal from the total volume (Table 1).

Table 1. Description of study plots.

Features	Site Plot										
	RATE	SATU	CAN1	DILM	CAN2	NIF1	NIF2	PINU	BRAS	CAN3	CAN4
Altitude (m a.s.l.)	300	295	370	700	325	305	320	665	505	325	340
Exposition	W	NW	N	SE	W	N	N	W	N	NW	W
Slope (°)	15	25	21	20	40	10	10	18	25	40	30
Soil substratum	marl	marl	marl	chalky sand stone	marl	clayey marl	marl	sands	loess	marl	marl
Soil type *	Cmeu	Cmeu	Cmeu	Cmeu-li	Cmeu	Cmeu	Cmeu	Cmeu-ll	Cmeu	Cmeu	Cmeu
Soil physiological thickness (cm)	42	46	48	32	45	45	54	50	47	47	48
Nitrogen in horizon A (%)	0.28	0.26	0.22	0.18	0.30	0.47	0.36	0.20	0.15	0.21	0.14
Carbon in horizon A (%)	2.95	3.76	2.05	2.37	3.52	7.03	5.08	2.78	1.80	3.21	1.63
Base saturation (%)	57.36	55.76	56.76	53.92	58.53	39.22	53.65	54.07	54.68	58.85	56.95
Cation exchange capacity in horizon A (0.01 meq g^{-1})	16.98	17.36	18.13	16.95	17.17	23.15	16.17	16.72	16.68	17.23	17.33
The year of plantation establishment	1992	1962	1962	1957	1952	1948	1935	1935	1947	1937	1937
Stand composition (%) **	90BP 10HD	100BP	70SP 20BP 10HD	100SP	70BP 20SP 10HD	90SP 10HD	100BP	90SP 10BP	70SP 30BP	80BP 20SP	100BP
Stand density (trees·ha^{-1})	1840	1680	1520	740	1100	1160	1080	680	1280	640	780
The management of plantations											
Stand age at intervention	-	31	32	39	43	25	-	-	42	59	59
Harvest intensity (% of removal)	-	3.2	3.8	2.2	7.0	6.4	-	-	21.1	1.5	3.2

* Cmeu: Eutric Cambisols; Cmeu-li: Leptic-eutric Cambisols; Cmeu-ll: Lamellic Cambisols. ** SP–Scots pine, BP–Black pine, HD–miscellaneous hardwoods [69].

Erosion was stopped in all the sampled sites. The main type of soils in this area is cambisoil, which makes up 48% of the total forestry soils in Romania [70] and whose features depend on the geomorphological units [71]. In the current state of site recovery, cambisoil exhibits lower volume, slight acidification of the mineral horizons, sandy-loamy texture in the mineral horizon A, and higher content of sand and humus. These are medium-deep soils (physiological thickness 45.8 ± 1.7 cm), poorly stony (the content of rock material 1–9%), high organic carbon (180–210 t/ha [72]), and medium supply of water and nitrogen available to the woody plants (0.25 ± 0.1% in the A horizon). The organic horizon is unbroken only in plantations older than 75 years [73].

2.3. Sampling Design and Data Compilation

Before sampling, we screened the degraded land areas afforested with pines in order to cover all the recovery levels of the local sites. Forty plots of 20 × 25 m^2 were established in the chosen sites, where we performed full inventories. For cross-dating, a master chronology was also used, from a 108 years old black pine stand grown in sites favorable for pine vegetation, at an altitude of 790–815 m.

The core sampling was drawn from 11 stands, mostly older than 40 years, which facilitated analysis of the long-term growth trend. For comparison, a younger stand was also chosen (Table 1). One sample 5 mm thick was taken from each tree, in the uphill breast height, with an increment borer. We opted for a single drilling direction, following a preliminary study in the CAN4 plot, where it was found that the differences in ring width between breast height radii were not statistically significant ($F = 0.482$, $p = 0.70$). A total of 30 trees in each plot were cored. Each tree was classified according to its top social position (in the Kraft classification), which became the KP variable. KP is a discrete variable with the following values: 1: dominant; 2: codominant; 3: subdominant; 4: suppressed; 5: dying [74].

The climatic data for the dendroecological study came from weather stations in Buzau (45°09′ N, 26°49′ E, 102 m a.s.l.) and Pătârlagele (45°19′ N, 25°91′ E, 390 m a.s.l.) in the research area. The continuous climatic time series have a length of 53 years (1961–2013).

2.4. Dendroecological Study

After seasoning, the samples were glued to wooden frames and sanded at a 60–240× granulation [75]. They were then scanned at 1200 dpi resolution using the WinDENDRO Density 2006c device from Régent Instruments [76]. We obtained raw time series for the 5 variables of the annual rings: annual ring width (RW), earlywood width (EW), latewood width (LW), and the corresponding proportions of earlywood (EP) and latewood (LP). For each individual series of RW, we subsequently calculated the range values, designated by the variable ring width range (RWR).

For cross-dating, which was also performed on the WinDENDRO, we used a reference series from the control stand. We kept the growth series with the Gleichläufigkeit nonparametric correlation coefficient with the control series [77] over 0.65.

The series of individuals from the same plot were averaged. At the far end, the average series was truncated at the depth of at least 3 trees [78]. The average series was converted by standardization in series of indices, performed again on WinDENDRO, where the smoothing filter was the spline curve and for which the value of −4 of the Lagrange parameter was chosen [79]. The signal thus extracted (a series of indices) was verified against the climate through the nonparametric test of signs [80]. Age-related and stand dynamics effects were quantified as the difference between the raw series and the standardized series of RW.

In order to identify the events from the average series of RW and the standardized indices, we calculated the moving growth change rates using a 4-year moving window, adapted from [81] and [82]:

$$GC_i = \frac{\sum_{i}^{i+3} X_i - \sum_{i-4}^{i-1} X_i}{\sum_{i-4}^{i-1} X_i} \times 100(\%) \quad (1)$$

where i is the current calendar year and X_i is the RW value of year i (from the raw series) or the value corresponding to the signal (standardized series) or the value of the noise series (the difference between the previous series). Consequently, three types of growth change rate were used. Change rates of radial growth (GC_RW) denote the level of the periodic changes in the annual ring width from the breast height. Change rates of standardized indices (GC_D) indicate the level of fluctuation in the signal. GC_N indicate the level of fluctuation in the noise series.

In order to separate the contribution of the climate from the logging effects, we similarly calculated the change rates for temperature and precipitation (TC and RC, respectively). TC expresses the relative change in average temperature from four consecutive calendar years after the current year as compared to the 4 previous years. The relationship of these climatic indices with the growth change rates was verified by nonparametric correlation.

For the diagnosis of the recent trees' vitality, we extracted the last 10 years from the growth time series [83], thus we obtained the average width of the last 10 years of the series (10 RW variable).

2.5. Statistical Analyses

To recognize the influences on growth, we chose a multifactorial design, wherein the dependents were the tree ring variables, the fixed factors were the species of trees and their social position, the covariates were the age of the trees and the current stand density, and the tree was random. To avoid age-related bias in the growth series, we chose analysis of covariance and partial correlation. Thus, to assess the involvement of species and social position in tree growth, independent of their age and of the density of plantation, analysis of variance was replaced by analysis of covariance (ANCOVA) [84]. The variables were previously checked to avoid multicollinearity. In ANCOVA, the tree species and social position were designated as categorically independent variables, the annual growth (RW, RWR, 10R, LP) as dependent, and the age of the trees and density of the stands as covariates. Similarly, in order to remove the age share in the relationship of stand density with tree growth, partial correlations were chosen. To measure the strength of the relationship between climate variables (as predictors) and growth (as dependent variables), multiple correlation was used. Time series synchronization was checked with cross correlation function [85]. Statistical data processing was carried out in Statistica 8.0.

3. Results

3.1. Differences in Tree Growth between the Two Pine Species

RW ranges from 0.026 to 21.11 mm·year^{-1}. At the breast height, the RW reaches its peak in the first 13 years, especially around the age of 7. RW together with LP allow the partition of radial growth into 5–18 juvenile wood rings, followed by a transition to mature wood, which can extend up to the age of 36. The differences between the sampled sites with regards to radial growth occur from the age of 11 years. Starting the age of 40, when all the trees are stucturally mature, the RW stabilizes around an average of 1.1 mm. The width of the annual rings in the last decade of trees growth is 50–80% of the average value over the entire series. In what concern the dominated trees over 40 years old, the width of the rings of the last decade did not reach 20% of the average value per series. Tree age, stand density, and tree social position are factors that influence RW size (Table 2). The contribution of species to growth differences between the trees seems to be small. At the same age, Scots pine grows on average 0.1–0.35 mm·year^{-1} faster than black pine. There are no differences between the two pine species with regards to the age when the annual rings become mature ($\chi^2 = 0.018$, $p = 0.89$, $df = 1$). Even though the radial growth is fairly similar for the two pines, Scots pine shows larger spreading of its annual values (higher values by about 1 mm of RWR). The radial growth from the last decade, which is placed in the decline, also does not distinguish between the two species.

Table 2. Analysis of covariance (ANCOVA) of the growth structural traits.

Dependent Variables	Median	Total Variance between Rings	Predictors (Fixed Effects)					
			Tree Age	Stand Density	Specie		Tree Social Position	
					Covariates			
					Tree Age	Stand Density	Tree Age	Stand Density
			p Values from *F* Test					
Annual ring width, mm	1.50	2.65	<0.01	<0.001	0.04	0.04	<0.001	<0.001
Annual range of ring width, mm	6.61	3.78	0.30	<0.001	<0.01	<0.01	0.05	0.08
Annual average of the last 10 years ring width, mm	0.97	0.34	<0.01	<0.001	0.53	0.57	<0.001	<0.001

The differences between species is increased for Scots pine, into more sparse stands (Figure 2).

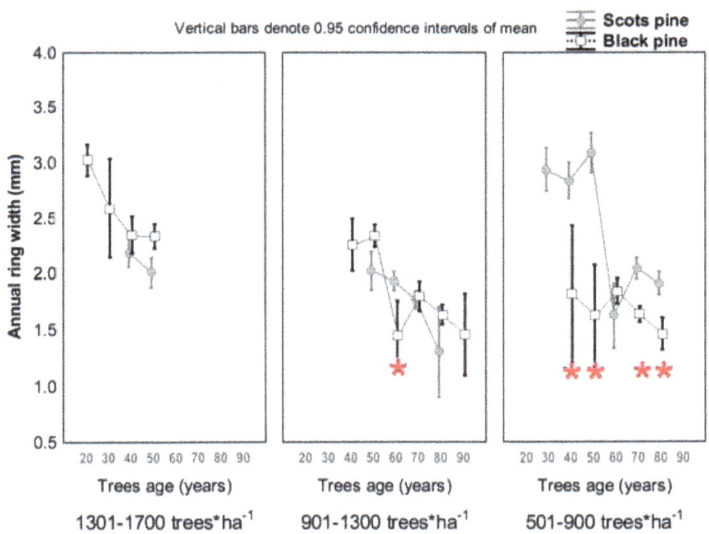

Figure 2. Specie x tree age x stand density interaction in pines radial growth at breast height (the asterisks mark significant differences ($p < 0.05$)).

The hierarchical tree position (KP) is an even more important source of annual growth variation. The understories (KP = 4 and 5) emphasize the differences between the two species of pine (Figure 3).

In older plantations, radial growth had a faster start in the Scots pine than in the black pine (Figure 4). Faster growth is supported longer by Scots pine, but a decline in growth usually occurs simultaneously after 39 years. The averages of RW were recorded simultaneously for the two pines without being influenced by the current stand density.

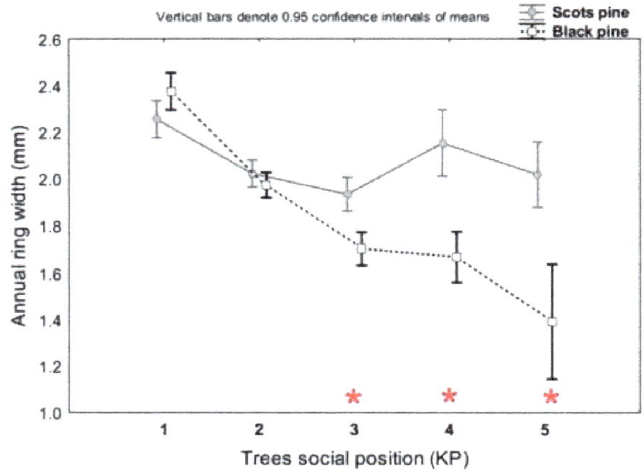

Figure 3. Distinguishing between social classes stand in terms of radial growth (1: dominant; 2: codominant; 3: subdominant; 4: suppressed; 5: dying. Stars mark significant differences: $p < 0.001$).

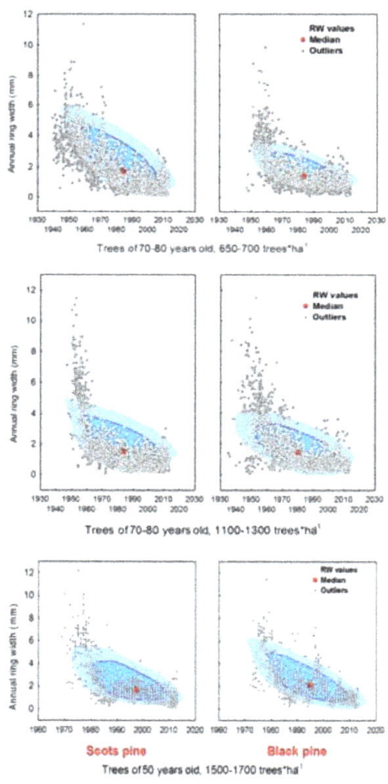

Figure 4. Ring width dynamic: comparisons between the black pine and the Scots pine for densities level.

In Table 3 we presented the density of the tree stands whose growth depression is in fact irreversible. The mixtures of black pine and Scots pine, rather than pure tree stands, survived better on degraded land.

Table 3. The current density of plantations with the age of more than 40 years.

Stand Composition	No of Trees·ha^{-1} (Mean ± Standard Deviation)
Scots pine stands	1193 ± 228
Black pine stands	1109 ± 468
Pine mixtures	1364 ± 344

3.2. Climate Share in Radial Growth

Multiple correlations of detrended growth change rate with climate change rate (rainfall together with air temperature act to separate the climate share from radial tree growth (Figure 5). We started from the assumption that any change in temperature and precipitation (shown by RC and TC) would cause a corresponding shift in growth (reflected by the GC from the signal). The hypothesis was verified in the sequence of recent and very narrow rings (with an average RW per plot <1 mm).

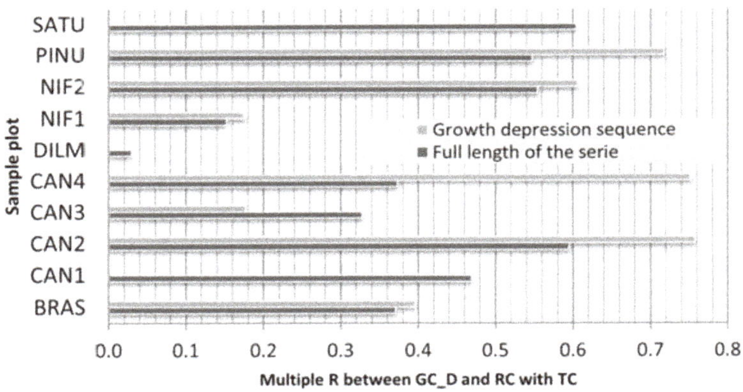

Figure 5. Verifying the growth response to the climate.

The multiple correlation coefficients, varying from one plot to another, between 0.175 and 0.757, show a sound affinity of the growth regime to the climate regime. Therefore, the contribution of climatic range to the latest narrow ring widths can be 57% at most. The differences among plots originate from the time lags between them. Improving correlation coefficients with temperature x precipitation in the growth suppression phase (Figure 5) argues that the climate maintains and accentuates the decline. The local climate trend over the past 50 years shows a warming of 0.24 °C per decade (R^2 = 0.26; $p < 0.001$)), more pronounced since 1994. The precipitations have suffered an insignificant decrease by 0.3 mm·year^{-1} (R^2 = 0.002, p = 0.77), being more consistent in the dry decade 1981–1990. By consequence, in the researched area, we cannot talk about the climate becoming arid which means that the decline of growth is not an argument for this.

Growth response to rainfall varies from one plot to another (Figure 6). The drought from the ninth decade of the last century did not seem to have any influence on the RW oscillations. An improvement in the water supply in 1998 was not able to revive the trees' growth. Crosscorrelations in time series analysis indicate a growth rate delay of 1–3 years versus rainfall change rate. It has been noted that the site with highest delay in growth as a response to rainfall (lag = 3 years) is distinguished by the highest carbon content and the lowest base saturation in the top soil layer. The correlation between growth change rate and temperature change rate is poor (simple nonparametric correlation coefficients which in the modulus do not exceed 0.198) and not significant ($p > 0.19$).

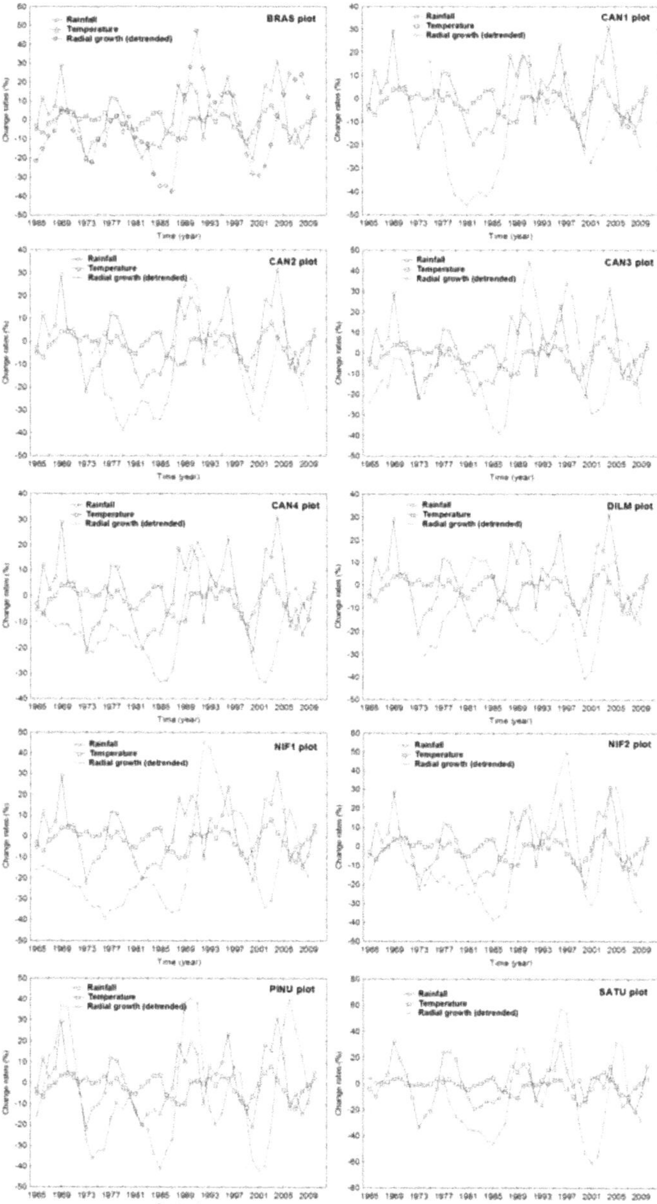

Figure 6. The dynamic of the change rates of detrended ring width, temperatures, and rainfall in the examined sample plots.

3.3. Logging Effects

Only logging from the BRAS and CAN3 stands seemed to stimulate radial growth (Table 4). In order to customize the logging effects, the nonclimatic noise series of radial growth was employed.

By comparing the GC values with the multiannual climatic regime, we noticed several aspects. The increased growth following the relative powerful thinning in the BRAS plot (GC = 294%) could have been caused, to some extent, by rainfall (RC = 10.5%). Indeed, the amount in the year prior to the intervention exceeded the multiannual average by 84.4 mm and was reached especially in May and June. In the following years, it oscillated a great deal, from 818.5 mm in 1991 to 461.7 mm in 1992. After 1989, the year of the intervention, the temperature increased again until 1994.

In order to delimit the contribution of climatic factors, we verified the multiple correlations between the detrended growth series and the temperature along with precipitation. For the eight-year sequence around the intervention, multiple R (0.774) shows a contribution to RW of up to 55.4% from temperature and rainfall from the period following the intervention. The sudden increase of GC_N was prepared in the previous year, when it became positive (Figure 7). Consequently, the revival of growth after 1989 in the BRAS plot was arranged by rainfall, but was actually allowed by thinning. The increase rate lasted only nine years, with GC subsequently returning to negative values.

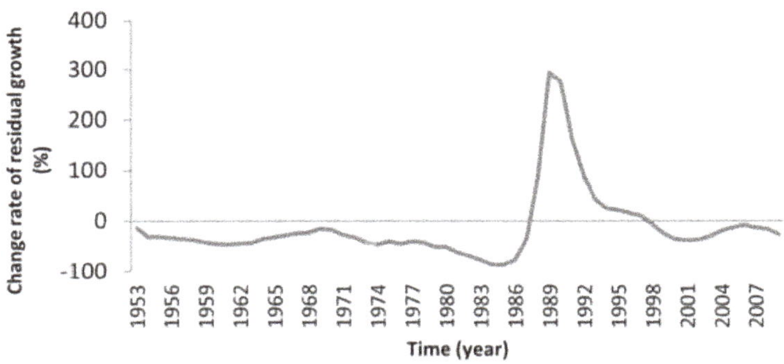

Figure 7. The variation of the growth change rate in the residual series of the ring width on the plot intensively managed.

The moderated logging on CAN2 seems to have had no bearing on growth (Table 4), despite the climatic incitement. The logging effects from the NIF1 plot on growth (Table 4) seem to be neutralized by temporary dryness. Actually, the RC value was on account of the low amount of rainfall during the year of logging, the surrounding seasons being normally supplied with water.

Table 4. The logging effects on the radial growth of trees.

The Managed Plots	The Growth Changes after the Intervention			
	Change Rate of Undetrended Ring Width (%)	Change Rate of Residual Serie of Ring Width (%)	Change Rate of Air Temperature (%)	Change Rate of Rainfall (%)
BRAS	+18.93	+294.23	−9.60	+10.46
CAN1	−7.03	−25.61	−0.90	+4.51
CAN2	−2.69	−15.98	+3.49	+10.16
CAN3	+16.47	7.72	+2.46	+23.56
CAN4	+2.37	+86.64	+2.46	+23.56
DILM	−12.18	−18.86	+2.46	+23.56
NIF1	−30.62	−36.74	+2.43	−21.63
SATU	−17.36	−26.65	+3.00	+8.49

4. Discussion

In the present study, the retrospective analysis of time series of the radial increments allowed the reconstruction of the behavior of pines planted on pluvial eroded land, at 80 years after the first afforestation. We have identified the sources of variation of the radial growth size, among which the age, the species and the fluctuations of rainfall, together with the temperature, have the greatest impact. The radial growth trends were analyzed in relation to the particular soil conditions. We diagnosed, at the same time, the effect of tending operations (cleanings and thinnings) on the rate of growth in relation to their intensity and climatic context.

Despite the adequate fitting and the relevant contribution of this research, we speculate that these multifactorial models could get more resolution with dendroecological data using individual tree models based on other studies [40]. For a better estimate of radial growth, the model should be strengthened with other variables related to vegetation, such as competition indices, spatial distribution indices, trees quality and health [86–92]. These should be tracked in dynamics, which extends research over time. The model could be improved with other climate variables as fixed effects, such as sunshine hours, soil temperature, relative air humidity, vapour pressure deficiency, as well rainfall and air temperature extremes [93,94].

4.1. Scots Pine vs. Black Pine in Terms of Radial Growth and Survival

The size and the dynamics of tree growth are specie, genotype and age specific. In previous studies, growth rate differences have been reported between the two pine species, which are usually in favor of the black pine (on carbonate soils [95], on rendzinic leptosols [96], on deep soils on limestone [97], on the sunny slopes and on superficial soils in steppe [10], and in the tree nursery [98]), sometimes in favor of the Scots pine (on degraded land in the dry areas [99]). In the natural tree stands vulnerable to fires from the Dinaric Alps [100], as well on the well sunny slopes or on superficial soils in the Carpathians [51], the differences of growth between the two pin species are insignificant. In the present study, the differences in growth between the two species became perceptible only after stratification of the values according to the social positioning of the trees and the actual density of the plantations (Table 2, Figures 2 and 3).

If we take as starting point the assumption of physiological similarities of pine species, especially the photosynthetic ones, then the differences between black pine and Scots pine in terms of net production are due to a different level of climatic sensitivities [101]. Out the investigations we carried so far, it has been demonstrated that black pine is more resistant to drought [97]. The Scots pine is more sensitive to heat and drought [102], even after irrigation [103], due to the more drastic limitation of photosynthesis and transpiration, in the soil water deficit [104], by a stronger stomatal control [97]. Pines resistance to drought is due to their isohydric behavior [105], to the reduction of crown conductivity [106,107] and to a more efficient way to use the water [108] that leads to a diminished water loss. In relation to the pines, the dryness leads to cell walls becoming thinner and to tracheid's lumen becoming larger, with positive impact on the conductivity of the water [97]. The accelerate juvenile growth of the Scots pine comparing to the black pine (Figure 4) it was noticed shortly after planting pine trees on the degraded lands [99].

In our sites, it may be the reaction of pines to soil conditions rather than their different climatic sensitivities. The sandy cambisoils, with a high level of humus and a small level of stone, in a climate where rainfalls sometimes exceed 800 m·year^{-1}, have a positive influence on Scots pine [109]. The Scots pine also makes better use of the favorable microclimatic conditions in the under storey (Figure 3). Limiting factors such as shading or heavily declivous slopes, as well as the reduced physiological thickness of soils affects the two pine species equally.

In plantations established on degraded lands, more the erosion is advanced, more the black pine turned out to be competitive at a higher level than the Scots pine, in terms of the survival [10]. In our sites, moderately affected by pluvial erosion, 80 years after planting, only 13–22% of the initial trees (5000 trees·ha^{-1}) survived. The mixture of black pine and Scots pine, rather than pure tree stands, survived better on degraded land (Table 3). Scots pine in Romanian plantations proved to be vulnerable to wind and breaks caused by the snow [51]. It is, moreover, one of the most important causes of the reduced survival rate (Table 3). In these plantations, Scots pine grew faster (Figure 2), also because it is more shade-intolerant than the black pine [109]. The trees that survived have a good health. When they occur, the defoliations do not exceed 25% of the crown and are more pronounced to Scots pine than to black pine [110].

4.2. Age-Dependency against Climate-Related Trends in Growth

The ontogenetic exponential decay of radial increments in sampled plantations (Figure 4) has the appearance of dynamics generated by the age of cambium. In Scots pine, the rapid decline in growth with age is a characteristic of trees with low wood quality of trunks [111], and the delay in mature wood formation is a result of difficult growing conditions [112]. Some sample plots from the present study showed a linear decline in radial growth, which is similar with a delay in maturation. The Scots pine stands from our plots are located on soils with very low level of nitrogen (0.15–0.20% nitrogen in mineral horizon A). The depletion of nitrogen occurred by sustaining an increased growth level of the trees for over 40 years. In fact, it is known that intensive land use with coniferous plantations, particularly pine, leads to alteration of soil fertility [113,114]. The length of the juvenile wood formation period in the examined trees corresponds to the values from the literature–between 6 and 16 years [115].

From 14–18 years, growth in the analyzed stands is declining. The decline of growth is primarily an aging process, physiologically and genetically controlled, which can be maintained by the environment [38].

The decline of radial growth is primarily an aging process, physiologically and genetically controlled, which can be supported by the environment [38]. Separating factors becomes even more difficult with maturation because the genetic control of wood formation gradually decreases [116]. The decline of growth closely followed the decline in the efficiency of water use [117]. Thus, it is not a random coincidence that the water intake in pine trees peaks at the age of 30–35 years [31], ie the age at which the transition to mature wood ends and the decline in growth is inevitable. In the internal silvosteppe, the decline of pine plantations is even faster at 25 years [21]. On degraded lands, the growth suppression phase ends with the die-back of Scots pine at 30 years, and black pine at 40–45 years [30].

Quantitatively, the climate (rainfall and temperature) contributed 3.1–57.3% to tree radial growth from the last decade variance in our plots. The contribution of rainfall seems to be greater than that of temperature (Figure 6), proving the xeric character of the sites. Temperature smooths the growth fluctuation with rainfall. The contribution of rainfall to black pine radial growth was estimated to be 54% [24], very similar to the cumulated contribution of rainfall and temperature to our BRAS plot. In black pine marginal populations, smaller contributions were reported (34% from rainfall and 27% from temperature) [118].

The temporal instability of the climate-growth link (Figure 5) was also noticed by Johnson et al. [64] and Pärn [119]. In an analysis of the red spruce's decline after 1960 in the eastern USA, this instability advocates the hypothesis that the causes of the recent auxologic regress are different from those in past centuries [38].

At the stage of decline, increased rainfall is no longer able to revive tree growth (Figure 6). Thus, the decline seems irreversible. In these conditions, the constancy of growth in the last decades seems to be mostly the physiological age effect, and the growth amount due to the effect of inhospitable soil conditions.

4.3. The Management of the Ecological Reconstruction on Land Degraded by Pluvial Erosion

The degraded lands are unable to recover spontaneously, requiring human interventions through ecological reconstruction [13]. In the ecological reconstruction of eroded land, pines are only a supportive solution, the vegetation makeup being directed toward restoring local biodiversity [13,53,120,121]. Moreover, in the pine plantations made on degraded lands on the outskirts of the Carpathians, we observed that, after the age of 40, the stands were invaded by rustic deciduous trees [51]. Pine plantations have managed to recover degraded stations (Figure 1) and remain models to follow for similar actions, especially in climate warming. On the soils well supplied with water, the Scots pine and the black pine mixes are better than monocultures (Table 3).

The first technical measure for achieving this should be thinning. Failing a remedial management response, denser crops are more vulnerable to drought [35] and very low in biodiversity [122], due to the physiological inhibition of the understorey [35]. However, the authors argued that the dense forest microclimate can mitigate climate excesses, which have escalated in the past 50 years [123]. The thinned stands can be subsequently cut for regeneration and hygiene [124], and thus gradually substituted by local mixtures that are more effective from a hydrological and protective point of view. The pine monocultures can become functionally and productively unfit after only 15–20 years, being exposed to breakage and windthrow [19].

The former recovery solutions for eroded land in Romania were directed toward optimizing anti-erosion protection with the gain of maximum biomass. Overstressing the forest productive potential cannot be a long-term solution [125]. Our previous inquiries [73] stated that pine sites with greater physiological thickness of soil are not the most productive (partial correlation of the soil's physiological thickness with the site class: -0.899, $p < 0.05$). We can therefore infer that the recovered sites we studied were not well exploited by pines. Thus, they represent a reserve of resources available to future late-successional vegetation. The yield availability of our crops does not look promising, since the vegetation successional trend toward the climax leads to reduced productivity [126].

The most influential management measures distribute the competition in the trees' layer. The density of pine plantations on degraded lands must be optimized to reduce runoff by avoiding water stress on individuals [35]. In the dense pine forests of Romania, the drought was felt more strongly, with the decline it caused stopped only by a radical intervention, heavy thinning [127]. All the examined plantations in this study had the same initial density (5000 seedlings$\cdot$ha^{-1}), and so the same competitive start. In the inhospitable sites where they were founded, the low survival demanded periodical gap-filling, which sometimes lasted up to 20 years. The behavior of the black pine understory (Figure 3) did not recommend it in gap-filling, which will use Scots pine or local hardwood, which is less light demanding.

4.4. Trees Response to Management

The tree growth answer to thinning is according to its intensity [128]. At Scots Pine, it was found that the thinnings stimulate the division of cambium [39]. In our stands, the moderate thinning had a slight effect on tree growth for nine consecutive years (Figure 6). The weak interventions were not able to counteract the temporary deficit of rainfall and stop the decline (Table 4). In our findings, the effects of the release of competition on growth are biased toward climate. According to the results obtained for other species, following heavy thinning, the responsiveness of the subjects to temperature increased [128]. For ponderosa pine, the competition does not affect climatic sensitivity in the overstory [78].

A forcible management on degraded land would be likely to be risky, especially in the warming of the local climate. As a matter of fact, in Scots pine heavy thinnings do not generate a surplus of growth compared to moderate ones, on the contrary [129]. Pine plantations from recovered sites can be managed, with moderate intensities, also after the age at which growth enters declining, depending on the state of tree health. For Scots pine, for example, the prolonged decline in radial growth precedes mortality by 15–40 years [130]. However, it needs to be taken into account that the longevity of pine trees on eroded land is lower [10] and postponing the renewal of the vegetation layer could compromise ecological reconstruction. Because the degraded sites in the area we studied have not heavily treated, the present study can provide only suggestions for forest management, requiring further experimental research.

5. Conclusions

In the sites moderately affected by sheet erosion from Buzau, Scots pine showed a slightly greater increase in radial growth than black pine.

Moderately eroded lands in Buzău under the Carpathians (Eastern Romania) have been systematically afforested for 80 years. A retrospective analysis of the radial growth time series revealed several differences between the two pine species used for afforestation. In comparison with black pine, Scots pine had (1) stronger youth growth, (2) a higher growth range over time, and (3) a significant growth advantage for the plantations with lower survival rates and trees from the understory.

A decline in growth occurs in both species after 39 years. Declining growth rates are similar for both pines. Climate (air temperature together with rainfall) has a global contribution that can reach 57% in tree growth. The contribution is higher in the growth decline sequence. The response to rainfall can be prompt or delayed by 1–4 years. By analyzing the reaction to drought and improving rainfall in individual sample plots, the conclusion is that climate has allowed the decline or even accentuated it, without causing it.

Management of the pine plantations on degraded land in Romania has been quite mild. Slight to moderate logging did not have a substantial impact on the growth of standing trees. The largest growth rate after logging occurred with rainfall assistance. In this case, the growth increase after logging lasted nine years.

Acknowledgments: The authors received no specific funding for this work.

Author Contributions: Conceived and designed the experiment: C.V.S.-G. and F.D. Performed the experiment: C.V.S.-G. Analyzed the data: F.D., R.E.E., V.E.C. and A.F.H. Interpreted the results: F.D., L.C.D., V.E.C. and G.S. Conceived the paper, wrote the first draft and edited the manuscript: F.D., R.E.E. and C.V.S.-G. Supervised the manuscript: G.S., L.C.D.

Conflicts of Interest: The authors declare no conflict of interest.

References

1. Roman-Amat, B. Les programmes d'amélioration: Résultats actuels et escomptés: Pin sylvestre (*Pinus sylvestris* L). *Rev. For. Fr.* **1986**, *38*, 132–134. [CrossRef]
2. Portefaix, C.; Roman-Amat, B. Les programmes d'amélioration: Résultats actuels et escomptés: Pins laricio de Corse et de Calabre (*Pinus nigra* Arb. ssp. *laricio* Poiret 1804) (Breeding programmes: Current and anticipated results: Corsican and Calabrian pine). *Rev. For. Fr.* **1986**, *38*, 129–131.
3. Schulze, E.-D.; Beck, E.; Müller-Hohenstein, K. *Plant Ecology*; Springer: Berlin/Heidelberg, Germany, 2005; p. 702. ISBN 9783540208334.
4. Vorob'ev, V.N. *Pine Forests: Utilization of Their Products*; Science Publishers: Enfield, CT, USA, 2007; p. 271. ISBN 9781578085941.
5. Critchfield, W.B.; Little, E.L. *Geographic Distribution of the Pines of the World*; U.S. Department of Agriculture, Forest Service: Washington, DC, USA, 1966; Volume 19, p. 116.
6. Nys, C.; Bonischot, R.; Gelhaye, D. Réponse d'un peuplement de pin laricio de corse (*Pinus nigra* espèce *laricio*) à la fertilisation en Sologne (The response to fertilizers of a Corsican pine (*Pinus nigra* var. *laricio*) stand in Sologne). *Rev. For. Fr.* **1985**, *37*, 482–486.
7. Schultz, R.P. Loblolly—The pine for the twenty-first century. *New For.* **1999**, *17*, 71–88. [CrossRef]
8. Baumann, K.; Schneider, B.U.; Marschner, P.; Hüttl, R.F. Seedling biomass and element content of *Pinus sylvestris* and *Pinus nigra* grown in sandy substrates with lignite. *Geoderma* **2006**, *136*, 573–578. [CrossRef]
9. Jolyet, A. Lemnul de pin negru (*Pinus austriaca*). *Rev. Păd* **1908**, *23*, 53–55.
10. Traci, C. *Împădurirea Terenurilor Degradate (Afforestation of Degraded Lands)*; Ceres Publishing House: Bucharest, Romania, 1985; p. 282.
11. Sanchez, L.G.; Prada, M.A. Los pinos como especies basicas de la restauracion forestal en el medio mediterraneo. *Ecología* **1993**, *7*, 113–126.
12. Madrigal, A. Problématica de la ordenación de masas artificiales en España. *Cuad. Soc. Esp. Cienc. For.* **1998**, *6*, 13–20.

13. Cortina, J.; Amat, B.; Castillo, V.; Fuentes, D.; Maestre, F.T.; Padilla, F.M.; Rojo, L. The restoration of vegetation cover in the semi-arid Iberian southeast. *J. Arid Environ.* **2011**, *75*, 1377–1384. [CrossRef]
14. Kuznetsova, T.; Tilk, M.; Pärn, H.; Lukjanova, A.; Mandre, M. Growth, aboveground biomass, and nutrient concentrationof young Scots pine and longepole pine in oil shale post-mining landscapes in Estonia. *Environ. Monit. Assess.* **2011**, *183*, 341–350. [CrossRef] [PubMed]
15. Ganatsas, P.; Tsitsoni, T.; Tsakaldimi, M.; Zagas, T. Reforestation of degraded Kermes oak shrublands with planted pines: Effects on vegetation cover, species diversity and community structure. *New For.* **2012**, *43*, 1–11. [CrossRef]
16. Ciortuz, I.; Păcurar, V. *Amelioraţii Silvice*; LuxLibris Publishing House: Braşov, Romania, 2004; p. 231. ISBN 9739458130.
17. Giurgiu, V. Consideraţii asupra stării pădurilor României I: Declinul suprafeţei pădurilor şi marginalizarea împădurilor. *Rev. Păd* **2010**, *125*, 3–16.
18. Untaru, E.; Constandache, C.; Nistor, S. Starea actuală şi proiecţii pentru viitor în privinţa reconstrucţiei ecologice prin împădurire a terenurilor degradate din România. (I). *Rev. Păd* **2012**, *127*, 28–34.
19. Constandache, C. Ameliorarea şi Refacerea Pinetelor Necorespunzătoare sub Raport Productiv şi Protectiv Instalate pe Terenurile Degradate Din Bazinul Hidrografic al Râului Putna. Ph.D. Thesis, Transilvania University of Braşov, Braşov, Romania, 2003.
20. Traci, C.; Untaru, E. *Comportarea şi Efectul Ameliorativ şi de Consolidare a Culturilor Forestiere de pe Terenurile Degradate din Perimetrele Experimentale*; Forest Research and Management Institute Research Paper; Forest Research and Management Institute: Bucharest, Romania, 1986.
21. Traci, C.; Muşat, I. Folosirea pinului negru şi a pinului silvestru la împăduririle terenurilor degradate. *Rev. Păd* **1955**, *70*, 211–217.
22. Bouriaud, O.; Popa, I. Comparative dendroclimatic study of Scots pine, Norway spruce and silver fir in the Vrancea Range, Eastern Carpathians Mountains. *Trees* **2009**, *23*, 95–106. [CrossRef]
23. Sander, D.H. *Height-Age Curves for Austrian Pine in Windbreaks on Loess Soils of Nebraska*; USDA Forest Service Research Note RM; Rocky Mountain Forest and Range Experiment Station: Fort Collins, CO, USA, 1963; p. 13.
24. Amodei, T.; Guibal, F.; Fady, B. Relationship between climate and radial growth in black pine (*Pinus nigra* Arnold ssp. *salzmanni* (Dunal) Franco) from the south of France. *Ann. For. Sci.* **2013**, *70*, 41–47.
25. Martín-Benito, D.; Cherubini, P.; del Río, M.; Cañellas, I. Growth response to climate and drought in *Pinus nigra* Arn. trees of different crown classes. *Trees* **2008**, *22*, 363–373. [CrossRef]
26. Ruiz-Labourdette, D.; Génova, M.; Schmitz, M.F. Summer rainfall variability in European Mediterranean mountains from the sixteenth to the twentieth century reconstructed from tree rings. *Int. J. Biometeorol.* **2014**, *58*, 1627–1639. [CrossRef] [PubMed]
27. Sangüesa-Barreda, G.; Camarero, J.J.; Linares, J.C.; Hernández, R.; Oliva, J.; Gazol, A.; González de Andrés, E.; Montes, F.; García-Martín, A.; de la Riva, J. Papel de los factores bióticos y las sequías en el decaimiento del bosque: Aportaciones desde la dendroecología (Role of biotic factors and droughts in the forest decline: Contributions from dendroecology). *Rev. Ecosist.* **2015**, *24*, 15–23. [CrossRef]
28. Rigling, A.; Bräker, O.; Schneider, G.; Schweingruber, F. Intra-annual tree-ring parameters indicating differences in drought stress of *Pinus sylvestris* forests within the Erico-Pinion in the Valais (Switzerland). *Plant Ecol.* **2002**, *163*, 105–121. [CrossRef]
29. Orwig, D.A.; Abrams, M.D. Variation in radial growth responses to drought among species, site and canopy strata. *Trees* **1997**, *11*, 474–484. [CrossRef]
30. Simionescu, A.; Traci, C.; Fraţian, A.; Popescu, T.; Mărcioiu, A. Despre uscarea unor arborete de pin şi măsuri de prevenire a extinderii acestui fenomen. *Rev. Păd* **1963**, *78*, 442–446.
31. Ceuca, G.; Constantinescu, N.; Drocan, R.; Georgescu, C.C.; Niţu, G.; Tomescu, A. Studiu privind condiţiile de vegetaţie ale arboretelor de pin cu fenomene de uscare. *Ann. For. Res.* **1957**, *18*, 204–249.
32. Untaru, E. Premise privind împădurirea terenurilor degradate în condiţiile schimbărilor climatice generate de încălzirea globală. *Rev. Păd* **2010**, *125*, 20–25.
33. Lebourgeois, F. Climatic signals in earlywood, latewood and total ring width of Corsican pine from western France. *Ann. For. Sci.* **2000**, *57*, 155–164. [CrossRef]
34. Waring, R.H. Characteristics of trees predisposed to die. *BioScience* **1987**, *37*, 569–574. [CrossRef]

35. Moreno-Gutiérrez, C.; Battipaglia, G.; Cherubini, P.; Saurer, M.; Nicolás, E.; Contreras, S.; Querejeta, J. Stand structure modulates the long-term vulnerability of *Pinus halepensis* to climatic drought in a semiarid Mediterranean ecosystem. *Plant Cell Environ.* **2012**, *35*, 1026–1039. [CrossRef] [PubMed]
36. Schweingruber, F.H. *Wood Structure and Environment*, 1st ed.; Springer: Berlin/Heidelberg, Germany, 2007; p. 279. ISBN 978-3-540-48548-3.
37. Bednarz, B.; Kochanowski, D. Radial growth response of Scots pine (*Pinus sylvestris* L.), black pine (*Pinus nigra* Arnold), and black alder (*Alnus glutinosa* (L.) Gaertn.) to the nun moth (*Lymantria monacha* L.) outbreak in the SŁowiński National Park and the Damnica Forest District (Northern Poland). *Electron. J. Pol. Agric. Univ.* **2010**, *13*, 1–12.
38. Cook, E.R.; Zedaker, S.M. The dendroecology of red spruce decline. In *Ecology and Decline of Red Spruce in the Eastern United States*; Eagar, C., Adams, M.B., Eds.; Springer: New York, NY, USA, 1992; pp. 192–234. ISBN 9780387977867.
39. Wodzicki, T.J. Natural factors affecting wood structure. *Wood Sci. Technol.* **2001**, *35*, 5–26. [CrossRef]
40. Montoro Girona, M.; Rossi, S.; Lussier, J.-M.; Walsh, D.; Morin, H. Understanding tree growth responses after partial cuttings: A new approach. *PLoS ONE* **2017**, *12*, e0172653. [CrossRef] [PubMed]
41. Sullivan, T.P.; Sullivan, D.S. Acceleration of old-growth structural attributes in lodgepole pine forest: Tree growth and stand structure 25 years after thinning. *For. Ecol. Manag.* **2016**, *365*, 96–106. [CrossRef]
42. Mäkinen, H.; Hynynen, J. Wood density and tracheid properties of Scots pine: Responses to repeated fertilization and timing of the first commercial thinning. *Forestry* **2014**, *87*, 437–447. [CrossRef]
43. Missanjo, E.; Kamanga-Thole, G. Effect of first thinning and pruning on the individual growth of *Pinus patula* tree species. *J. For. Res.* **2015**, *26*, 827. [CrossRef]
44. Goudiaby, V.; Brais, S.; Berninger, F.; Schneider, R. Vertical patterns in specific volume increment along stems of dominant jack pine (*Pinus banksiana*) and black spruce (*Picea mariana*) after thinning. *Can. J. For. Res.* **2012**, *42*, 733–748. [CrossRef]
45. Oktaba, J.; Paschalis, P.; Staniszewski, P. Selected indicators of pine and spruce wood technical quality from the forest being under the impact of industrial pollution. *Folia For. Pol. Ser. A For.* **2002**, *44*, 77–86.
46. Valor, T.; Piqué, M.; López, B.C.; González-Olabarria, J.R. Influence of tree size, reduced competition, and climate on the growth response of *Pinus nigra* Arn. *salzmannii* after fire. *Ann. For. Sci.* **2013**, *70*, 503–513.
47. Tikhonova, I.V.; Tarakanov, V.V.; Knorre, A.A. Contributions of genotypic and meteorological factors to variation of annual tree increment in clonal Scots pine plantations. *Russ. J. Ecol.* **2012**, *43*, 179–184. [CrossRef]
48. Dobbertin, M. Tree growth as indicator of tree vitality and of tree reaction to environmental stress: A review. *Eur. J. For. Res.* **2005**, *124*, 319–333. [CrossRef]
49. Hâruţa, O.; Fodor, E.; Teuşdea, A. Boli complexe la *Pinus nigra* Arnold în defileul Crişului Repede (Complex diseases in *Pinus nigra* Arnold situated along Crisul Repede river gorge). *Ann. For. Res.* **2007**, *50*, 169–184.
50. Maestre, F.T.; Cortina, J. Are *Pinus halepensis* plantations useful as a restoration tool in semiarid Mediterranean areas? *For. Ecol. Manag.* **2004**, *198*, 303–317. [CrossRef]
51. Constandache, C. Cercetări privind regenerarea sub masiv şi introducerea la adăpostul masivului a unor specii autohtone valoroase, în arborete apropiate de exploatabilitate, de pe terenuri degradate. *Ann. For. Res.* **2004**, *47*, 63–81.
52. Barbero, M.; Loisel, L.; Médail, F.; Quézel, P. Signification biogéographique et biodiversité des forêts du bassin méditerranéen. *Bocconea* **2001**, *13*, 11–25.
53. Vallauri, D.R.; Aronson, J.; Barbero, M. An analysis of forest restoration 120 years after reforestation on badlands in the southwestern Alps. *Restor. Ecol.* **2002**, *10*, 16–26. [CrossRef]
54. Constandache, C.; Untaru, E.; Ivan, V. Cercetări privind refacerea-ameliorarea arboretelor necorespunzătoare de pe terenuri degradate din Vrancea (Research regarding the reconstruction of destructed stand on claimed lands in Vrancea County). *Ann. For. Res.* **2001**, *44*, 168–173.
55. Greavu, M. Cercetări Privind Împădurirea Terenurilor Erodate, Ravenate şi Stâncoase Din Podişul Dobrogei de Nord. Ph.D. Thesis, Transilvania University of Braşov, Braşov, Romania, 2003.
56. Niţă, M.D.; Clinciu, I.; Popa, B. Evaluation of stream bed dynamics from vidas torrential valley using terrestrial measurements and gis techniques. *Environ. Eng. Manag. J.* **2015**, *15*, 1387–1395.
57. Dresner, S.; Ekins, P.; McGeevor, K.; Tomei, J. Forests and climate change—Global understandings and possible responses. In *Forestry and Climate Change*; Freer-Smith, P.H., Broadmeadow, M.S.J., Lynch, J.M., Eds.; CAB International: Wallingford, UK, 2007; pp. 38–48. ISBN 9781845932954.

58. Spencer, T.; Douglas, I.; Greer, T.; Sinun, W. Vegetation and fluvial geomorphic processes in South-east Asian tropical rainforests. In *Vegetation and Erosion: Processes and Environments*; Thornes, J.B., Ed.; John Wiley: Chichester, UK, 1990; pp. 451–469. ISBN 0471926302.
59. Velcea, V.A.; Savu, A. *Geografia Carpaţilor şi Subcarpaţilor*; Didactică si Pedagogică Publishing House: Bucharest, Romania, 1982; p. 300.
60. Lefter, R. Studiul terenurilor degradate din Podişul Moldovei şi ameliorarea lor prin culturi forestiere (The study of the Moldavian Plateau eroded lands and their improvement by means of the forest crops). *Rev. Păd* **1966**, *81*, 570–576.
61. Martínez-Vilalta, J.; Piñol, J. Drought-induced mortality and hydraulic architecture in pine populations of the NE Iberian Peninsula. *For. Ecol. Manag.* **2002**, *161*, 247–256. [CrossRef]
62. Martínez-Vilalta, J.; Piñol, J.; Sala, A.M. The hydraulic architecture of Pinaceae—A review. *Plant Ecol.* **2004**, *171*, 3–13. [CrossRef]
63. Rundel, P.W.; Yoder, B.J. Ecophysiology of *Pinus*. In *Ecology and Biogeography of Pinus*; Richardson, D.M., Ed.; Cambridge University Press: Cambridge, UK, 1998; pp. 296–323. ISBN 978-0521789103.
64. Johnson, A.H.; Cook, E.R.; Siccama, T.G. Climate and red spruce growth and decline in the northern Appalachians. *Proc. Natl. Acad. Sci. USA* **1988**, *85*, 5369–5373. [CrossRef] [PubMed]
65. Dinulică, F. *Lemnul de Compresiune la Brad (Compression Wood from Silver Fir)*; Ceres Publishing House: Bucharest, Romania, 2012; p. 296. ISBN 978-973-40-0981-7.
66. Badea, L.; Băcăuanu, V.; Posea, G. Relieful României. In *Geografia României I: Geografia Fizică*; Badea, L., Gâştescu, P., Velcea, V., Eds.; Romanian Academy Publishing House: Bucharest, Romania, 1983; pp. 64–194.
67. Ţîştea, D.; Stoenescu, Ş.M.; Dissescu, C.; Donciu, C.; Topor, N.; Fetov, V. Date climatologice. In *Clima Republicii Populare Române*; Institutul Meteorologic: Bucharest, Romania, 1961; Volume 2, pp. 242–249.
68. Bogdan, O.; Ţîştea, D. Clima României. In *Geografia României*; Badea, L., Gâştescu, P., Velcea, V., Eds.; Romanian Academy Publishing House: Bucharest, Romania, 1983; Volume 1, pp. 195–292.
69. Food and Agriculture Organization (FAO). *World Reference Base for Soils Resources*; World Soil Research Report No. 84; FAO: Rome, Italy, 1998; Available online: http://www.fao.org/docrep/W8594E/W8594E00.htm (accessed on 2 April 2017).
70. Dincă, L.; Spârchez, G.; Dincă, M. Romanian's forest soil GIS map and database and their ecological implications. *Carpath. J. Earth Environ.* **2014**, *9*, 133–142.
71. Spârchez, G.; Dincă, L.; Marin, G.; Dincă, M.; Enescu, E.R. Variation of eutric cambisols' chemical properties based on altitudinal and geomorphological zoning. *Environ. Eng. Manag. J.* **2017**, *16*, 2911–2918.
72. Dincă, L.; Dincă, M.; Vasile, D.; Spârchez, G.; Holonec, L. Calculating organic carbon stock from forest soils. *Not. Bot. Hort Agrobot.* **2015**, *43*, 568–575. [CrossRef]
73. Silvestru-Grigore, C.V.; Enescu, R.E.; Spârchez, G. Specificul ecologic al staţiunilor plantate cu pin pe terenuri degradate din Subcarpaţii Buzăului. *Rev. Păd* **2015**, *130*, 98–107.
74. Kraft, G. *Beiträge zur Lehre von den Durchforstungen, Schlagstellungen und Lichtungshieben (Contributions to the Theory of Thinnings, Distance between Trees and Natural Regeneration of Stands)*; Klindworth's Verlag: Hannover, Germany, 1884; p. 147.
75. Guay, R.; Gagnon, R.; Morin, H. A new automatic and interactive tree ring measurement system based on a line scan camera. *For. Chron.* **1992**, *68*, 138–141. [CrossRef]
76. WinDENDRO. *WinDENDROTM2006 for Tree-Ring Analysis, Manual of Exploitation*; Régent Instruments Inc.: Québec City, QC, Canada, 2007; p. 133.
77. Pilcher, J.R. Sample preparation, cross-dating and measurement. In *Methods of Dendrochronology*; Cook, E.R., Kairiukstis, L.A., Eds.; Kluwer Academic Publishers: Dordrecht, The Netherlands, 1989; pp. 40–51. ISBN 0-7923-0586-8.
78. Kerhoulas, L.P.; Kane, J.M. Sensitivity of ring growth and carbon allocation to climatic variation vary within ponderosa pine trees. *Tree Physiol.* **2011**, *32*, 14–32. [CrossRef] [PubMed]
79. Cook, E.R.; Peters, K. The smoothing spline: A new approach to standardizing forest interior tree-ring width series for dendroclimatic studies. *Tree-Ring Bull.* **1981**, *41*, 45–53.
80. Fritts, H.C.; Guiot, J.; Gordon, G.A. Verification. In *Methods of Dendrochronology*; Cook, E.R., Kairiukstis, L.A., Eds.; Kluwer Academic Publishers: Dordrecht, The Netherlands, 1989; pp. 178–185. ISBN 0-7923-0586-8.
81. Cropper, J.P. Tree-ring skeleton plotting by computer. *Tree-Ring Bull.* **1979**, *39*, 47–54.

82. Schweingruber, F.H.; Eckstein, D.; Serre-Bachet, F.; Bräker, O.U. Identification, presentation and interpretation of the event years and pointer years in dendrochronology. *Dendrochronologia* **1990**, *8*, 9–38.
83. Tsoumis, G.; Panagiotidis, N. Effect of growth conditions on wood quality characteristics of Black pine (*Pinus nigra* Arn.). *Wood Sci. Technol.* **1980**, *14*, 301–310. [CrossRef]
84. Dytham, C. *Choosing and Using Statistics: A Biologist's Guide*, 3rd ed.; Wiley-Blackwell: Oxford, UK, 2011; pp. 171–172. ISBN 978-1-4051-9838-7.
85. Box, G.E.P.; Jenkins, G.M.; Reinsel, G.R. *Time Series Analysis: Forecasting and Control*; John Wiley & Sons, Inc.: Hoboken, NJ, USA, 2008; pp. 474–489. ISBN 978-0-470-27284-8.
86. Bruchwald, A.; Kliczkowska, A. Kształtowani się bonitacji dla drzewostanów sosnowych Polski (Characteristics of pine stand quality class in Poland). *Prace Instytutu Badawczego Leśnictwa* **1997**, *836/842*, 63–73.
87. Crecente-Campo, F.; Pommerening, A.; Rodríguez-Soalleiro, R. Impacts of thinning on structure, growth and risk of crown fire in a *Pinus sylvestris* L. plantation in northern Spain. *For. Ecol. Manag.* **2009**, *257*, 1945–1954. [CrossRef]
88. Sławski, M. Ilościowa charakterystyka zróżnicowania struktury borów sosnowych różnego wieku (Quantitative characteristic of structure in Scots pine stands of various age). *Sylwan* **2012**, *156*, 349–359.
89. Condés, S.; del Río, M. Climate modifies tree interactions in terms of basal area growth and mortality in monospecific and mixed *Fagus sylvatica* and *Pinus sylvestris* forests. *Eur. J. For. Res.* **2015**, *134*, 1095–1108. [CrossRef]
90. Newton, P.F. Simulating the potential effects of a changing climate on black spruce and jack pine plantation productivity by site quality and locale through model adaptation. *Forests* **2016**, *7*, 223. [CrossRef]
91. Riofrío, J.; Río, M.; del Bravo, F. Mixing effects on growth efficiency in mixed pine forests. *Forestry* **2017**, *90*, 381–392. [CrossRef]
92. Strimbu, V.C.; Bokalo, M.; Comeau, P.G. Deterministic Models of Growth and Mortality for Jack Pine in Boreal Forests of Western Canada. *Forests* **2017**, *8*, 410. [CrossRef]
93. Xenakis, G.; Ray, D.; Mencuccini, M. Effects of climate and site characteristics on Scots pine growth. *Eur. J. For. Res.* **2012**, *131*, 427–439. [CrossRef]
94. Redmond, M.D.; Kelsey, K.C.; Urza, A.K.; Barger, N.N. Interacting effects of climate and landscape physiography on piñon pine growth using an individual-based approach. *Ecosphere* **2017**, *8*, e01681. [CrossRef]
95. Šeho, M.; Kohnle, U.; Albrecht, A.; Lenk, E. Wachstumsanalysen von vier Schwarzkiefer-Provenienzen (*Pinus nigra*) auf trockenen Standorten in Baden-Württemberg (Growth analyses of four provenances of European Black Pine (*Pinus nigra*) growing on dry sites in southwest Germany (Baden-Wuerttemberg)). *Allg. Forst. Jagdztg.* **2010**, *181*, 104–116.
96. Eilmann, B.; Rigling, A. Tree-growth analyses to estimate tree species' drought tolerance. *Tree Physiol.* **2012**, *32*, 178–187. [CrossRef] [PubMed]
97. Martin-Benito, D.; Anchukaitis, K.J.; Evans, M.N.; del Río, M.; Beeckman, H.; Cañellas, I. Effects of Drought on Xylem Anatomy and Water-Use Efficiency of Two Co-Occurring Pine Species. *Forests* **2017**, *8*, 332. [CrossRef]
98. Öner, N.; Eren, F. The comparisons between root collar diameter and height growth of black pine (*Pinus nigra* Arnold.) and Scots pine (*Pinus sylvestris* L.) seedlings in Bolu forest nursery. *J. Appl. Biol. Sci.* **2008**, *2*, 7–12.
99. Haralamb, A. Specii de tranziţie în lucrările de fixarea terenurilor degradate. *Rev. Păd* **1935**, *48*, 319–326.
100. Poljanšek, S.; Levanič, T.; Ballian, D.; Jalkanen, R. Tree growth and needle dynamics of *P. nigra* and *P. sylvestris* and their response to climate and fire disturbances. *Trees* **2015**, *29*, 683–694.
101. Rundel, P.W.; Richardson, D.M. Pines. In *Encyclopedia of Forest Sciences*; Burley, J., Evans, J., Youngquist, J.A., Eds.; Elsevier Ltd.: Oxford, UK, 2004; pp. 1430–1441. ISBN 0-12-145160-7.
102. Lévesque, M.; Rigling, A.; Brang, P. Réponse à la sécheresse de conifères indigènes et exotiques: Une étude dendroécologique (Drought response of native and non-native conifers: A dendroecological study). *Schweiz. Z. Forstwes.* **2015**, *166*, 372–379. [CrossRef]
103. Feichtinger, L.M.; Eilmann, B.; Buchmann, N.; Rigling, A. Growth adjustments of conifers to drought and to century-long irrigation. *For. Ecol. Manag.* **2014**, *334*, 96–105. [CrossRef]

104. Lévesque, M.; Siegwolf, R.; Saurer, M.; Eilmann, B.; Rigling, A. Increased water-use efficiency does not lead to enhanced tree growth under xeric and mesic conditions. *New Phytol.* **2014**, *203*, 94–109. [CrossRef] [PubMed]
105. Leo, M.; Oberhuber, W.; Schuster, R.; Grams, T.E.; Matyssek, R.; Wieser, G. Evaluating the effect of plant water availability on inner alpine coniferous trees based on sap flow measurements. *Eur. J. For. Res.* **2014**, *133*, 691–698. [CrossRef]
106. Yakoto, A.; Takahara, K.; Akashi, K. Water stress. In *Physiology and Molecular Biology of Stress Tolerance in Plants*; Madhava Rao, K.V., Raghavendra, A.S., Reddy, K.J., Eds.; Springer: Berlin/Heidelberg, Germany, 2006; pp. 15–39. ISBN 978-1-4020-4225-6.
107. Wieser, G.; Leo, M.; Oberhuber, W. Transpiration and canopy conductance in an inner alpine Scots pine (*Pinus sylvestris* L.) forest. *Flora* **2014**, *209*, 491–498. [CrossRef] [PubMed]
108. Petrucco, L.; Nardini, A.; Von Arx, G.; Saurer, M.; Cherubini, P.; Tognetti, R. Isotope signals and anatomical features in tree rings suggest a role for hydraulic strategies in diffuse drought-induced die-back of *Pinus nigra*. *Tree Physiol.* **2017**, *37*, 523–535. [PubMed]
109. Șofletea, N.; Curtu, L. *Dendrologie (Dendrology)*; Transilvania University Press: Brașov, Romania, 2007; pp. 68–79.
110. Silvestru-Grigore, C.V.; Spârchez, G.; Dinulică, F. Starea de sănătate a arboretelor de pin instalate pe terenuri degradate din Subcarpații Buzăului (The health condition of pine stands installed on degraded lands in Buzau under Carpahians). *Rev. Păd* **2016**, *131*, 7–18.
111. Pazdrowski, W. Technological value of Scots pine (*Pinus sylvestris* L.) wood depending on the quality of tree stems in final crops (in Polish). *Roczniki Akad. Rol. Pozn.* **1988**, *170*, 72.
112. Kärenlampi, P.P.; Riekkinen, M. Pine heartwood formation as a maturation phenomenon. *J. Wood Sci.* **2002**, *48*, 467–472. [CrossRef]
113. Evans, J. The productivity of second and third rotations of pine in the Usutu Forest. *Swazil. Commonw. For. Rev.* **1986**, *65*, 205–214.
114. Sheppard, K.R. *Plantation Silviculture*; Martinus Nijhoff Publishers: Dordrecht, The Netherlands, 1986; p. 322.
115. Zobel, B.J.; Sprague, J.R. *Juvenile Wood in Forest Trees*; Springer: Berlin, Germany, 1998; p. 300. ISBN 978-3-642-72128-1.
116. Zobel, B.J.; Jett, J.B. *Genetics of Wood Production*; Springer: Berlin/Heidelberg, Germany; New York, NY, USA, 1995; p. 337.
117. Keenan, R.J.; van Dijk, A.I.J.M. Planted forests and water in perspective. *For. Ecol. Manag.* **2007**, *251*, 1–9.
118. Génova, M.; Moya, P. Dendroecological analysis of relict pine forests in the centre of the Iberian Peninsula. *Biodivers. Conserv.* **2012**, *21*, 2949–2965. [CrossRef]
119. Pärn, H. Hariliku männi puistute radiaalkasvu ja kliimategurite vaheliste seoste ajalisest varieeruvusest (Temporal variability in the relationships between the radial growth of Scots pine stands and the climate). *Metsanduslikud Uurim.* **2004**, *40*, 65–79.
120. Barčić, D.; Španjol, Ž.; Rosavec, R. Utjecaj na stanište i razvoj šumskih kultura crnoga bora (*Pinus nigra* J. F. Arnold) na krškom submediteranskom području (Impact on site and development of black pine (*Pinus nigra* J.F. Arnold) forest cultures in the submediterannean karst area). *Croat. J. For. Eng.* **2011**, *32*, 131–139.
121. Li, Y.T.; Lo, Y.H.; Lin, Y.C.; Guan, B.T.; Blanco, J.A.; You, C.H. Bringing the natives back: Identifying and alleviating establishment limitations of native hardwood species in a conifer plantation. *Forests* **2018**, *9*, 4. [CrossRef]
122. Lebreton, P.; Choisy, J.P. Avifaune et altérations forestières. III. Incidences avifaunistiques des aménagements forestières: Substitution *Quercus/Pinus* en milieu subméditerranéen. *Bul. D'écologie* **1991**, *22*, 213–220.
123. Birsan, M.-V.; Dumitrescu, A.; Micu, D.M.; Cheval, S. Changes in annual temperature extremes in the Carpathians since AD 1961. *Nat. Hazards* **2014**, *74*, 1899–1910. [CrossRef]
124. Rey, F.; Berger, F. Management of Austrian black pine on marly lands for sustainable protection against erosion (Southern Alps, France). *New For.* **2006**, *31*, 535–545. [CrossRef]
125. Burger, J.A. Management effects on growth, production and sustainability of managed forest ecosystems: Past trends and future directions. *For. Ecol. Manag.* **2009**, *258*, 2335–2346. [CrossRef]
126. Abrams, M.D.; Orwig, D.A. Structure, radial growth dynamics and recent climatic variations of a 320-year-old *Pinus rigida* rock outcrop community. *Oecologia* **1995**, *101*, 353–360. [CrossRef] [PubMed]

127. Greavu, M.; Untaru, E.; Filat, M. Cercetări privind îngrijirea și conducerea arboretelor instalate pe terenuri degradate. *Ann. For. Res.* **1995**, *43*, 31–38.
128. Pérez-de-Lis, G.; García-González, I.; Rozas, V.; Arévalo, J.R. Effects of thinning intensity on radial growth patterns and temperature sensitivity in *Pinus canariensis* afforestations on Tenerife Islands, Spain. *Ann. For. Sci.* **2011**, *68*, 1093–1104. [CrossRef]
129. Primicia, I.; Artázcoz, R.; Imbert, J.; Puertas, F.; Traver, M.; Castillo, F. Influence of thinning intensity and canopy type on Scots pine stand and growth dynamics in a mixed managed forest. *For. Syst.* **2016**, *25*, e057. [CrossRef]
130. Hereş, A.M.; Martínez-Vilalta, J.; Claramunt López, B. Growth patterns in relation to drought-induced mortality at two Scots pine (*Pinus sylvestris* L.) sites in NE Iberian Peninsula. *Trees* **2012**, *26*, 621–630. [CrossRef]

forests

MDPI

Article

Secondary Invasions Hinder the Recovery of Native Communities after the Removal of Nonnative Pines Along a Precipitation Gradient in Patagonia

Agostina Torres [1], Pablo A. E. Alarcón [2], Mariano A. Rodríguez-Cabal [1,†] and Martín A. Nuñez [1,*,†]

1 Grupo de Ecología de Invasiones, INIBIOMA, Universidad Nacional del Comahue, CONICET, Avenida de los Pioneros 2350, San Carlos de Bariloche, Río Negro, Argentina; torresa@comahue-conicet.gob.ar or torresa@agro.uba.ar (A.T.); rodriguezcabal@gmail.com (M.A.R.C.)

2 Grupo de Investigaciones en Biología de la Conservación, INIBIOMA, Universidad Nacional del Comahue, CONICET, Quintral 1250, San Carlos de Bariloche, Río Negro, Argentina; pabloalarcon@comahue-conicet.gob.ar or paealarcon@gmail.com

* Correspondence: nunezm@gmail.com; Tel.: +54-294-444-2676

† Senior Authors.

Received: 20 May 2018; Accepted: 29 June 2018; Published: 4 July 2018

Abstract: The removal of nonnative species can lead to re-invasion by nonnative species, especially in communities with multiple co-occurring invaders. Biotic and abiotic conditions shape community structure, reducing the predictability of nonnative management. We evaluated plant community recovery after the removal of nonnative pines with an emphasis on the effect of environmental conditions on the nonnative species response. We compared clearcuts (where pine plantations were removed), pine plantations, and native communities along a precipitation gradient in Patagonia. Nonnative richness and cover were higher in clearcuts compared to native communities along nearly the entire precipitation gradient, with the exception of the harshest sites. Compared to native communities, invasion resistance was lower in clearcuts in the wetter sites. Native richness and cover were lower in clearcuts relative to native communities along the gradient. Species composition in clearcuts diverged in similarity from native communities towards the wetter sites. Plantations showed an extremely lower richness and cover compared to both clearcuts and native communities. Our study highlights that clearcutting is an ineffective strategy to manage nonnatives aimed at restoring native communities and elucidates the importance of environmental context in management approaches. Taken together, our findings reinforce the important consideration of both the biotic and abiotic context of nonnative management.

Keywords: Argentina; invasive species; management; environmental gradient; Pinaceae; restoration; silviculture

---

1. Introduction

The management of nonnative species is a current challenge for ecological restoration [1–3], whose major goal is recovering the characteristics of an ecosystem that were prevalent before invasion, such as increasing biodiversity and restoring ecological functions [4]. The outcome of nonnative species management is highly unpredictable and the recovery of community structure and ecosystem functioning are hardly ever achieved or even evaluated [2,5]. An increasingly reported problem is that after the removal of a dominant nonnative species, other nonnative species invade the area, a process called secondary invasions [2,6,7]. Yet, most studies addressing the management of nonnative species focus on the management of a single-invader, without considering their community context [2].

The fact that many ecosystems are invaded by multiple co-occurring nonnative species, and that many factors can modulate nonnative species establishment, both contribute towards the unpredictability of nonnative species management [6]. Thus, a better understanding of the conditions that promote nonnative species invasion after the removal of a dominant nonnative can help predict management outcomes, as well as improve the allocation of management efforts [8].

Under certain conditions, the removal of nonnative species has led to successful outcomes [9,10]. However, it is not clear yet which conditions favored successful or unsuccessful restoration outcomes. Abiotic conditions can shape the variability in relative abundances and overall plant species composition in post-removal areas, which increases the unpredictability of nonnative species management [6]. Particularly, harsh environmental conditions (i.e., sites with extreme resource limitations where the occurrence of abiotic conditions that create rapid plant mortality is common, such as frost, extreme heat, and drought) may act as a strong filter for nonnative species. In fact, it has been found that harsh environments have a lower number of invaders than favorable environments [11–15]. For example, Sorte et al. [16] found that drought favored native over nonnative species. Secondary invasions in harsh sites will likely depend on nonnative species adapted to harsh conditions being present and able to respond rapidly under poor growing conditions [6,17]. Therefore, secondary invasions should take place less often under harsh environmental conditions than in more benign conditions.

Biotic conditions may also shape the community response in different ways to environmental conditions. For instance, the diversity-invasibility hypothesis posits that more diverse communities exhibit greater resistance to invasions than less diverse communities [18,19]. More diverse communities have fewer unexploited resources reducing invasions via resource competition [20]. Further, resource competition may be stronger in more benign environments, consequently promoting biotic resistance [21]. Similarly, interactions among nonnative species can influence secondary invasions [7,22]. Competition among nonnative species can determine that the removal of a dominant nonnative releases other sub-dominant nonnatives from competition, thus favoring secondary invasions [7,22]. Additionally, an indirect positive interaction among nonnatives can drive the accumulation of nonnative species in the community, an interaction mediated by the reduction of native species abundance [22]. This positive interaction promotes secondary invasions after the removal of dominant nonnative species [6]. In this context, there is a need to develop general principles regarding invader interactions across varying environmental conditions so that secondary invasions can be anticipated and managers can allocate efforts toward pre- or post- removal actions [6].

Nonnative Pinaceae species (hereafter pines) have been planted in several regions of the southern hemisphere (e.g., New Zealand, Australia, South Africa, and South America) for forestry purposes and have subsequently invaded native habitats [23]. Both pine plantations and invasions produce a wide spectrum of changes in native ecosystems [23]. For example, pines have changed vegetation structure and fuel loads in Patagonian treeless ecosystems, which increase the intensity and frequency of fires [24]. As a consequence, changes in fire regimes reduce the recovery of nonnative species and promote further nonnative invasions [25,26]. Additionally, below-ground impacts may be more difficult to reverse, giving rise to both biotic and abiotic soil legacy effects (e.g., changes in soil nutrients, soil biota, or soil seed bank) that can drive changes in subsequent plant community structure and ecosystem processes [27,28].

Removal of nonnative pines (both planted and invasive) is a common management strategy around the world aimed at passively recovering native ecosystems, yet little is known about its efficacy [23]. For example, in Patagonia, many pine plantations are harvested for timber but not replanted owing to current bans on planting nonnative tree species. Moreover, many plantations are clearcut with the goal of restoration to native communities, especially in forest ecosystems [29]. However, it is well known that passive restoration to pre-existing states can be a challenge [30,31]. Removing nonnative pines can lead to undesired invasions of other nonnative species, halting the recovery of native ecosystems [30,32], and leaving vast areas with low timber productivity or conservation values. Therefore, assessing the effect of nonnative pine removal on plant community restoration, and understanding conditions that promote nonnative species in areas previously occupied by pines, is critical to properly managing pine plantations after timber harvest and ultimately restoring native communities.

The objective of our study was to evaluate plant community recovery after the removal of nonnative pine plantations and whether the effects of clearcutting varied with environmental conditions. We hypothesize that the previous presence of pines favors the establishment of nonnative over native species due to soil legacy effects (e.g., changes in soil nutrients, depletion of soil seed bank, changes in mycorrhizal communities) [23,30] and disturbance effects (e.g., an increase in resource availability, mainly light) [33]. We also hypothesize that the strength of these effects changes along a precipitation gradient, where they are weaker under harsher environments (i.e., drier areas) than in more benign environments (i.e., wetter areas) [6,14]. Additionally, we expect that steppe native species will better respond to clearcut conditions than forest native species, as light conditions in clearcuts are more similar to those of steppes than forests. Overall, we predict that secondary invasions should be higher and native community recovery lower in clearcuts in more benign sites. Since pine plantations are the prior state of clearcuts, we also evaluated nonnative species invasions and the similarity of plantations in comparison with clearcuts and native communities. These comparisons allow us to control for the effect of initial conditions (plantation understory) on clearcut community structure and to evaluate the impact of this land-use change on native communities, respectively.

2. Materials and Methods

2.1. Study Region

We conducted this study in Northwestern Patagonia, Argentina. This region is characterized by a steep west–east natural precipitation gradient caused by the rain shadow effect of the Andes, which acts as a barrier to the moist air coming from the Pacific Ocean [34]. Rainfall is concentrated between April and September and decreases from ca. 3000 to 500 mm per year over 100 km [35,36]. In this study, mean annual precipitation decreased from 1270 mm per year in the most mesic sites to 630 mm in the driest sites. Precipitation data for each site was obtained from Fetch Climate Web [37]. Mean annual temperature is 7.9 °C, with maximum temperatures occurring during January and February [35]. Vegetation shifts as mean annual precipitation decreases. Along this gradient, the wettest sites are temperate forests, dominated by *Nothofagus* spp. that are first replaced by *Austrocedrus chilensis* forest and matorral vegetation type in the forest-steppe ecotone, and finally by semi-arid grasslands or shrublands in the dry steppe ecosystem [38].

Three distinct physiognomic units occur from west to east: forests, shrublands, and steppes [34]. In the western area of the region, the Patagonian-Andean forest dominates; a vegetation unit dominated by deciduous, evergreen, and mixed forests [39]. The deciduous forest is mainly dominated by *N. pumilio* and *N. Antarctica*, which are restricted to the wettest and highest elevations of the gradient [39]. *N. Antarctica* also dominates stumpy forests in the driest and eastern part of the gradient [40]. Between 37.8° and 47° S, there are also forests of *N. dombeyi*, *N. obliqua* and *A. chilensis* [39], which are the most represented in the region spanned by our study sites. In the northern portion of this region, *A. araucana* appears as a subdominant species in these forests. The following trees and shrubs are also common: *Lomatia hirsuta*, *Maytenus boaria*, *Schinus patagonicus*, *Azara microphylla*, *Aristotelia chilensis*, *Chusquea culeou*, and *Berberis* sp. [39]. In the extra-Andean portion, shrubs increase and grasses decrease as mean annual precipitation decreases [34]. In this ecotone, we find the grained steppe that enters into the eastern sector of the deciduous forests, shaping a mosaic of both vegetation types. The vegetation cover is relatively high (64%) and it is dominated by *Festuca pallescens* and accompanied by *Rytidosperma pictum*, *Lathyrus magellanicus*, and some shrubs such as *Senecio sericeonites* and *Azorella prolifera* [39]. In the driest portion of the gradient, the typical vegetation is the grained-shrubby steppe where the typical vegetation is dominated by *Pappostipa speciosa*, *Pappostipa humilis*, *Poa ligularis*, and *Poa lanuginosa*, as well as by the shrubs *Adesmia volckmannii* and *Berberis microphylla*. This vegetation type has many variants according to the subdominant species [39].

In this region, many nonnatives have been introduced since European occupation. For example, several tree species from the Pinaceae family are spreading in the southern hemisphere, including Argentina [23]. In this region, *Pinus contorta* and *Pseudotsuga menziesii* are the main invaders of the native communities. Although conifers are naturally represented in these communities by two native trees (*A. chilensis* and *A. araucana*), there is no native species from the Pinaceae family in this region (e.g., all pines are nonnative). Besides pines, there is a high richness of nonnative species in the region [41,42]. Some species are only casual but others are highly invasive, such as *Rosa rubiginosa* and *Cytisus scoparius* [43,44].

2.2. Study Design

To evaluate if the effect of clearcutting on secondary plant invasion and community structure varied with environmental conditions, we surveyed 16 sites (Table S1) along a precipitation gradient in the 2016–2017 growing season. The mean distance between sites was ~20 km. At each site, we selected three land-use types: (1) Clearcut: communities assembled after pine plantation removal. Clearcuts were considered to be different from others when previous pine species were different or when clearcuts had different ages. Clearcut age varied from two to eight years. It is well known that time since clearcutting is an important factor determining native vegetation recovery, and this could produce a bias in our results if there was a correlation between clearcut age and precipitation. We evaluated this and we did not find a correlation between the age of the clearcut and precipitation ($r = -0.013$, p-value = 0.96). Therefore, we did not find evidence of a possible bias in our results regarding clearcut age co-varying with precipitation; (2) Plantation: pine plantations that represented the situation previous to clearcut. All pine plantations surveyed were at a mature stage as our purpose was to represent the ecosystem state previous to clearcuts; (3) Native communities: areas dominated by native vegetation, with low levels of anthropogenic disturbance that represent a reference community. Within each land-use type, we randomly placed three observational plots (4 m^2 each) to assess plant community structure. In each plot, we recorded plant species composition and abundance (i.e., percent aerial cover per species) (Table S2). Species were classified by origin as native or nonnative following Zuloaga et al. [36].

2.3. Data Analyses

We evaluated the interactive effect of land-use type and precipitation on different descriptors of community structure: (1) native and nonnative species richness; (2) native and nonnative species cover; (3) proportion of nonnative species; (4) proportion of nonnative cover; and (5) Shannon diversity index based on species-specific foliar cover.

We tested the interactive effect of land-use type and precipitation by fitting a set of Bayesian hierarchical linear models. We modeled each community structure descriptor separately, and all models included land-use type and precipitation as predictors. To capture the hierarchical structure in the data (where plots were nested into sites), we set the land-use type variable as a categorical plot-level predictor and the precipitation variable as a continuous site-level predictor. While these models varied in their probability distributions, all of them were represented using similar deterministic functions that can be summarized as follows:

Plot-level model:

$$\text{Response variable}_{[j]} = \alpha_{N[j]} + \alpha_{C[j]} * \text{clearcut}_{[i]} + \alpha_{P[j]} * \text{plantation}_{[i]} \quad (1)$$

Site-level model:

$$\alpha_{N[j]} = \beta_{0N} + \beta_{1N} * \text{precipitation}_{[j]} \quad (1.1)$$

$$\alpha_{C[j]} = \beta_{0C} + \beta_{1C} * \text{precipitation}_{[j]} \quad (1.2)$$

$$\alpha_{P[j]} = \beta_{0P} + \beta_{1P} * \text{precipitation}_{[j]} \quad (1.3)$$

In Equation (1), α_N represents the effect of native communities at each j precipitation level, whereas α_C and α_P are the analogous effect (i.e., effect size) of clearcut and plantation communities, respectively, compared to native communities (our reference community). As the model was fitted at two levels, parameters at the plot-level were allowed to vary with precipitation. Thus, β_{1N}, β_{1C}, and β_{1P} are the slopes of the site-level linear regression models (Equations (1.1)–(1.3)) for the native, clearcut, and plantation land-use type, respectively, and represent the rate at which the effect of land-use type changed with precipitation.

To describe variability around the above deterministic pattern, we used different probability distributions depending on the nature of the response variable (i.e., on the values that it can theoretically take). To model the species richness (count data), we assumed that the response variable drew a Poisson distribution [45,46]. The Poisson parameter (λ) was modeled as a linear function of community type and precipitation by means of a log link function (Code S1). To model species cover, we assumed that the response variable drew a Gaussian distribution, and modeled the parameter μ as a linear function of the same predictors (Code S2). To evaluate whether or not the richness and cover changed with species origin, we included this variable in the above models (Code S1 and S2). To model the nonnative richness and cover proportion (varying from 0 to 1), we assumed a Binomial distribution for the response variable [45,46]. We modeled the Binomial parameter ρ as a linear function of the predictors using a logit link function (Code S3 and S4). Finally, to model the Shannon diversity index, we assumed that the response variable drew a Gaussian distribution, where the parameter μ was a linear function of land-use and precipitation (Code S5). The response variables with continuous positive values (i.e., Shannon Index and cover) were modeled using Gaussian distributions as preliminary models employing log-normal distributions failed to converge. All these models were implemented in JAGS via the R package '*jagsUI*'. We ran three chains with 10,000 iterations each discarding the first 5000 as burn-in.

To evaluate shifts in species' composition among land-use types, we performed Non-Metrical Multidimensional Scaling (NMDS). The ordination reduced the dimensionality of the distance matrix, and provided a first step for visualizing community dissimilarities [47]. We carried out a meta-NMDS from the R '*vegan*' package [48] that generated an ordination of the Bray-Curtis distance matrix. Bray-Curtis distances represent how dissimilar two communities are, not only taking into account species composition (the list of species), but also the cover per species [47]. Bray Curtis distances were obtained with the '*vegan*' package from a matrix in which the abundances per species registered at each of the three plots surveyed at each land-use type at each site were averaged. Additionally, we performed a permutational analysis of variance with the adonis function implemented in the '*vegan*' package [48]. This non-parametric test allowed us to evaluate the interactive effect of precipitation and land-use type on the dissimilarity among communities. Finally, to evaluate if more invaded communities such as clearcuts tended to have less species turnover along the precipitation gradient, we estimated the Simpson beta-diversity index (B_{SIM}) for each land-use type along the gradient. If species invading clearcuts were the same along the precipitation gradient, we would expect that nonnative species composition would be less variable than total species composition along the gradient. For each land-use type, we estimated B_{SIM} for total species composition and B_{SIM} for nonnative species composition. Beta diversity indexes were estimated from the R package '*betapart*' [49]. All analyses were conducted in R 3.4.3 (R Core Team, R Foundation for Statistical Computing, Vienna, Austria) [50].

3. Results

At the regional scale, we recorded 130 plant species (85 natives and 45 nonnatives) across all land-use types. We found 92 species (68 native and 24 nonnative species) in native communities, 83 species (46 native and 37 nonnative species) in clearcuts, and 31 species (19 native and 12 nonnative species) in pine plantations. The modeled native richness was lower in clearcut and plantation communities compared to the native communities at all precipitation levels (Figure S1B,C). In contrast, our model showed that the effect of clearcutting and plantation on nonnative richness depended on the precipitation level (Figures 1D–F and S1E,F). Yet the modelling of the proportion of nonnative richness resulted in higher values in clearcuts and plantations than in native communities throughout the precipitation gradient (Figures 1G–I and S1H,I). In comparison with clearcuts, plantations harbored the lowest species richness, regardless of plant species' origin (Figure 1B,C,E,F).

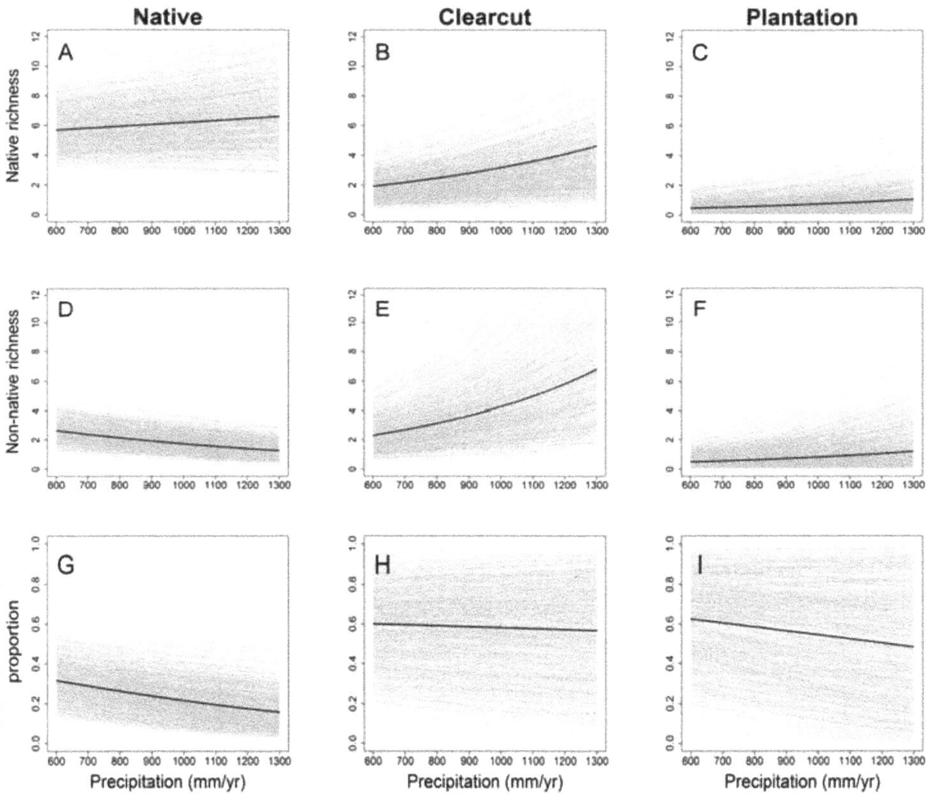

Figure 1. Estimates of native richness (**A–C**), nonnative richness (**D–F**), and proportion of nonnative richness (**G–I**) along the precipitation gradient for each land-use type (native, clearcut, and plantation), resulting from the hierarchical linear model. The black curve represents the mean estimate of the model that regressed richness/proportion with origin (only for richness), land-use type, and precipitation. Gray lines correspond to 1000 simulations from the posterior distributions of the estimated parameters.

According to our models, species richness and proportion of nonnative species depended on the precipitation level in all land-use types, as reflected by the non-zero slopes in Figure 2. While native richness did not vary significantly with precipitation, nonnative richness increased as precipitation increased in native communities (Figures 1A,D and 2A). This resulted in a lower proportion of nonnative richness in more benign (wetter) sites compared to harsher (drier) sites (Figures 1G and 2B), as our model showed. In clearcut communities, both native and nonnative species increased towards more benign sites (Figures 1B,E and 2A). Modeled nonnative richness was, on average, ~3-fold higher in more benign sites (higher precipitation) than in harsher sites (lower precipitation) in this land-use type. Thus, the modeled nonnative richness was higher in clearcuts compared to native communities along nearly the entire precipitation gradient, with the exception of the drier sites (Figures 1D,E and S1E). However, the proportion of nonnative species in clearcuts did not change with precipitation (slope close to zero) although the slope was marginally different compared to native communities (Figures 1H and 2B). Our model indicated that, compared to native communities, plantations had fewer nonnative species in harsher sites but the difference in native and nonnative richness was diluted in more benign sites (Figures 1D,F and S1F).

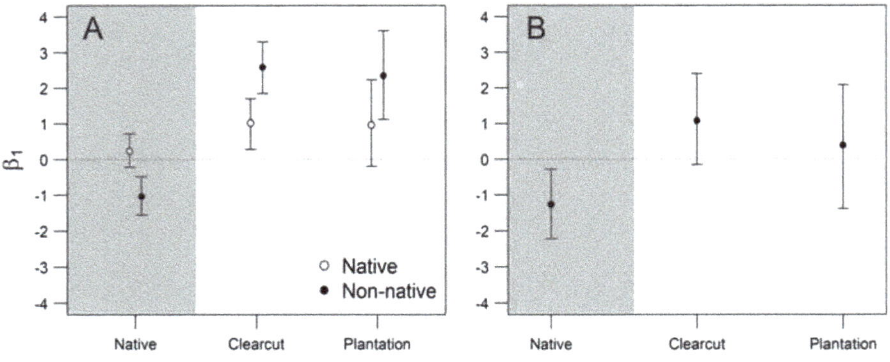

Figure 2. Modeled rate of change of native and nonnative richness (**A**), and proportion of nonnative species (**B**) along the precipitation gradient. Points represent the mean estimated slope of the linear model that regressed richness/proportion with origin (only for richness), land-use type, and precipitation. Vertical lines are 95% confidence intervals (95 CI) of the posterior distribution. Slopes in control communities represent the rate of change of species richness/proportion with precipitation. Responses were considered different between native and nonnative species and among land-use types when 95 CI did not overlap with each other or with zero, respectively. Slopes in clearcut and plantation communities are relative to native communities (i.e., effect size). A positive slope indicates that species richness/proportion increased at a higher rate than in native communities, while a negative slope means that species richness/proportion decreased at a lower rate than native communities.

The modeled native plant cover was lower in clearcuts and plantations in comparison to native communities along the precipitation gradient (Figures 3A–C and S2B,C). Instead, our model showed that nonnative cover increased with precipitation (Figure A1A), following a similar pattern as nonnative richness. In the harsher sites, nonnative cover in clearcuts was similar to native communities, but in clearcuts, it tended to increase towards more benign sites (Figures 3D,E and S2E). Proportion of nonnative cover decreased as sites became wetter in native and clearcut communities (Figures 3G,H and A1B). In plantations, the modeled native and nonnative cover was close to zero along the precipitation gradient, but nonnative cover increased as sites became more benign compared to native communities (Figures 3C,F and S2F).

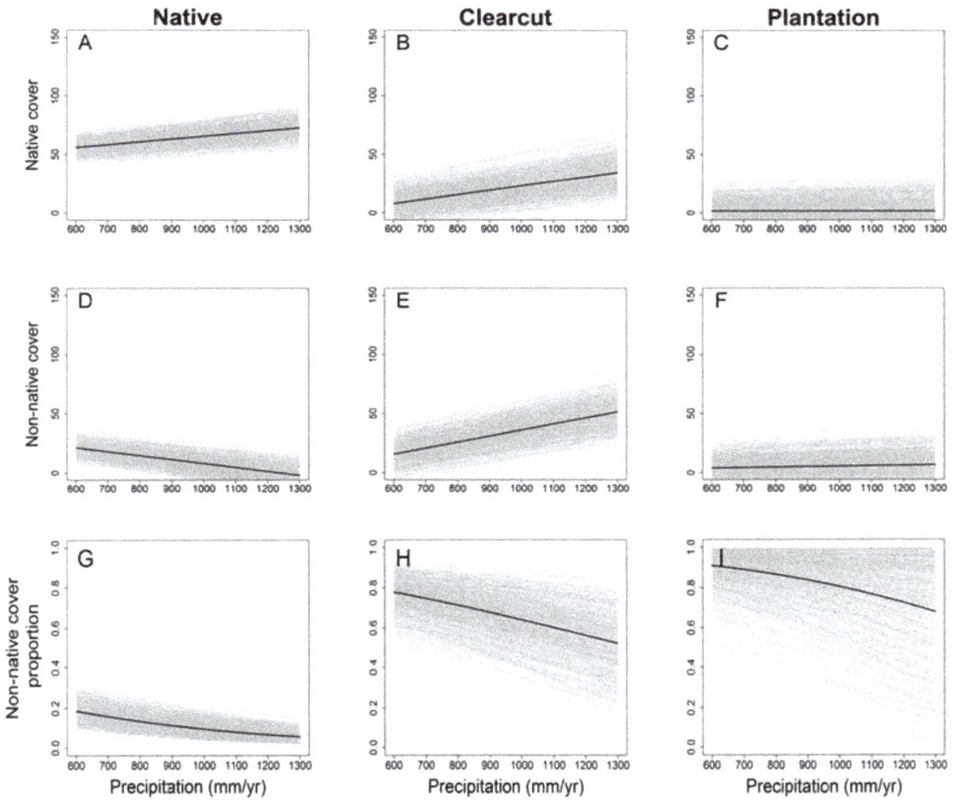

Figure 3. Estimates of native cover (**A–C**), nonnative cover (**D–F**), and proportion of nonnative cover (**G–I**) along a precipitation gradient for each land-use type (native, clearcut, and plantation), resulting from the hierarchical linear model. The black curve represents the mean estimate of the model that regressed cover/proportion with origin (only for cover), land-use type, and precipitation. Gray lines correspond to 1000 simulations from the posterior distributions of the estimated parameters.

Clearcut and native communities were significantly different in terms of species composition and species' relative abundance (F = 3.06, *p*-value = 0.001; Figure 4). The stress value obtained from the NMDS was 0.15, suggesting that the ordination was a good representation of the observed distances in the reduced dimensions. The differences among clearcuts and native communities were high (Bray Curtis distances higher than 0.5 in all cases) along the precipitation gradient. As the adonis test shows, the differences in community structure among land-use types varied with level of precipitation (F = 1.74, *p*-value = 0.001). Pairwise comparisons indicated greater differences between clearcuts and native communities in the more benign sites relative to the harsher sites (F = 1.83, *p*-value = 0.006), as suggested by NMDS (Figure 4). Additionally, the diversity of clearcuts increased with precipitation (Figures 5 and A2); clearcuts were less diverse than native communities in the drier sites and became more diverse in the wetter sites (Figures 5 and S3B). Likewise, the dissimilarity among plantations and native communities was also affected by the precipitation gradient (F = 1.80, *p*-value = 0.01). Finally, species turnover along the precipitation gradient (i.e., Simpson beta-diversity index (B_{SIM})) was high for overall species composition ($B_{SIM\text{-}control}$ = 0.876, $B_{SIM\text{-}clearcut}$ = 0.851, $B_{SIM\text{-}plantation}$ = 0.713) and for nonnative species composition ($B_{SIM\text{-}control}$ = 0.871, $B_{SIM\text{-}clearcut}$ = 0.814, $B_{SIM\text{-}plantation}$ = 0.692). In all land-use types, the identity of overall species composition and nonnative species alike varied along the gradient [40].

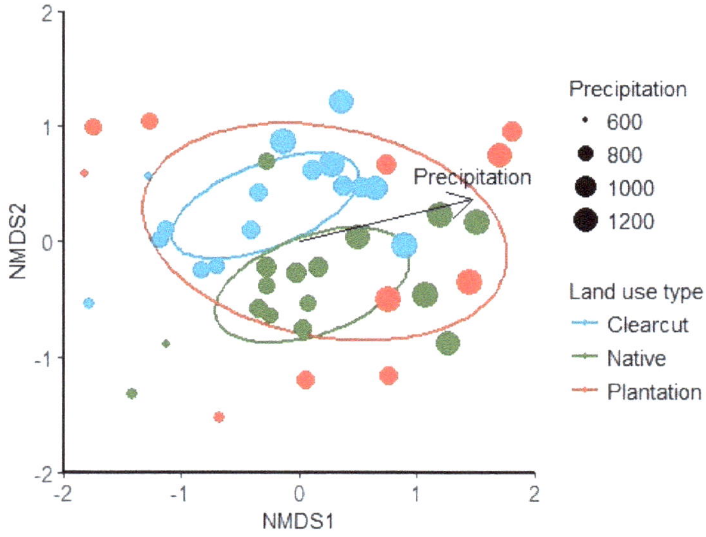

Figure 4. Non-metric multidimensional scaling (NMDS) ordination plot of communities in two-dimensional scales. Each point represents the ordination score of a community, and the distance between any two points represents the difference between those two communities according to Bray Curtis distances. Communities that are closer together are more similar in composition, while communities that are farther apart are less similar. Ellipses represent 95% confidence intervals around the centroid of each land-use type. Colors indicate different land-use types: green for native communities, blue for clearcut communities, and red for plantation communities. Arrows point to the sites with higher precipitation. Point size indicates the mean annual precipitation of each site, where the larger the size of the point, the wetter the site.

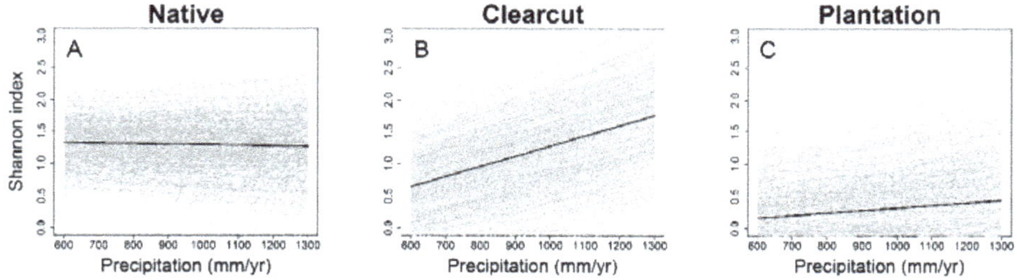

Figure 5. Estimates of the Shannon diversity index along the precipitation gradient for each land-use type: native (**A**), clearcut (**B**), and plantation (**C**), resulting from the hierarchical linear model. The black curve represents the mean estimates of the model that regressed Shannon with land-use type and precipitation. Gray lines correspond to 1000 simulations from the posterior distributions of the estimated parameters.

Changes among land uses and along the gradient on the community descriptors were accompanied by changes in species dominance. Overall, clearcuts had greater relative cover of annual and perennial herbs and lower relative cover of shrubs and trees than native communities (Table A1). In clearcuts, the native tree *Aristotelia chilensis* and the non-natives *Holcus lanatus* (annual herb), *Pseudotsuga menziesii* (tree), *Cirsium vulgare* (annual/biannual herb), *Pinus ponderosa*, and *Rumex acetosella* had the highest cover values (Table S2). In contrast, in native communities, the most abundant species were the native perennial herb *Chusquea culeou*, the native trees *Nothofagus antarctica*, *Schinus patagonicus*, and *Maytenus boaria*, and the native shrubs *Berberis microphylla* and *Colletia hystrix* (Table S2). Plantations showed lower total cover of both native and nonnative species but higher relative cover of trees than clearcuts and native communities (Table A1). The most abundant species in plantations were the non-native tree *Pseudotsuga menziesii*, and the rest of the species were notably less abundant than in the other community types (Table S2). Furthermore, we found a high level of turnover where few nonnative species occurred in more than half the sites (Table S2).

4. Discussion

Our results support the hypothesis that clearcut communities are more invaded by nonnative species than native communities. Proportion of nonnative richness and cover were higher in clearcut communities along the entire precipitation gradient. These results are similar to other cases previously reported where the removal of mature nonnative pines led to secondary invasions [10,30,32,51], but see [52]. However, Pauchard and Alaback [53] did not find high levels of invasions after pine clearcutting in the native range of the pine species. Disturbance generated by the removal of nonnative pines may increase resource availability and favor nonnative species, as suggested by the fluctuating resource hypotheses [54]. This occurs either by reducing resource uptake [33,54,55] or by increasing resource supply through residual biomass of the harvested trees [30,56]. However, these effects are more likely to occur immediately after clearcutting [56], which suggests that long-lasting legacy effects of pines may influence secondary invasions. In fact, pines produce below-ground impacts that can indirectly affect post-removal above-ground communities [57]. For instance, pines can reduce soil nutrient pools [58], decomposition rates [59], and soil pH [60,61]. Pines can also affect soil biota and native mutualisms [31,62,63]. Thus, ecological legacies of pines can indirectly promote the performance of nonnatives while hindering native species. For example, in New Zealand, the nonnative *Pinus contorta* altered biogeochemical cycles and increased ectomycorrhizal inoculum, which consequently generated a no-analog assemblage of species dominated by nonnative grasses and herbs after pine removal [30].

Native and nonnative species richness in clearcuts was higher at sites with the highest amount of precipitation. This suggests that native-rich communities tended to have more nonnative species than native-poor communities in clearcuts [12,13]. One possible explanation is that in drier sites, there would be fewer nonnative species adapted to the harsher conditions to be able to rapidly establish [6,14]. This would likely be due to introduction biases that altered nonnative species pools; nonnative species adapted to harsher conditions may be underrepresented in the nonnative species pool compared to nonnative species adapted to high-resource levels [64]. Instead, in the wettest sites, more benign conditions would not filter out the stress-tolerant species. Therefore, clearcut communities would have higher native and nonnative species richness and abundance [12]. The opposite occurred in the native communities where nonnative species richness decreased in wetter sites. This led us to hypothesize that under benign conditions, biotic resistance in native communities is higher than in clearcuts. In undisturbed conditions, an increase in biotic resistance in native communities may explain the lower nonnative cover and richness. Biotic resistance in native communities may be mediated by an increase in native cover. Previous studies have discussed the role of native species enhancing resistance to invasion owing to negative interactions among native and nonnative species [18–20]. Overall, young clearcuts showed greater nonnative invasions (had lower resistance to invasion) in more benign sites compared to native communities.

Regardless of secondary invasion, the success of passive restoration depends on the capacity of native species to (1) survive underneath pines and grow after removal or (2) recolonize the site from the soil seed bank or seed rain [9]. Here, we found that plantations had a negative effect on native richness and cover along the gradient, as has been previously reported in Patagonia [65,66]. This suggests that plantations are not harboring native species in either the harsh or in the more benign environments and that it is unlikely that they determine native community composition in clearcuts. However, under more benign environmental conditions than evaluated in our study, plantations can provide habitats for native species [67,68] and may accelerate passive restoration. For example, in more benign environments (1855 mm of mean annual precipitation) in New Zealand, Brockerhoff et al. [69] found a similar understory cover of native species in plantations and in the native forests. Seed banks allow new species that do not occur in the understory vegetation to occur after the removal of pines. Although our study did not directly address the role of seed banks influencing species composition, evidence suggests that soil seed banks are mainly dominated by nonnative species in pine-invaded communities around the world [10,70] and in other disturbed communities in Patagonia [71]. However, the importance of seed banks determining native vegetation dynamics in Patagonian communities is low [72,73] and variable along the precipitation gradient [73].

As we expected, young clearcut communities converge with native communities in the harshest sites. We hypothesized that the current high-light environment of the clearcuts favors the dominant native shade-intolerant species found in the harsher sites (steppe) and hinders native shade-tolerant species commonly found in the more benign sites (forest). Patterns of diversity suggest that the increase in native and nonnative richness in clearcut communities is driven by an increase in species abundance. Thus, the differences among clearcuts and native communities along the gradient could likely be explained by the higher proportion of nonnatives found in clearcuts across the gradient. Moreover, clearcuts presented a higher relative cover of annual herbs and lower relative cover of shrubs and trees in comparison with native communities, especially in the more benign sites. As native communities are mainly dominated by shrubs and trees, native species in clearcuts are notably different from those from native communities, especially in the wettest sites. We also found high species turnover along the gradient. Nonnative species in the harsher sites may not be a subgroup of those established in the more benign sites. Overall, one possibility is that clearcuts would need more time for passive restoration to succeed, particularly if the stage dominated by annual herbs is transient or facilitates the establishment of longer-live species typical of native communities. A more pessimistic scenario is that clearcut trajectories diverge from native communities, leading to alternative states, which can occur when plantations are burned [26]. Additionally, differences among plantation and native communities can be due to the extremely low diversity and understory cover found in plantations.

If the goal is to restore native communities following invasion, clearcut practices may not be an ideal technique to manage pines in Patagonia. Alternative practices, such as selective logging or techniques that leave dead pines standing (e.g., through girdling or herbicide application), may alleviate abiotic conditions (e.g., light, moisture) and promote native species compared to nonnative species if native propagules are not limiting [51,53,69]. Moreover, the management of current plantations (e.g., opening canopy) may increase understory biodiversity and accelerate clearcut restoration [65]. Additionally, management timing can influence restoration outcomes [23,66]. It has been found that the removal of pines allowed the regeneration of native communities in early stages of invasion. However, larger legacy effects appeared in later stages of invasion, hindering passive restoration [9,10,30]. In a more pessimistic scenario, return to the original native community may require additional interventions such as the re-introduction of locally extinct native species and their mutualists, or the modification of habitat conditions to make them more suitable for native species establishment [31]. It is important to note that secondary invasions may also generate economic problems by hindering the growth of desired species. For example, in its native range, *Pseudotsuga menziesii* (a species of interest for forestry worldwide) was negatively affected by the previous presence of the invasive nitrogen fixer *Cytisus scoparius* [74], a common nonnative species in clearcut communities in our study area.

Our work provides empirical evidence that furthers our understanding of the response of native and nonnative species composition to management under different environmental conditions. Thus, it may contribute to improve management approaches towards nonnative species and help the development of a theoretical framework for biological invasions [2]. Our results highlight the need to consider environmental context in the management of nonnative species. In harsh environments, a lower number of nonnative species respond positively to the removal of a primary invader compared to more benign environments, which also may have lower population and individual growth rates, as suggested by lower nonnative cover. Therefore, managers may have more time to manage or control the secondary invaders in harsher sites in comparison to more benign ones [2,6]. However, to achieve conclusive results, further studies should increase the sampling effort and time elapsed since nonnative removal, as well as conduct experimental studies that address the mechanisms underlying secondary invasions' patterns. Beyond that, based on our findings, it is not likely that clearcut communities that are already invaded by multiple nonnative species will recover and resemble native communities through passive restoration.

5. Conclusions

Removal of nonnative pines drives the secondary invasion of multiple nonnative species, altering plant species composition relative to those of native community assemblages. However, nonnative species richness and cover were higher in more benign (wetter) sites and clearcut communities were more similar to native communities in the harshest (drier) sites. The results of our work highlight the inefficacy of clearcutting to manage nonnative pines and restore native communities, especially in the wettest sites. Our conclusions also draw attention to the importance of the environmental context of management and reinforce recent arguments [2,6] that account for the biotic and abiotic context of nonnative species management as crucial.

Supplementary Materials: The following are available online at http://www.mdpi.com/1999-4907/9//394/s1, Table S1: Site locations (coordinates) with mean annual precipitation (mm) and time since clearcutting (yr) at each site, Table S2: Species cover (%) in 4 m^2 at each land-use type at each site; Figure S1: Effect size of each land-use type at each precipitation level for native richness, nonnative richness and proportion of nonnative species, Figure S2: Effect size of each land-use type at each precipitation level for native cover, nonnative cover and proportion of nonnative cover, Figure S3: Effect size of each land-use type at each precipitation level for Shannon diversity index, Code S1: Hierarchical linear regression model for richness, Code S2: Hierarchical linear regression model for cover, Code S3: Hierarchical linear regression model for proportion of nonnative richness, Code S4: Hierarchical linear regression model for proportion of nonnative cover, Code S5: Hierarchical linear regression model for Shannon diversity index.

Author Contributions: A.T., M.A.-N. and M.A.R.-C. conceived and designed the study. A.T. collected the data and led the writing of the manuscript. A.T. and P.A.E.A. designed and carried out data analysis. All authors participated in data interpretation and revised the manuscript.

Funding: This research was funded by FONCyT grant number PICT 2014-0662 and PICT 2016-1412.

Acknowledgments: We are grateful to Cecilia Ezcurra and Laura Ventura for assistance in species determination, and to Ariel Mayoral and Pablo Alvear for field assistance. We thank the land owners who facilitated our access to the study sites. We also thank Teresa Moran Lopez and Florencia Tiribelli for helpful ideas for data analysis and Mariana C. Chiuffo for insightful discussions. We are grateful to Lara Souza and Sara Kuebbing for revising the manuscript. This manuscript greatly benefited from the thorough revision of four anonymous reviewers.

Conflicts of Interest: The authors declare no conflict of interest.

Appendix A

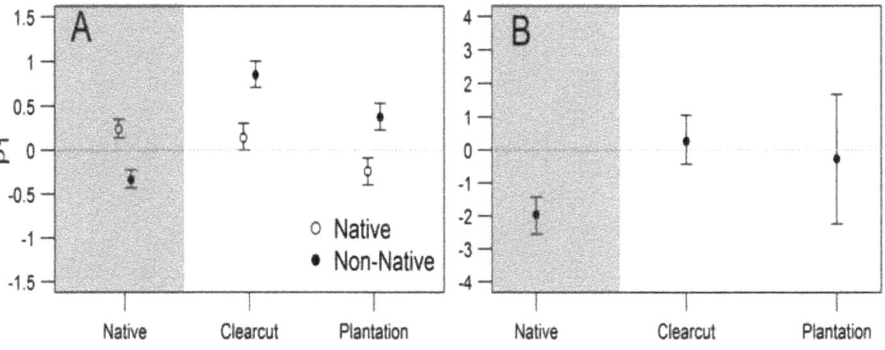

Figure A1. Modeled rate of change of native and nonnative cover (**A**), and proportion of nonnative cover (**B**) along the precipitation gradient. Points represent the mean estimated slope of the hierarchical linear model that regressed cover/proportion with origin (only for cover), land-use type, and precipitation. Vertical lines are 95% confidence intervals (95 CI) of the posterior distribution. Responses were considered different between native and nonnative species and among land-use types when 95 CI did not overlap with each other or with zero, respectively. Slopes in clearcut and plantation communities are relative to native communities (i.e., effect size). A positive slope means that cover/proportion increased at a higher rate than in native communities, while a negative slope means that cover/proportion decreased at a lower rate than native communities.

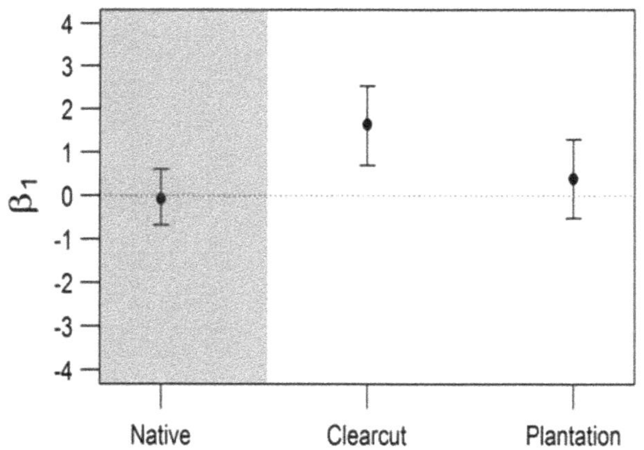

Figure A2. Estimated rate of change of the Shannon diversity index along the precipitation gradient. Points represent the mean estimate of the slope of the hierarchical linear model that regressed Shannon with land-use type and precipitation. Vertical lines are credible intervals of 95% (95 CI) of the posterior distribution. Responses were considered to be different when 95 CI did not overlap with zero. For clearcut and plantation communities, the slopes represent the changes in the effect size of the treatment (i.e., relative to native communities) with precipitation. A positive slope means that Shannon diversity increased at a higher rate than in native communities, while a negative slope means that Shannon diversity decreased at a lower rate than native communities.

Table A1. Mean total cover (%) and mean cover (%) according to life forms (perennial herbs, annual herbs, shrubs, and trees) for each land-use type at each site.

Sites	Land-Use Type	Precipitation	Total Cover	Perennial Herbs	Annual Herbs	Shrubs	Trees
AA 2	Native	619	79.1	25.9	0.0	53.2	0.0
AA 2	Plantation	619	1.3	0.3	0.0	0.0	1.0
AA 2	Clearcut	619	15.5	0.9	0.2	1.7	12.8
AA 1	Native	648	70.4	0.0	0.0	10.8	59.7
AA 1	Plantation	648	3.2	1.2	0.0	2.0	0.0
AA 1	Clearcut	648	0.5	0.5	0.0	0.0	0.0
MC	Native	805	117.4	30.5	15.6	25.5	45.8
MC	Plantation	805	0.3	0.0	0.0	0.0	0.3
MC	Clearcut	805	31.8	20.8	9.1	1.7	0.3
SJ 2	Native	824	65.3	10.4	3.6	28.5	22.8
SJ 2	Plantation	824	0.0	0.0	0.0	0.0	0.0
SJ 2	Clearcut	824	59.4	28.8	0.3	0.0	30.4
SJ 3	Native	827	33.5	5.5	3.1	20.7	4.3
SJ 3	Plantation	827	1.0	0.0	0.0	0.0	1.0
SJ 3	Clearcut	827	52.8	42.5	8.7	0.0	1.7
SJ 1	Native	846	108.3	24.2	37.4	31.3	15.4
SJ 1	Plantation	846	2.0	0.0	0.0	0.7	1.3
SJ 1	Clearcut	846	27.6	16.1	1.1	0.0	10.4
CON 1	Native	863	67.3	10.8	10.0	5.8	40.7
CON 1	Plantation	863	3.0	0.0	0.0	0.3	2.7
CON 1	Clearcut	863	62.3	10.3	48.1	0.5	3.5
CON 2	Native	876	66.5	10.1	1.1	50.4	4.8
CON 2	Plantation	876	2.2	1.4	0.0	0.0	0.9
CON 2	Clearcut	876	52.6	19.0	16.8	2.1	14.8
AMU 1	Native	973	106.3	27.8	0.3	24.3	53.8
AMU 1	Plantation	973	45.7	2.4	0.9	0.1	42.3
AMU 1	Clearcut	973	20.7	2.5	17.8	0.0	0.4
AMU 3	Native	983	96.3	11.3	3.6	18.0	63.4
AMU 3	Plantation	983	0.0	0.0	0.0	0.0	0.0
AMU 3	Clearcut	983	108.1	8.9	41.3	1.7	56.2
AMU 2	Native	995	97.6	7.6	0.6	25.5	63.9
AMU 2	Plantation	995	14.9	0.0	0.0	0.0	14.9
AMU 2	Clearcut	995	94.8	0.9	44.1	3.1	46.7
PM	Native	1181	20.6	0.0	0.0	6.0	14.4
PM	Plantation	1181	0.3	0.0	0.0	0.0	0.3
PM	Clearcut	1181	68.2	32.3	17.7	18.2	0.0
VM	Native	1216	38.2	23.3	0.3	8.3	6.4
VM	Plantation	1216	9.0	0.1	0.0	0.0	9.0
VM	Clearcut	1216	65.8	15.5	9.6	3.6	37.2
QQ3	Native	1262	111.3	76.8	0.0	0.0	34.5
QQ3	Plantation	1262	3.7	3.4	0.0	0.2	0.1
QQ3	Clearcut	1262	88.3	17.5	1.1	2.1	67.6
CF	Native	1265	56.5	1.0	0.0	1.9	53.5
CF	Plantation	1265	21.5	2.6	0.4	11.7	6.8
CF	Clearcut	1265	48.4	17.8	20.4	9.0	1.2
QQ2	Native	1274	59.5	54.6	0.0	0.0	4.9
QQ2	Plantation	1274	0.0	0.0	0.0	0.0	0.0
QQ2	Clearcut	1274	99.0	47.3	47.3	3.8	0.6

References

1. Kettenring, K.M.; Adams, C.R. Lessons learned from invasive plant control experiments: A systematic review and meta-analysis. *J. Appl. Ecol.* **2011**, *48*, 970–979. [CrossRef]
2. Pearson, D.E.; Ortega, Y.K.; Runyon, J.B.; Butler, J.L. Secondary invasion: The bane of weed management. *Biol. Conserv.* **2016**, *197*, 8–17. [CrossRef]

3. Zavaleta, E.S.; Hobbs, R.J.; Mooney, H.A. Viewing invasive species removal in a whole-ecosystem context. *Trends Ecol. Evol.* **2001**, *16*, 454–459. [CrossRef]
4. Benayas, J.M.R.; Newton, A.C.; Diaz, A.; Bullock, J.M. Enhancement of biodiversity and ecosystem services by ecological restoration: A meta-analysis. *Science* **2009**, *325*, 1121–1124. [CrossRef] [PubMed]
5. Prior, K.M.; Adams, D.C.; Klepzig, K.D.; Hulcr, J. When does invasive species removal lead to ecological recovery? Implications for management success. *Biol. Invasions* **2018**, *20*, 267–283. [CrossRef]
6. D'Antonio, C.M.; Ostertag, R.; Cordell, S.; Yelenik, S. Interactions Among Invasive Plants: Lessons from Hawai 'i. *Annu. Rev. Ecol. Evol. Syst.* **2017**, *48*, 521–541. [CrossRef]
7. Kuebbing, S.E.; Nuñez, M.A. Negative, neutral, and positive interactions among nonnative plants: Patterns, processes, and management implications. *Glob. Chang. Biol.* **2015**, *21*, 926–934. [CrossRef] [PubMed]
8. Lodge, D.M.; Williams, S.; MacIsaac, H.J.; Hayes, K.R.; Leung, B.; Reichard, S.; Mack, R.N.; Moyle, P.B.; Smith, M.; Andow, D.A.; et al. Biological invasions: Recommendations for US policy and management. *Ecol. Appl.* **2006**, *16*, 2035–2054. [CrossRef]
9. Cuevas, Y.A.; Zalba, S.M. Recovery of native grasslands after removing invasive pines. *Restor. Ecol.* **2010**, *18*, 711–719. [CrossRef]
10. Galloway, A.D.; Holmes, P.; Gaertner, M.; Esler, K. The impact of pine plantations on fynbos above-ground vegetation and soil seed bank composition. *S. Afr. J. Bot.* **2017**, *113*, 300–307. [CrossRef]
11. Funk, J.L. The physiology of invasive plants in low-resource environments. *Conserv. Physiol.* **2013**, *1*. [CrossRef] [PubMed]
12. Stohlgren, T.J.; Barnett, D.T.; Kartesz, J.T. The rich get richer: Patterns of plant invasions in the United States. *Front. Ecol. Environ.* **2003**, *1*, 11–14. [CrossRef]
13. Stohlgren, T.J.; Barnett, D.T.; Jarnevich, C.S.; Flather, C.; Kartesz, J. The myth of plant species saturation. *Ecol. Lett.* **2008**, *11*, 313–322. [CrossRef] [PubMed]
14. Zefferman, E.; Stevens, J.T.; Charles, G.K.; Dunbar-Irwin, M.; Emam, T.; Fick, S.; Morales, L.V.; Wolf, K.M.; Young, D.J.; Young, T.P. Plant communities in harsh sites are less invaded: A summary of observations and proposed explanations. *AoB Plants* **2015**, *7*. [CrossRef] [PubMed]
15. D'Antonio, C.M.; Levine, J.; Thomsen, M. Ecosystem resistance to invasion and the role of propagule supply: A California perspective. *J. Mediterr. Ecol.* **2001**, *2*, 233–246.
16. Sorte, C.J.; Ibáñez, I.; Blumenthal, D.M.; Molinari, N.A.; Miller, L.P.; Grosholz, E.D.; Diez, J.M.; D'Antonio, C.M.; Olden, J.D.; Jones, S.J.; et al. Poised to prosper? A cross-system comparison of climate change effects on native and non-native species performance. *Ecol. Lett.* **2013**, *16*, 261–270. [CrossRef] [PubMed]
17. Harms, R.S.; Hiebert, R.D. Vegetation response following invasive tamarisk (*Tamarix* spp.) removal and implications for riparian restoration. *Restor. Ecol.* **2006**, *14*, 461–472. [CrossRef]
18. Elton, C.S. *The Ecology of Invasions by Animals and Plants*; University of Chicago Press: London, UK, 1958; ISBN 0-226-20638-6.
19. Levine, J.M. Species diversity and biological invasions: relating local process to community pattern. *Science* **2000**, *288*, 852–854. [CrossRef] [PubMed]
20. Levine, J.M.; Adler, P.B.; Yelenik, S.G. A meta-analysis of biotic resistance to exotic plant invasions. *Ecol. Lett.* **2004**, *7*, 975–989. [CrossRef]
21. Bertness, M.D.; Callaway, R. Positive interactions in communities. *Trends Ecol. Evol.* **1994**, *9*, 191–193. [CrossRef]
22. Kuebbing, S.E.; Nuñez, M.A. Invasive non-native plants have a greater effect on neighbouring natives than other non-natives. *Nat. Plants* **2016**, *2*, 16134. [CrossRef] [PubMed]
23. Nuñez, M.A.; Chiuffo, M.C.; Torres, A.; Paul, T.; Dimarco, R.D.; Raal, P.; Policelli, N.; Moyano, J.; García, R.A.; van Wilgen, B.W.; et al. Ecology and management of invasive Pinaceae around the world: progress and challenges. *Biol. Invasions* **2017**, *19*, 3099–3120. [CrossRef]
24. Paritsis, J.; Landesmann, J.B.; Kitzberger, T.; Tiribelli, F.; Sasal, Y.; Quintero, C.; Dimarco, R.D.; Barrios-García, M.N.; Iglesias, A.L.; Diez, J.P.; et al. Pine Plantations and Invasion Alter Fuel Structure and Potential Fire Behavior in a Patagonian Forest-Steppe Ecotone. *Forests* **2018**, *9*, 117. [CrossRef]
25. Franzese, J.; Raffaele, E. Fire as a driver of pine invasions in the Southern Hemisphere: A review. *Biol. Invasions* **2017**, *19*, 2237–2246. [CrossRef]
26. Nuñez, M.A.; Raffaele, E. Afforestation causes changes in post-fire regeneration in native shrubland communities of northwestern Patagonia, Argentina. *J. Veg. Sci.* **2007**, *18*, 827–834. [CrossRef]

27. Cuddington, K. Legacy effects: the persistent impact of ecological interactions. *Biol. Theory* **2011**, *6*, 203–210. [CrossRef]
28. Suding, K.N.; Stanley Harpole, W.; Fukami, T.; Kulmatiski, A.; MacDougall, A.S.; Stein, C.; Putten, W.H. Consequences of plant-soil feedbacks in invasion. *J. Ecol.* **2013**, *101*, 298–308. [CrossRef]
29. Silvicultural Management Project for the Gradual Eradicaion of Exotic Forest Species of Isla Victoria. In *Agreement between the National Park Administration and the Province of Río Negro*; NPA (National Park Administration): Buenos Aires, Argentina, 1988.
30. Dickie, I.A.; St John, M.G.; Yeates, G.W.; Morse, C.W.; Bonner, K.I.; Orwin, K.; Peltzer, D.A. Belowground legacies of Pinus contorta invasion and removal result in multiple mechanisms of invasional meltdown. *AoB Plants* **2014**, *6*. [CrossRef] [PubMed]
31. Wardle, D.A.; Peltzer, D.A. Impacts of invasive biota in forest ecosystems in an aboveground-belowground context. *Biol. Invasions* **2017**, *19*, 3301–3316. [CrossRef]
32. Sample, M.W. Exotic Conifer Removal in Northern Argentina: Ecological Outcomes and Management Evaluation. Master's Thesis, Northern Arizona University, Flagstaff, AZ, USA, December 2016.
33. Jauni, M.; Gripenberg, S.; Ramula, S. Non-native plant species benefit from disturbance: A meta-analysis. *Oikos* **2015**, *124*, 122–129. [CrossRef]
34. Paruelo, J.M.; Beltran, A.; Jobbagy, E.; Sala, O.E.; Golluscio, R.A. The climate of Patagonia: General patterns and controls on biotic processes. *Ecol. Austral* **1998**, *8*, 85–101.
35. Suarez, L.M.; Kitzberger, T. Differential effects of climate variability on forest dynamics along a precipitation gradient in northern Patagonia. *J. Ecol.* **2010**, *98*, 1023–1034. [CrossRef]
36. Jobbágy, E.G.; Paruelo, J.M.; León, R.J. Estimación del régimen de precipitación a partir de la distancia a la cordillera en el noroeste de la Patagonia. *Ecol. Austral* **1995**, *5*, 47–53.
37. Grechka, D.A.; Berezin, S.B.; Emmott, S.; Lyutsarev, V.; Smith, M.J.; Purves, D.W. Universal, easy access to geotemporal information: FetchClimate. *Ecography* **2016**, *39*, 904–911. [CrossRef]
38. Cabrera, Á.L.; Willink, A. *Biogeografía de América Latina*, 1st ed.; Organization of American States: Washington, DC, USA, 1973; p. 120.
39. Oyarzabal, M.; Clavijo, J.; Oakley, L.; Biganzoli, F.; Tognetti, P.; Barberis, I.; Maturo, H.M.; Aragón, R.; Campanello, P.I.: Prado, D.; et al. Unidades de vegetación de la Argentina. *Ecol. Austral* **2018**, *28*, 40–63. [CrossRef]
40. Oliva, G.; González, L.; Rial, P. El ambiente en la Patagonia Austral. In *Ganadería Ovina Sustenable en la Patagonia Austral. Tecnología de Manejo Extensivo*; Borelli, P., Oliva, G., Eds.; INTA: Santa Cruz, Argentina, 2001; Chapter 2; pp. 17–80.
41. Ezcurra, C.; Brion, C. *Plants of the Nahuel Huapi: Catalogue of the Vascular Flora of the Nahual Huapi National Park*; Universidad Nacional del Comahue: San Carlos de Bariloche, Argentina, 2005; p. 70.
42. Zuloaga, F.O.; Morrone, O.; Belgrano, M.J. *Catálogo de las Plantas Vasculares del Cono Sur (Argentina, sur de Brasil, Chile, Paraguay y Uruguay)*; Missouri Botanical Garden: St Louis, MI, USA, 2008; ISBN 978-19-307-2370-2.
43. Damascos, M.A.; Gallopin, G.G. Ecología de un arbusto introducido (*Rosa rubiginosa* L.= *Rosa eglanteria* L.): riesgo de invasión y efectos en las comunidades vegetales de la región andino-patagónica de Argentina. *Rev. Chil. Hist. Nat.* **1992**, *65*, 395–407.
44. Simberloff, D.; Relva, M.A.; Nuñez, M. Introduced species and management of a Nothofagus/Austrocedrus forest. *Environ. Manag.* **2003**, *31*, 0263–0275. [CrossRef] [PubMed]
45. Gelman, A.; Hill, J. *Data Analysis Using Regression and Multilevel/Hierarchical Models*; Cambridge University Press: New York, NY, USA, 2006; ISBN 1-139-46093-5.
46. Crawley, M.J. *The R Book*, 2nd ed.; John Wiley & Sons: Chichester, UK, 2013; ISBN 1-118-44896-0.
47. Legendre, P.; Legendre, L. *Numerical Ecology*; Elsevier: Amsterdam, The Netherlands, 2012; ISBN 978-0-444-53869-7.
48. Oksanen, J.; Blanchet, F.G.; Kindt, R.; Legendre, P.; O'hara, R.; Simpson, G.L.; Solymos, P.; Stevens, M.H.H.; Wagner, H. vegan: Community Ecology Package. R Package Version 2.4-6. 2017. Available online: https://CRAN.R-project.org/package=vegan (accessed on 10 October 2017).
49. Baselga, A.; Orme, C.D.L. betapart: An R package for the study of beta diversity. *Methods Ecol. Evol.* **2012**, *3*, 808–812. [CrossRef]
50. Team, R.C. *R: A Language and Environment for Statistical Computing, Version 3.4.3*; Foundation for Statistical Computing: Vienna, Austria, 2017; Available online: http://www.R-project.org (accessed on 10 October 2017).

51. McAlpine, K.G.; Howell, C.J.; Wotton, D.M. Effects of tree control method, seed addition, and introduced mammal exclusion on seedling establishment in an invasive Pinus contorta forest. *N. Z. J. Ecol.* **2016**, *40*, 302–309. [CrossRef]
52. Alday, J.G.; Etxeberria, E.; Ametzaga, I. Conversion of Pinus radiata plantations to native forest after harvest operations: A north Iberian Peninsula case study. *Eur. J. For. Res.* **2017**, *136*, 801–810. [CrossRef]
53. Pauchard, A.; Alaback, P.B. Edge type defines alien plant species invasions along Pinus contorta burned, highway and clearcut forest edges. *For. Ecol. Manag.* **2006**, *223*, 327–335. [CrossRef]
54. Davis, M.A.; Grime, J.P.; Thompson, K. Fluctuating resources in plant communities: A general theory of invasibility. *J. Ecol.* **2000**, *88*, 528–534. [CrossRef]
55. Pearson, D.E.; Ortega, Y.K.; Villarreal, D.; Lekberg, Y.; Cock, M.C.; Eren, Ö.; Hierro, J. The fluctuating resource hypothesis explains invasibility, but not exotic advantage following disturbance. *Ecology* **2018**, *99*, 1296–1305. [CrossRef] [PubMed]
56. Grove, S.; Parker, I.M.; Haubensak, K.A. Persistence of a soil legacy following removal of a nitrogen-fixing invader. *Biol. Invasions* **2015**, *17*, 2621–2631. [CrossRef]
57. Corbin, J.D.; D'Antonio, C.M. Gone but not forgotten? Invasive plants' legacies on community and ecosystem properties. *Invasive Plant Sci. Manag.* **2012**, *5*, 117–124. [CrossRef]
58. Hess, L.J.; Austin, A.T. Pinus ponderosa alters nitrogen dynamics and diminishes the climate footprint in natural ecosystems of Patagonia. *J. Ecol.* **2014**, *102*, 610–621. [CrossRef]
59. Araujo, P.I.; Austin, A.T. A shady business: Pine afforestation alters the primary controls on litter decomposition along a precipitation gradient in Patagonia, Argentina. *J. Ecol.* **2015**, *103*, 1408–1420. [CrossRef]
60. Dehlin, H.; Peltzer, D.A.; Allison, V.J.; Yeates, G.W.; Nilsson, M.C.; Wardle, D.A. Tree seedling performance and below-ground properties in stands of invasive and native tree species. *N. Z. J. Ecol.* **2008**, *32*, 67–79.
61. Reich, P.B.; Oleksyn, J.; Modrzynski, J.; Mrozinski, P.; Hobbie, S.E.; Eissenstat, D.M.; Chorover, J.; Chadwick, O.A.; Hale, C.M.; Tjoelker, M.G. Linking litter calcium, earthworms and soil properties: A common garden test with 14 tree species. *Ecol. Lett.* **2005**, *8*, 811–818. [CrossRef]
62. Nuñez, M.A.; Dickie, I.A. Invasive belowground mutualists of woody plants. *Biol. Invasions* **2014**, *16*, 645–661. [CrossRef]
63. Reinhart, K.O.; Callaway, R.M. Soil biota and invasive plants. *New Phytol.* **2006**, *170*, 445–457. [CrossRef] [PubMed]
64. Buckley, Y.M.; Catford, J. Does the biogeographic origin of species matter? Ecological effects of native and non-native species and the use of origin to guide management. *J. Ecol.* **2016**, *104*, 4–17. [CrossRef]
65. Paritsis, J.; Aizen, M.A. Effects of exotic conifer plantations on the biodiversity of understory plants, epigeal beetles and birds in Nothofagus dombeyi forests. *For. Ecol. Manag.* **2008**, *255*, 1575–1583. [CrossRef]
66. Franzese, J.; Urrutia, J.; García, R.A.; Taylor, K.; Pauchard, A. Pine invasion impacts on plant diversity in Patagonia: Invader size and invaded habitat matter. *Biol. Invasions* **2017**, *19*, 1015–1027. [CrossRef]
67. Pawson, S.M.; Ecroyd, C.E.; Seaton, R.; Shaw, W.B.; Brockerhoff, E.G. New Zealand's exotic plantation forests as habitats for threatened indigenous species. *N. Z. J. Ecol.* **2010**, *34*, 342–355.
68. Poch, T.J.; Simonetti, J.A. Insectivory in Pinus radiata plantations with different degree of structural complexity. *For. Ecol. Manag.* **2013**, *304*, 132–136. [CrossRef]
69. Brockerhoff, E.G.; Ecroyd, C.E.; Leckie, A.C.; Kimberley, M.O. Diversity and succession of adventive and indigenous vascular understorey plants in Pinus radiata plantation forests in New Zealand. *For. Ecol. Manag.* **2003**, *185*, 307–326. [CrossRef]
70. Drake, D.R. Relationships among the seed rain, seed bank and vegetation of a Hawaiian forest. *J. Veg. Sci.* **1998**, *9*, 103–112. [CrossRef]
71. Varela, S.A.; Gobbi, M.E.; Laos, F. Banco de semillas de un bosque quemado de Nothofagus pumilio: Efecto de la aplicación de compost de biosólidos. *Ecol. Austral* **2006**, *16*, 63–78.
72. Urretavizcaya, M.F.; Defossé, G.E. Soil seed bank of *Austrocedrus chilensis* (D. Don) Pic. Serm. *et* Bizarri related to different degrees of fire disturbance in two sites of southern Patagonia, Argentina. *For. Ecol. Manag.* **2004**, *187*, 361–372. [CrossRef]

73. Raffaele, E.; Gobbi, M. Seed bank composition and variability in *Austrocedrus chilensis* forest sites in Patagonia, Argentina. *Int. J. Ecol. Environ. Sci.* **1996**, *22*, 59–72.
74. Grove, S.; Parker, I.M.; Haubensak, K.A. Do impacts of an invasive nitrogen-fixing shrub on Douglas-fir and its ectomycorrhizal mutualism change over time following invasion? *J. Ecol.* **2017**, *105*, 1687–1697. [CrossRef]

forests

MDPI

Article

Stem Circadian Phenology of Four Pine Species in Naturally Contrasting Climates from Sky-Island Forests of the Western USA

Emanuele Ziaco and Franco Biondi *

DendroLab, Department of Natural Resources and Environmental Science, University of Nevada, Reno, NV 89557, USA; eziaco@unr.edu

* Correspondence: fbiondi@unr.edu; Tel.: +1-775-784-6921

Received: 12 June 2018; Accepted: 29 June 2018; Published: 4 July 2018

Abstract: We identified intra-annual climatic drivers of stem water relationships and growth in foundation conifers at a subalpine site in the Great Basin Desert and at a montane site in the Mojave Desert (Nevada, USA). Sites were instrumented to represent naturally different levels of heat and drought stress as part of NevCAN (the Nevada Climate and ecohydrological Assessment Network). We analyzed three years (2013–2015) of sub-hourly dendrometer records for limber (*Pinus flexilis*) and bristlecone pine (*Pinus longaeva*) at the subalpine site, and for ponderosa (*Pinus ponderosa*) and pinyon pine (*Pinus monophylla*) at the montane site. Multiple logistic regression was used to identify relationships with environmental variables measured in-situ. At both sites, stem expansion occurred during the early morning and late afternoon, and irreversible stem increment was concentrated in the early morning hours. In subalpine species, stem growth started in late spring and continued until August, while at the montane site stem growth was episodic, peaking during summer monsoonal rainstorms. Circadian maximum and minimum stem size during the growing season were reversed during the dormant season at the colder/wetter subalpine site but not at the warmer/drier montane one. Knowledge of intra-annual tree-water relationships and stem growth helps understand how sky island forests grow under highly diverse climatic conditions, including severe drought and heat stress.

Keywords: arid environments; tree water deficit; point dendrometers; hydroclimate; north American monsoon; stem water dynamics; *Pinus longaeva*; *Pinus flexilis*; *Pinus ponderosa*; *Pinus monophylla*

1. Introduction

The functioning and health of forest ecosystems depend on the adaptation of dominant tree species to varying environmental conditions at multiple timescales [1]. Annual to multi-decadal climate-growth relationships have been investigated in a broad range of biomes for a number of conifer and broadleaf species using tree-ring proxies [2]. In order to understand how forest ecosystems will fare under future global warming scenarios, it has become increasingly important to investigate the physiological basis of individual plant adaptation to different levels of environmental stress [3–5]. Attention has therefore been driven towards identifying the mechanisms underlying patterns of hourly to sub-hourly stem growth [6,7] and in particular how such patterns are affected by meteorological and environmental variables under natural conditions [8,9].

Daily variations of woody stem size are the consequences of reversible phases of shrinkage and swelling, reflecting reversible dynamics of stem water storage, as well as irreversible stem growth [10]. Stem contraction occurs when water uptake from the soil does not compensate for water losses from canopy evapotranspiration [11], forcing the plant to recall water from the elastic tissues of the stem (i.e., phloem, bark) into the sapwood [12]. Contrarily, when absorption from the roots is greater than the amount of water lost to the atmosphere, expansion of the stem is observed [13]. However, under hyperarid conditions a tree may undergo prolonged periods of stem shrinkage [14]. Recently an innovative methodological approach was proposed to separate water-deficit induced stem shrinkage from irreversible radial growth [15], assuming that no growth is occurring during stem shrinkage because of limited turgor pressure in the cambium cells [7].

The amount of water stored in tree stems and its contribution to daily water fluxes depends on the tree species and individual characteristics, in particular tree size [16]. Stems, especially the sapwood, hold the largest reserve of water in trees, and larger individuals hold greater reserves of water available to fulfill daily physiological processes [17]. Larger proportions of sapwood may also facilitate water movement through the stem under lower tensions compared to species with less sapwood under similar evapotranspiration demands [18]. Even if the proportion of sapwood over total basal area is not linked to environmental conditions [19], climatic stressors (such as drought) may deplete internal storage by increasing water demand for transpiration [20] or by preventing the replenishment of reserves [21].

Monitoring stem-radius changes using automated dendrometers is a widely used method to define stem water dynamics [11,20,22,23] and tree growth response to short-term climatic factors [24]. Daily oscillations define the stem circadian cycle, which is composed by a contraction phase (i.e., depletion of water reserves), an expansion phase (i.e., replenishment of water reserve), and occasionally a stem increment phase [22]. While diurnal (or diel) cycles depend on evapotranspiration, ultimately balancing root uptake with canopy losses under normal conditions [12], prolonged periods of stem swelling or shrinking may result in long cycles exceeding 24 h [4]. Positive/negative fluctuations of stem radius resulting from diurnal and long cycles can then be used to highlight correlations with environmental variables and their effect on tree growth and stem water dynamics ([10] and references therein).

The Great Basin and Mojave Deserts of the western USA are characterized by arid and semi-arid conditions, and water availability is generally the main limiting factor for natural systems [25]. Both deserts are characterized by pronounced elevation gradients in precipitation and temperature, so that mountain ranges, depending on their elevation and topography, can be dominated by sky-island conifer forests that experience dry and hot summers as well as cold and snowy winters [26]. Latitudinal variability in climate also exists, as the northern areas tend to be characterized by winter and spring precipitation, with associated snowpack dynamics, whereas the southern areas tend to be more affected by summertime rainfall brought by the northwest edge of the North American Monsoon system [27]. Tree species growing in these regions are able to survive under a climatic regime characterized by periods of hyperarid conditions [28], so that greater understanding of their ecohydrological dynamics may help predict how tree species will fare under the warmer and drier world forecasted by global warming scenarios in the western US.

We assessed tree-water relationships and the environmental drivers of stem circadian cycles in four "foundation" [29] conifer species from the Great Basin and Mojave deserts of North America. We analyzed three years (2013–2015) of continuous dendrometer records from two contrasting locations: a subalpine site in the central Great Basin and a montane site at the border between the Mojave and Great Basin deserts. Our study objectives were: i) to characterize daily and seasonal stem circadian cycles in conifers experiencing different levels of heat and drought stress; ii) to characterize seasonal dynamics and patterns of tree water deficit and irreversible radial growth; and iii) to investigate the climatic drivers of stem size variations in order to assess species-specific levels of adaptation to local conditions and hypothesize their responses to future warming scenarios.

2. Materials and Methods

2.1. Study Area

Two sites separated by 2.5° latitude were instrumented for this study as part of the Nevada Climate-ecohydrological Assessment Network (NevCAN; Figure 1a). Automated sensors were installed between 2010 and 2013 to measure climatic and environmental variables at sub-hourly time intervals [30]. The Snake Subalpine West site (38°54′22″ N, 114°18′32″ W; 3355 m a.s.l.) is located on the western flank of the Snake Range in central-eastern Nevada. Vegetation is a mixed-conifer stand dominated by bristlecone pine (*Pinus longaeva* D. K. Bailey), limber pine (*Pinus flexilis* E. James), and Engelmann spruce (*Picea engelmannii* Parry ex Engelm.). The Sheep Montane site (36°35′25″ N, 115°12′51″ W; 2320 m a.s.l.) is located in southern Nevada, about 50 km north of Las Vegas. The canopy is composed of ponderosa pines (*Pinus ponderosa* var. *scopulorum* Engelm.) and pinyon pines (*Pinus monophylla* Torr. & Frém.). The shrub layer is absent or extremely sporadic at both sites. Parent rock material at both sites is composed by a colluvium derived from limestone and dolomite. Soils are loamy-skeletal Lithic Cryorthens at the Snake Range location, and loamy-skeletal Aridic Lithic Haplustolls at the Sheep Range site [31].

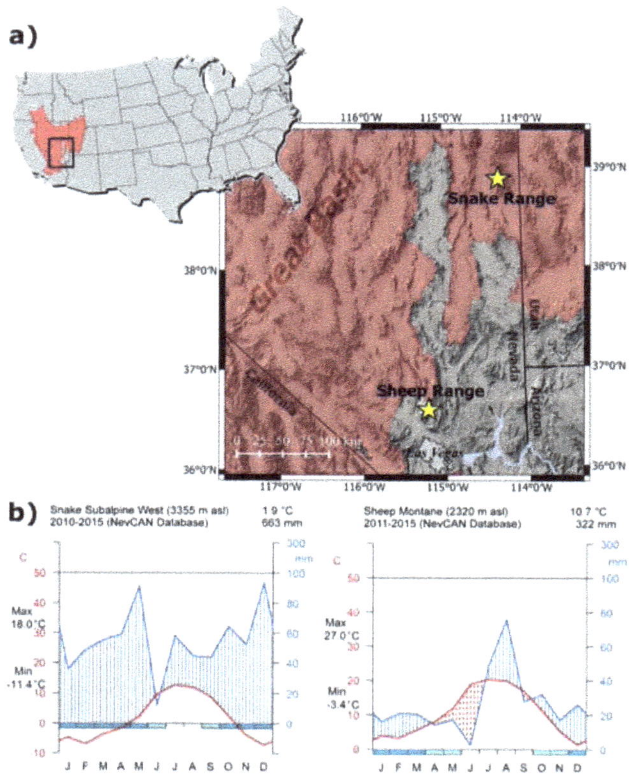

Figure 1. (a) Geographic location of the two study sites with respect to the hydrographic boundary of the Great Basin (red shaded area). (b) Walter-Lieth climatic diagrams for the two study sites with temperature (red labels) on the left y axis and total precipitation (blue labels) on the right one. Bars below the x-axes indicate frost season (dark blue), probable frost season (light blue), and frost-free season (no bars).

The Snake Range, which falls within the hydrographic Great Basin, is characterized by cool-season precipitation and associated winter snowpack dynamics. Average annual temperature during the first five years of NevCAN operation (2010–2015) was 1.9 °C, and average total annual precipitation was 663 mm (Figure 1b). The Sheep Range is located near the hydrographic boundary between the Great Basin and the Mojave Desert, one of the driest regions influenced by the North American Monsoon [32]. Based on NevCAN data for 2011–2015, the mean annual temperature was 10.7 °C, and the mean total annual precipitation was 322 mm, mostly concentrated in July and August (Figure 1b).

2.2. Dendrometer Data

Automated point dendrometers manufactured by Agricultural Electronics Corporation (Tucson, Arizona, USA) were installed on two pine species at each of the two sites. Observational noise caused by the hydration and de-hydration of the bark is minimized by the mm^2-area of the sensor tip and by shaving off the outer bark before installing the sensor [33]. Stem radius is quantified by measuring the horizontal linear displacement of a sensing rod, which is then turned into an electric signal with a resolution of 4 μm over a range of 15,000 μm [34]. At the Snake Range, a total of 12 dendrometers were installed on 6 bristlecone pines (*Pinus longaeva*, abbreviated as PILO), and 7 dendrometers were installed on 4 limber pines (*Pinus flexilis*, PIFL) (on one tree it was possible to install only one dendrometer). At the Sheep Range, a total of 8 dendrometers were installed on 4 ponderosa pines (*Pinus ponderosa*, PIPO) and 8 dendrometers were installed on 4 pinyon pines (*Pinus monophylla*, PIMO). Trees equipped with two dendrometers at different height along the stem, usually between the ground level and about 1.5 m, were used to test that dendrometer traces did not depend on sensor location. For each species, there was at least one dendrometer located in one of the four cardinal directions.

Instrumented trees were selected to represent the variability of stem sizes under natural conditions. At the Snake Range the average diameter of instrumented trees was 41 ± 15 cm for PILO and 21 ± 9 cm for PIFL, and the average height was 8 ± 2 m and 5 ± 2 m, respectively. Average diameter at the Sheep Range was 53 ± 8 cm for ponderosa pine (mean height = 11 ± 2 m) and 30 ± 7 cm for pinyon pine (mean height = 6 ± 2 m). Bristlecone pines instrumented at the Snake Range exceeded 240 years of age [35], even though one of them was large enough that its stem age exceeded multiple centuries, while limber pines usually did not exceed 100 years of age [14]. At the Sheep Range, age of instrumented ponderosa and pinyon pines reached a maximum greater than 250 years.

2.3. Climatic Data

Climatic data recorded by the NevCAN stations were obtained from the Western Regional Climate Center (www.wrcc.dri.edu/SRtransect/; www.wrcc.dri.edu/GBtransect/) and used to produce hourly summaries for atmospheric and soil variables [35]. Correlation analysis was used to assess the degree of predictor multicollinearity, and seven variables were selected for further study (Table 1): mean air temperature (AirT), mean soil temperature at 50 cm belowground (SoilT), soil volumetric water content at 17 cm (VWC_{17}) and 32 cm (VWC_{32}), total precipitation (PPT), dew point (Dew), and Vapor Pressure Deficit (VPD). Dew point is directly related to atmospheric moisture content [36] and its importance in hydrologic modeling is well recognized in the western US [37,38]. Dew point synthetizes four atmospheric variables (relative humidity, barometric pressure, saturated and ambient vapor pressure), which are required for its computation. Relative humidity (%) and barometric pressure (mb) were acquired from the NevCAN station, while vapor pressures (hPa) were computed according to standard formulas (equations (1) and (2), [34]) and then used to calculate dew point (°C; equation (A.4) [39]).

Table 1. Summary of climatic variables included in this study. Hourly means and range (minimum/maximum) are shown for the June-September period during 2013–2015 (average climate conditions for the whole year are shown in Figure 1b).

Variable	Code	Unit	Snake Range Subalpine West	Sheep Range Montane
Air temperature	AirT	°C	10.4 (−12.8/25.7)	18.8 (−5.8/36.6)
Soil temperature (50 cm)	SoilT	°C	13.7 (−0.6/21.4)	19.0 (8.6/23.6)
Soil volumetric water content (17 cm)	VWC_{17}	%	7.0 (2.4/22.1)	6.9 (4.4/13.9)
Soil volumetric water content (32 cm)	VWC_{32}	%	8.8 (4.1/18.4)	8.3 (6.6/10.2)
Dew point	Dew	°C	−2.6 (−34.6/8.9)	0.7 (−25.1/17.0)
Vapor Pressure Deficit	VPD	hPa	7.7 (0.2/21.7)	15.5 (0.2/42.2)
Total precipitation	PPT	mm	179 (0/9)	171 (0/51)

2.4. Extraction of Stem-Cycle Phases

All dendrometer records were processed using *dendrometeR* [40], a task-specific, open-source, statistical package for analyzing sub-daily dendrometer data within the R software environment [41]. Half-hourly individual dendrometer measurements from January 1st 2013 to December 31st 2015 were converted into daily stem increments by subtracting the previous day maximum from the current day maximum, and then transformed into cumulative series by adding daily stem increments. Quality control was performed to identify data gaps due to malfunctioning of dendrometers or data loggers. Graphical and numerical data summaries were used to detect errors, and values > ± 2 standard deviations from the daily mean were set as missing. No data gaps >12 h were found in our records, and any shorter gaps were filled using an ARIMA model, as it is typically done in these studies [42]. To enhance the signal-to-noise ratio at the stand level while reducing possible effects of tree size on individual measurements, corrected half-hourly records were combined in a site composite by species [23]. Daily and monthly mean circadian cycles were computed for each species by aggregating half-hourly measurements of stem radius from 2013 to 2015.

Xylogenesis studies conducted through repeated microcoring at the Snake Range on both *P. longaeva* [35] and *P. flexilis* [14] found that the production of new xylem cells begins in early June and continues until the end of August. At the Sheep Range, cellular measurements conducted on *P. ponderosa* in 2015 and 2016 have shown that cambial phenology is driven by moisture availability, with the formation of new tracheids occurring earlier in the spring after wet winters, but delayed to late June-early July when water is supplied through monsoonal precipitation [43]. Therefore we included in our analysis circadian cycles that started after June 1st and ended before September 30th.

Since daily cycles of stem size variations do not cover exactly a 24-h period [23], we distinguished diurnal cycles (duration <28 h) from long cycles (duration ≥28 h) [4]. Long cycles from June to September exceeded the average duration of diurnal cycles by two standard deviations or more. For each species, we identified three different phases in circadian cycles [22]: 1) contraction (i.e., period between a maximum in stem radius and the following minimum); 2) expansion (i.e., period between a daily minimum and the following maximum); 3) stem increment (i.e., that portion of the expansion phase starting when stem radius exceeds the previous cycle's maximum). For each phase and for each whole cycle, defined as contraction + expansion + increment (when present), we computed the onset (hour of the day), the duration (hours) and the magnitude (μm), i.e., the difference between maximum and minimum stem radius.

2.5. Modeling Relationships between Climate, Tree Water Deficit (TWD), and Real Growth (RG)

Stem water dynamics were analyzed separately from irreversible stem radial growth for each species. To investigate daily patterns of stem water depletion and replenishment, we pooled together expansion and increment phases to indicate "stem swelling" (SS), which indicates recharging stem water reserves. Stem increment phases (*sensu* [22]) are often not associated with a "real" radial increment, especially when they occur as isolated peaks during prolonged phases of stem shrinkage [44]. Therefore, to identify radial growth causing an actual increment in stem diameter, we first created hourly time series of tree water deficit (TWD) according to [15]. Briefly, TWD is a measure of stem shrinkage caused by water deficit, and it is measured in the same unit as stem radius (either μm or mm). TWD is usually <0, indicating that the current stem radius is below the previous highest recorded radius. When TWD equals 0, stem radius is increasing, stem water potential is near saturation, and therefore irreversible real growth (RG) is observed [15]. Hourly values of stem radius measured at TWD = 0 were used to compute curves of cumulative RG from June 1st to September 30th.

Because dendrometers readings can be affected by tree size, we used a probabilistic approach to assess how the presence/absence of circadian phases, rather than their magnitude, was controlled by climatic drivers. The relationship between environmental variables, stem swelling/shrinkage, and real growth was investigated by species and for different time lags using logistic regression. Summaries for each climatic variable were calculated up to a 24-h lag with a 3-h interval (i.e., t_0, t_{-3}, t_{-6}, ..., t_{-24}), producing a total of nine time lags for each explanatory variable (i.e., AirT, PPT, SoilT, VWC_{17}, VWC_{32}, VPD and Dew), and a total of 63 predictors. Each hourly observation of stem size was converted to a binomial variable, with "0" for shrinkage and "1" for swelling. Analogously, each hourly record was converted to a binomial variable depending on the presence/absence of RG, with "1" for growth and "0" for no growth. A blockwise variable selection was performed to select within each group of variables the time lag that best fitted the binary dataset of stem swelling and real growth using simple logistic regression:

$$Logit\ (\pi_x) \ln\left(\frac{\pi_x}{1-\pi_x}\right) = \beta_0 + \beta_1 x$$

with π_x being the probability of stem swelling (or real growth) for a given value of climatic parameter x, and β_0 and β_1 the intercept and slope of the regression. Goodness of fit for each model was assessed using a Hosmer-Lemeshow test [45], and significance of individual parameters was tested with Wald's χ^2 [46]. If two or more variables, at any time lag, passed the Hosmer-Lemeshow test, then those variables were combined into a multiple logistic regression [45]:

$$Logit\ (\pi_x) = \ln\left(\frac{\pi_x}{1-\pi_x}\right) = \beta_0 + \beta_1 x + \beta_2 y + \ldots + \beta_n z$$

where x, y, and z are the n selected climatic variables, β_0 the intercept of the regression, and β_1, β_2, and βn the coefficients for the n climatic variables. The Hosmer-Lemeshow test was used again to evaluate model calibration, while model discrimination capacity (i.e., the model's ability to predict an event correctly) was assessed measuring the area under a Receiver Operating Characteristic (ROC) curve [47] computed using the R package *pROC* [48]. In order to account for time-of-day influence on stomatal conductance and xylem water potentials, models of stem swelling and real growth were computed separately for night (19:00–6:00) and day hours (7:00–18:00). Since more than 90% of the seasonal cumulative real growth was completed by the end of August, the relationship between climate, TWD, and RG was analyzed for the June-August period.

3. Results

3.1. Climate

Climatic data from 2013 to 2015 highlight differences between the two study sites (Figure 2), in particular for seasonal dynamics of precipitation and soil temperature. Average air and soil temperature during the summer at the Sheep Range site were higher, respectively by 8.4 °C and 5.3 °C, than at the Snake Range site (Table 1). At the Sheep Range, soil temperature at 50-cm depth was close to 0 °C only occasionally between December and January (Figure 2), while at the Snake Range soil temperature at 50-cm depth was constantly near 0 °C from November to May because of snowpack presence. From June to August, daily variability of soil temperature was higher at the Snake Range (northern site) than at the Sheep Range (southern site) because of larger thermal excursion between night and day. The average volumetric water content of both shallow (VWC_{17}) and deep (VWC_{32}) soil was similar at the two sites, despite a much broader range of soil moisture from June to August at the Snake Range (Table 1).

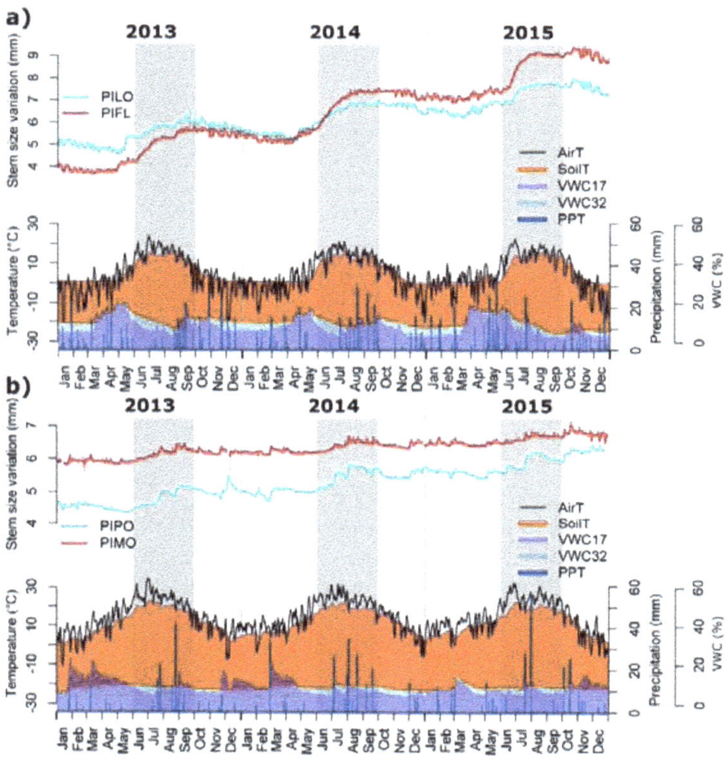

Figure 2. Site-wide composites of stem size variation during 2013–2015 computed for: (**a**) *Pinus longaeva* (PILO) and *Pinus flexilis* (PIFL) at the Snake Range Subalpine West site and (**b**) for *Pinus ponderosa* (PIPO) and *Pinus monophylla* (PIMO) at the Sheep Range Montane site. Daily time series of mean air temperature (AirT), soil temperature at 50 cm belowground (SoilT), volumetric water content at 2–17 cm (VWC_{17}) and 17–21 cm belowground (VWC_{32}), and total precipitation (PPT) are also shown. Gray shaded areas represent the 4-month period (June–September) considered for the analysis of circadian stem size phenology.

At both sites, no significant correlation emerged between VWC_{17} and VWC_{32} within a time lag of ±3–4 days. Total June–September precipitation was similar between the two sites (Table 1), even though at the Sheep Range precipitation was mostly concentrated in the July–August period. In fact, during 2013–2015 the fraction of total annual precipitation falling in July–August was only 15% at the Snake Range, but about 41% at the Sheep Range because of the North American Monsoon system. Consequently, during June–September average dew point was higher at the Sheep Range (0.7 °C) than at the Snake Range (−2.6 °C), even though the average vapor pressure deficit during the same period was almost twice as large at the Sheep Range than at the Snake Range (Table 1).

3.2. Seasonal and Daily Dynamics of Stem-Size Variation

Dendrometer traces (Figure 2) recorded at the Snake Range on PILO and PIFL (the northern species) displayed stem increment starting in May–June, concurrent with increasing deep soil temperature, and lasting until the end of August, with a subsequent gradual shrinkage during the winter (Figure 2a). At the Sheep Range, stem-size variation for both PIPO and PIMO (the southern species) was smaller, with major stem enlargement occurring in July with the arrival of monsoonal precipitation (Figure 2b). On average, stem-radius increase was 3.1 ± 1.1 mm for PILO and 5.2 ± 1.3 mm for PIFL between June 1st 2013 and September 30th 2015 (Figure 3a,b), while in the same period PIPO and PIMO showed lower size increases, respectively 2.2 ± 0.7 mm and 1.4 ± 0.6 mm (Figure 3c,d).

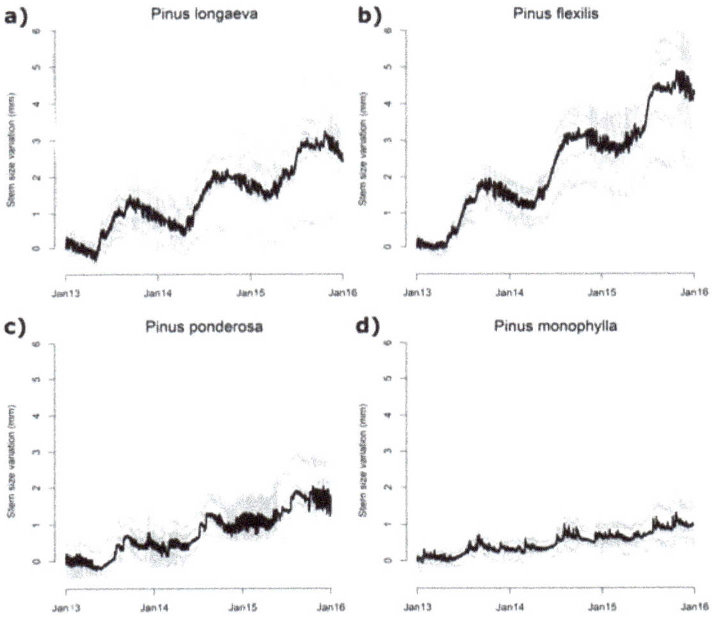

Figure 3. Half-hourly individual dendrometer tracks of stem size variation (gray lines) for PILO (**a**), PIFL (**b**), PIPO (**c**), and PIMO (**d**) from 1 January 2013 to 31 December 2015. Black bold lines represent the site average. Species codes are the same as in the Figure 2 caption.

Circadian cycles showed seasonal variability and substantial differences between the northern and southern pine species. At the Snake Range circadian cycles occurring during the cold season (November–May) were opposite in phase than those occurring during the warm season and the early fall (June–October) (Figure 4a). During the winter, for both PILO and PIFL, minimum stem radius occurred in the morning (7:00–7:30), while maxima were in the afternoon (15:30–16:00). During the summer and the early autumn, stem radius reached the maximum in the first hours of the morning in PILO (5:30) and a couple of hours later in PIFL (8:00). Then daily shrinkage began, culminating in the afternoon, first in PIFL (15:30) and then in PILO (17:00). At the Sheep Range no differences emerged between cold- and warm-season stem cycles in PIPO and PIMO (Figure 4b). Maximum stem radius occurred during the morning, first in PIMO (6:30–7:30) and then in PIPO (8:00–8:30). Minimum radial dimension was observed in the afternoon (16:30–17:00) during the warmest months, and about two hours later (18:00–18:30) during the cold season.

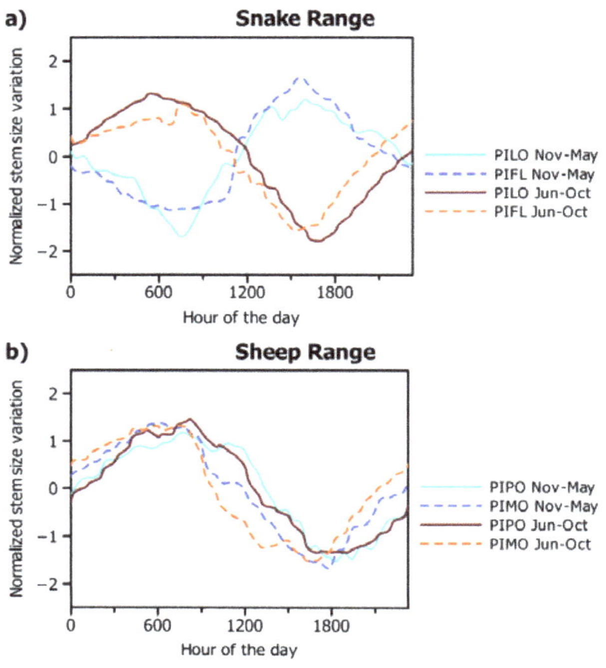

Figure 4. Daily circadian cycles calculated over three consecutive years (from January 2013 to December 2015) and averaged by species for the winter/spring (November–May) and for the summer/early autumn (June–October) at (**a**) the Snake Range, and (**b**) the Sheep Range. Species codes are the same as in the Figure 2 caption.

The average number of full cycles completed between June 1st and September 30th during the three years of study was 99 for PILO and 105 for PIFL at the Snake Range, and slightly less at the Sheep Range for PIPO (87) and PIMO (97). PIMO showed the highest amount of long cycles (60, representing 21% of total cycles), while in PIFL only 16% of all cycles were long (49). The duration of diurnal cycles was similar between species and sites, ranging between 23.7 and 23.8 hours (Table 2).

The onset of contraction started earlier at the Snake Range (05:20 in PILO and 04:29 in PIFL) than at the Sheep Range (06:49 in PIPO and 05:40 in PIMO). On average each diurnal cycle caused a positive increment of stem radius, as the stem size increase caused by the replenishment of stem water reserves (i.e., swelling) was larger than the reduction associated with water depletion (i.e., shrinkage). Average daily stem increment was about 1 μm for PIMO and 6 μm for PIPO at the Sheep Range, while it was about 4 μm in PILO and 11 μm in PIFL at the Snake Range. Long cycles, which showed very high variability, were longer at the Sheep Range (57.7 h in PIPO and 52.3 h in PIMO) compared to the Snake Range (52.1 h in PILO and 46.1 h in PIFL), but lower in magnitude (Table 2). The average increment of stem radius associated to long cycles was highest in PIPO (65 μm), and lowest in PIMO (38 μm).

Stem increment phases were mainly associated with long cycles: on average 68% of long cycles showed a stem increment phase against 48% of diurnal cycles, without any site-related difference between species. Long cycles produced higher stem increments, reflected in magnitudes that were three–four times larger than in diurnal cycles (Table 2). The duration of each phase was directly linked to its magnitude in all species, in particular for the increment, with similar correlations for PILO (Pearson's correlation coefficient $r = 0.89$, $p < 0.001$), PIFL ($r = 0.80$, $p < 0.001$) and PIPO ($r = 0.89$, $p < 0.001$) but lower in PIMO ($r = 0.68$, $p < 0.001$) (Figure S1).

Table 2. Summary statistics (mean ± 1 standard deviation) by species and phase at our study sites for diurnal and long cycles of stem variation occurring from June to September (see text for details). Onset = hour of day; Duration = hours; Magnitude = μm.

			Snake Range		Sheep Range	
Phase	**Type**	**Feature**	*Pinus longaeva*	*Pinus flexilis*	*Pinus ponderosa*	*Pinus monophylla*
Contraction	Diurnal	Onset	5:20 ± 1 h 51 m	4:29 ± 1 h 31 m	6:46 ± 2 h 16 m	5:40 ± 1 h 51 m
		Duration	11.3 ± 2.0	9.9 ± 1.6	10.8 ± 2.5	9.6 ± 2.2
		Magnitude	85 ± 40	69 ± 28	35 ± 50	60 ± 44
Expansion	Diurnal	Onset	16:16 ± 1 h 36 m	14:11 ± 1 h 33 m	17:05 ± 2 h 23 m	15:14 ± 1 h 30 m
		Duration	10.6 ± 2.3	10.2 ± 3.6	11.6 ± 3.0	12.6 ± 3.2
		Magnitude	75 ± 39	59 ± 27	28 ± 13	48 ± 24
Increment	Diurnal	Onset	4:52 ± 7 h 22 m	13:39 ± 9 h 01 m	5:19 ± 5 h 07 m	7:38 ±7 h 30 m
		Duration	3.5 ± 2.1	6.0 ± 3.4	3.1 ± 2.4	4.5 ± 4.4
		Magnitude	14 ± 14	21 ± 18	13 ± 73	13 ± 22
Whole cycle	Diurnal	Onset	5:20 ± 1 h 51 m	4:29 ± 1 h 31 m	6:46 ± 2 h 16 m	5:40 ± 1 h 51 m
		Duration	23.7 ± 1.9	23.8 ± 1.6	23.7 ± 1.9	23.8 ± 1.7
		Magnitude	95 ± 43	87 ± 27	44 ± 69	72 ± 61
Contraction	Long	Onset	6:58 ± 4 h 59 m	5:34 ± 4 h 40 m	7:39 ± 3 h 26 m	7:07 ± 4 h 41 m
		Duration	22.6 ± 14.3	14.9 ± 8.3	21.8 ± 19.2	19.8 ± 21.6
		Magnitude	73 ± 69	61 ± 38	35 ± 24	48 ± 34
Expansion	Long	Onset	15:25 ± 3 h 08 m	14:26 ± 1 h 59 m	17:42 ± 2 h 45 m	15:33 ± 1 h 47 m
		Duration	14.2 ± 11.6	15.3 ± 9.9	21.1 ± 19.4	23.7 ± 16.1
		Magnitude	45 ± 36	51 ± 34	30 ± 18	39 ± 29
Increment	Long	Onset	9:09 ± 7 h 04 m	12:33 ± 7 h 55 m	11:14 ± 8 h 05 m	11:58 ± 6 h 52 m
		Duration	24.4 ± 17.8	19.5 ± 14.0	25.4 ± 27.6	15.8 ± 15.6
		Magnitude	71 ± 54	53 ± 44	70 ± 101	47 ± 56
Whole cycle	Long	Onset	7:05 ± 4 h 57 m	5:24 ± 4 h 33 m	7:43 ± 3 h 26 m	7:10 ± 4 h 43 m
		Duration	52.1 ± 23.5	46.1 ± 16.6	57.7 ± 32.3	52.3 ± 27.8
		Magnitude	122 ± 90	111 ± 56	84 ± 94	85 ± 63

3.3. Climatic Influences on Tree Water Deficit and Real Growth

Time series of tree water deficit (TWD) showed different seasonal patterns of stem shrinkage/swelling between the Snake (Figure 5) and the Sheep Range (Figure 6). In general, all species experienced periods of water stress, particularly in the second part of the growing season (i.e., August). PILO and PIFL experienced lower values of TWD compared to PIPO and PIMO during June and part of July. At the Snake Range TWD values reached zero almost every day, at least for a few hours, resulting in a slow but constant real growth. A similar pattern was observed at the Sheep Range in June 2013 and 2014, but not later in the growing season, when TWD dropped for both PIPO and PIMO to approximately −0.25 mm, and RG occurred as episodic pulses driven by the onset of monsoonal precipitation (Figure 6). In 2015, after a severely dry winter, real growth for both species at the monsoonal site was concentrated in just three pulses during three major rainstorms: first in mid-June, second in mid-July, and third in early August. At the Snake Range, the species characterized by smaller individuals (PIFL) completed 25%, 50%, and 75% of total June–September radial increment usually 1–2 weeks before the species with larger average stems (PILO). At the Sheep Range radial growth was simultaneous between the two species, with no size effect. Stem swelling between June 1st and September 30th occurred prevalently during the night, between 18:00 and 06:00, especially at the Snake Range, whereas at the Sheep Range stem swelling was observed until 08:00 (Figure 7a,b). Real growth took place only in the morning hours, between 02:00 and 08:00 at the Sheep Range (with a drop between 06:00 and 07:00; Figure 7c) and between 04:00 and 05:00 at the Snake Range (Figure 7d).

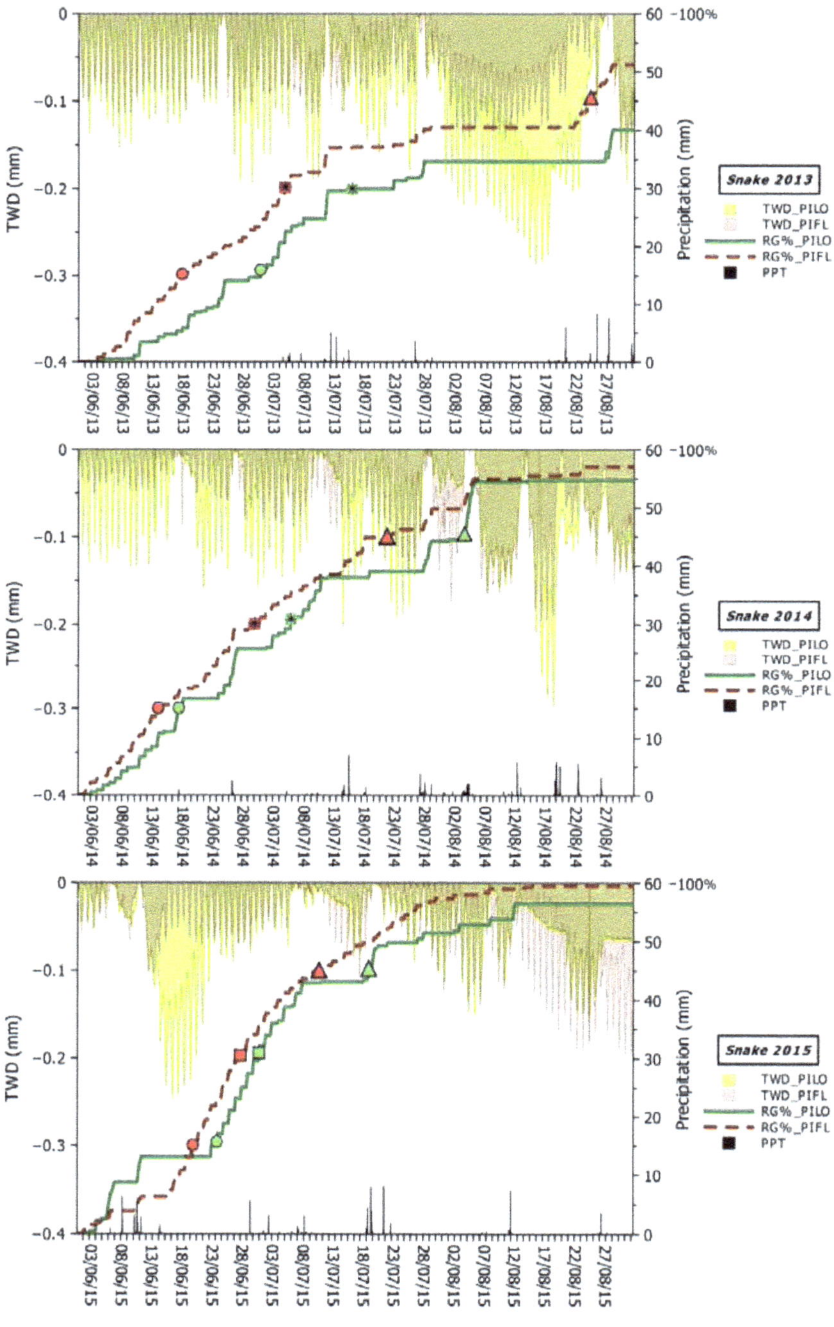

Figure 5. Tree water deficit (TWD) and real growth (RG) time series in 2013–2015 for PILO and PIFL (Snake Range). Symbols represent quartiles of real growth observed between June and September (circles = 25%; squares = 50%; triangles = 75%). Vertical bars represent total hourly precipitation. (Right y axes also represent 100% of RG). For a definition of TWD and RG see main text. Species codes are the same as in the Figure 2 caption.

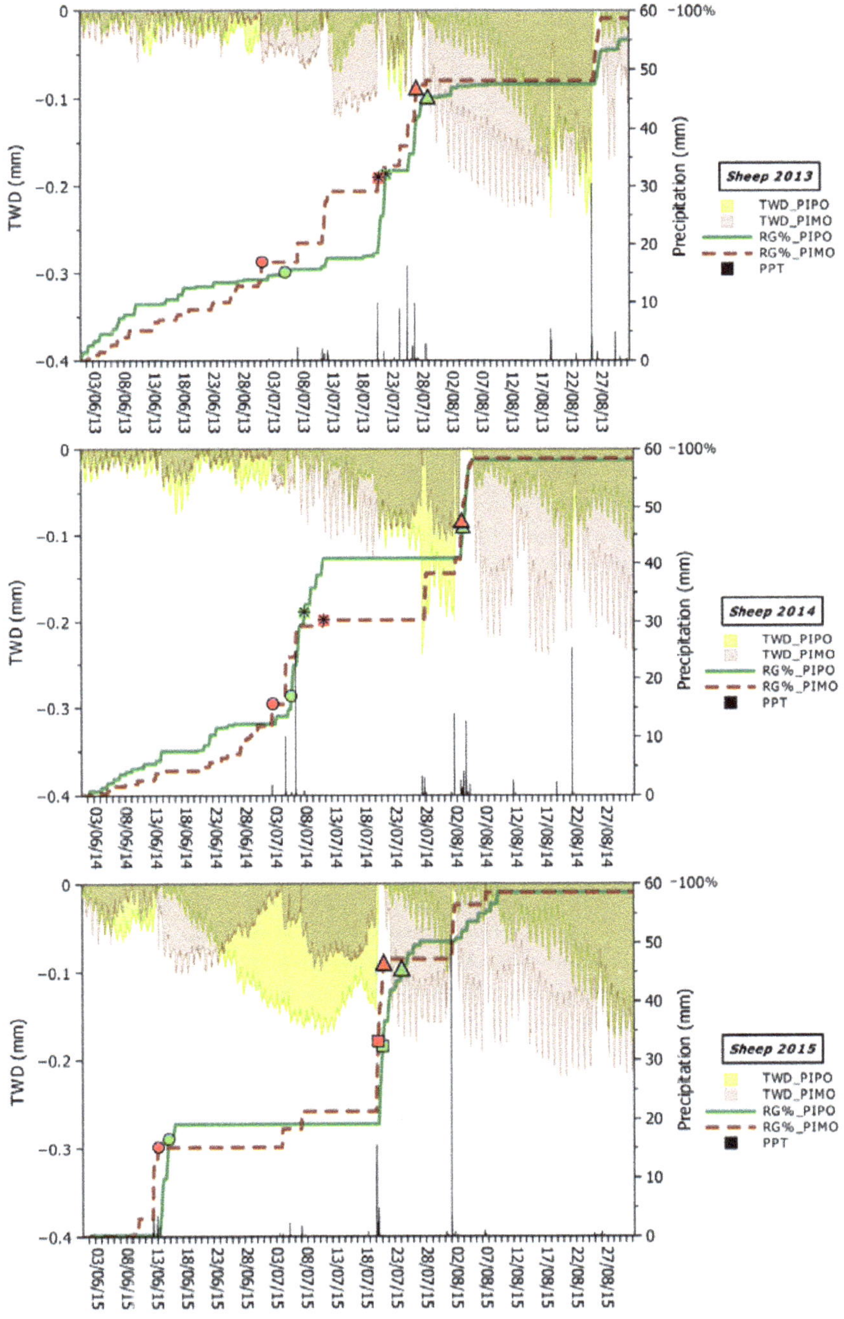

Figure 6. Tree water deficit (TWD) and real growth (RG) time series in 2013–2015 for PIPO and PIMO (Sheep Range). Symbols represent quartiles of real growth observed between June and September (circles = 25%; squares = 50%; triangles = 75%). Vertical bars represent total hourly precipitation. (Right y axes also represent 100% of RG). For a definition of TWD and RG see main text. Species codes are the same as in the Figure 2 caption.

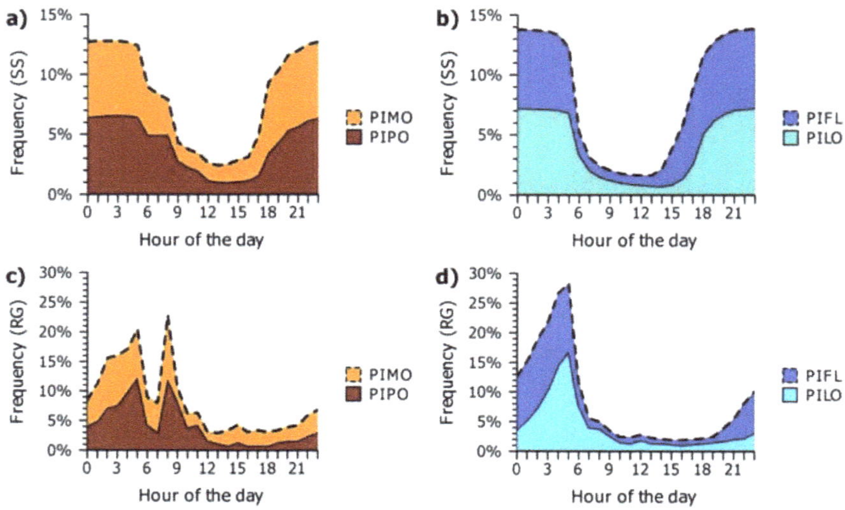

Figure 7. Average daily cycles of stem swelling (SS) and real growth (RG). Graphs (**a**) and (**c**) are for the Sheep Range; graphs (**b**) and (**d**) are for the Snake Range. Species codes are the same as in the Figure 2 caption.

Precipitation was the main predictor of stem swelling (SS) and radial growth (RG), appearing in all logistic models developed for day (Table S1) and night (Table S2) hours, although with different significance. Precipitation increased the odds of stem swelling and radial increment, and species response was less than six hours. During the day, a 1-mm increase in total precipitation in the previous 3 h increased the odds of SS by 2 and 2.5 times in PILO and PIFL respectively. At the Sheep Range, PIPO responded immediately to PPT (lag0), while the odds of SS in PIMO were affected by total precipitation in the previous 6 hours (Table S1). Precipitation in the previous hour was a significant predictor of RG in PIMO during both day- and night-time, and it increased the odds of RG by 1.2 times in PILO as well. Air temperature, vapor pressure deficit, and soil temperature were the other variables that emerged as significant predictors of SS and RG, although at different time lags depending on the species (Table S1). In general, AirT had a negative effect on the probability of SS and RG in all species during the day. At the Snake Range, stem swelling was favored by cooler mean temperatures in the previous hour in PILO and 24 h in PIFL. Within these time frames, a 1-°C increase in air temperature would make stem swelling 0.8 and 0.9 times less likely to happen respectively in PILO and PIFL. During the night, stem swelling was promoted by warm temperatures even at the Sheep Range (Table S2), where AirT in the previous 3 h significantly reduced the odds of RG, respectively 0.8 and 0.6 times in PIPO and PIMO. Vapor pressure deficit negatively affected both SS and RG in all species, but in particular it decreased the odds of irreversible growth during the day at the Snake Range, and the odds of stem swelling during the night at the Sheep Range. Warm soils favored SS at the Snake Range and RG of PIMO at the Sheep Range, despite limited daily variability of soil temperature.

All logistic regressions passed the Hosmer-Lemeshow test at the Sheep Range but not at the Snake Range (Tables S1 and S2). Model performances assessed through the area under the ROC curve varied by species and by process (i.e., stem swelling vs. real growth). At the Snake Range the daily dynamics of stem water replenishment were better captured for PILO, where multiple logistic regression successfully predicted 76.1% of events, while at the Sheep Range daily stem swelling dynamics were less well captured by the models identified for PIPO (65.8%) and PIMO (53.1%) (Table S1 and Figure S2a). During the night, logistic regressions correctly predicted events of SS 65.8% and 55.3% of times in PIPO and PIMO (Table S2 and Figure S2b). Within species and sites, real growth was in general better modeled than SS, with models for PIMO and PIPO predicting, respectively, 88.9% and 74.1% of RG events during the day (Table S1 and Figure S2c) and 74.9% and 75.8% during the night (Table S2 and Figure S2d). At the Snake Range, irreversible radial increment during the day was successfully predicted 96.2% in PIFL, but less well during the night (79.5% and 56.4% of events captured in PILO and PIFL respectively).

4. Discussion

4.1. Seasonal Dynamics of Circadian Cycles

The mechanisms underlying plant-water relationships and tree hydraulic status have been investigated on a large number of species under controlled conditions, but much less research has focused on understanding how such mechanisms are modulated by environmental and climatic factors in natural conditions [8,9,23]. Albeit not including all co-existing tree species, our two study areas are both representative of natural sky-island forests in the Great Basin and Mojave Deserts [30], and differ remarkably for the seasonal dynamics of total precipitation and soil temperature. Therefore they provide a unique opportunity to investigate daily and seasonal relationships between stem size variations, tree water status, and their environmental drivers. At the subalpine Snake Range site, lower air and soil temperatures together with the seasonal dynamics of the snowpack affect tree growth, especially in the spring and early fall. At the Sheep Range monsoonal rainfall occurring in July and August determines a temporary improvement of water availability, with higher soil and atmospheric moisture, alternating with hyperarid periods. The similar duration of circadian stem cycles at the two locations points toward a strong climatic control over the timing of replenishment (i.e., stem expansion) and depletion (i.e., stem contraction) of tree water reserves (Figure S3). Onset of stem water replenishment is much more variable than the onset of stem contraction in all species, regardless of diurnal or long cycles. This pattern suggests that even if environmental parameters driving transpiration (in particular temperature) play a primary role in stem-water dynamics [11,20,49], replenishment of stem water reserves in these arid environments may start anytime during the day if additional moisture is supplied [50]. The expansion phase began 1–2 h earlier in the afternoon in species characterized by smaller individuals, i.e., PIFL at the Snake Range and PIMO at the Sheep, compared to PILO and PIPO, suggesting that tree size, including stem diameter and crown dimension, could also influence stem rehydration.

The average yearly number of cycles was similar between species, and so was the proportion of long cycles, which ranged between 15 and 20%. The magnitude of both diurnal and long cycles is lower at the Sheep Range, even though the duration of long cycles, especially in PIPO, is greater. This is likely related to the fact that precipitation at the Sheep Range was episodic and concentrated in few abundant rainstorms that greatly raised soil moisture. The temporary improvement of moisture conditions triggered long cycles even though the severe evapotranspiration demands of the Sheep Range limited the magnitude of the radial increment. The importance for PIPO and PIMO of monsoonal water supplied in July and August is evident from the dynamics of stem size variation occurring after minor precipitation events (i.e., <5 mm/day). These atmospheric events are not strong enough to trigger a long cycle, and moderate stem increments associated with the temporary improvement of moisture conditions are usually followed by a sudden stem shrinkage driven by high transpiration related to summer air temperature. Most likely, under hyperarid conditions, once stem water has been refilled, environmental factors controlling transpiration determine whether stem water fulfils transpiration demands or can be diverted to the formation of new xylem cells [51].

Average monthly circadian cycles at the Sheep Range suggest that seasonality of soil freezing/thawing has little effect on stem water dynamics. Since mean soil and air temperature rarely drop below freezing, the risk of frost-induced cavitation for species growing at the southern site is limited if not absent. On the other hand, the reversal of stem cycle phases in winter compared to summer/early fall at the Snake Range indicates that extra-cellular water freezes during the cold season, creating a withdrawal of water from the bark and the outer stem tissues that leads to frost-induced stem shrinkage, reaching its peak in the coldest hours of the morning [52]. Reversed-phase circadian cycles between winter and summer months are commonly observed in mountain regions [20] but not in species experiencing mild winter temperatures [53].

4.2. Climatic Drivers of Stem Swelling and Real Growth

Cumulative dendrometer series showed that radial increment was more sustained in the northern species compared to the southern ones. More importantly, they captured site- and species-specific strategies to cope with seasonal heat and drought stress, reflecting a high level of phenological plasticity and adaptation to local conditions, in particular at the Sheep Range. PILO and PIFL present a well-defined period of stem enlargement, starting in April–May, continuing as a steep monotonic increase from June to August, and finally reaching a plateau in September. This pattern is typical of conifer species in cold environments [22,54] or from regions characterized by winter dormancy [55,56], and reflects the potential limitations imposed by frozen soil water during the first phases of xylogenesis [57]. At the Sheep Range, stem enlargement is not continuous from June to September but rather episodic, with peaks in radial increment occurring during summer monsoonal rainstorms, as particularly evident in 2015.

Logistic regression models were better at predicting irreversible radial growth than stem swelling (i.e., replenishment of water reserves), especially during daytime. Cellular measurements of wood formation conducted on PIPO at the Sheep Range in 2015 have linked stem rehydration driven by summer precipitation with enhanced cellular division, leading to the rapid formation of new layers of xylem, especially after dry winters, starting in mid-July [43]. These findings align with previous studies from Northern Arizona, where stem radial growth and net photosynthetic rates in PIPO were found to peak in July and August [58], when water stress is relieved by monsoonal precipitation [59]. Studies on wood formation conducted in 2013 and 2014 at the Snake Range on PILO [35] and PIFL [14] have highlighted seasonal patterns of xylem production comparable to radial growth dynamics recorded by point dendrometers, with new cell formation starting in June and ending in late August/early September.

A relatively limited number of variables (i.e., precipitation, air temperature, vapor pressure deficit, and soil temperature) affected stem circadian phenology at both sites. The recurrent presence of precipitation and air temperature as predictors for the presence/absence of stem swelling and irreversible increment suggests that patterns of radial growth reflect site-specific dynamics of evapotranspiration and interactions between moisture conditions and heat. All four pine species experienced prolonged periods of intense water deficit, characterized by no irreversible expansion. During such periods, the recharge of stem water reserves is incomplete, implying that water potential in the cambium remains negative and no radial growth takes place [44]. The range of tree water deficit was similar between sites, but subalpine species in the first part of the growing season presented daily variation of stem size larger than the montane ones. Nevertheless, PILO and PIFL produced regular radial increment, reducing stem water deficit at least for a few hours. PIPO and PIMO showed limited water deficit in June, but then entered a long phase of severe water deficit interrupted only by summer rainstorms.

Climatic response varied by site and species. At the Snake Range, soil temperature emerged as an important driver of stem water replenishment. We then infer that species growing at high elevations, like PILO and PIFL, benefit from increasing soil temperature even in semi-arid environments because in warmer soils water viscosity increases, enhancing root water supply and growth [60], whereas low soil temperature also decreases root hydraulic conductivity by inhibiting aquaporin activity [61]. The shifting sign in the relationship between air temperature and SS (negative during the day, positive at night), is consistent with the high transpiration demands of both sites which largely affect the daily circadian phenology for all species. On the other hand, the effect of air temperature on the odds of irreversible growth seems to be equally fast in both PIPO and PIMO, but much different in PILO and PIFL, with this last species responding to air temperature over a much longer period (12 h).

This finding highlights the role of canopy architecture in controlling transpiration rates, as larger trees with more portions of crown exposed to incoming solar radiation tend to transpire more than smaller ones [62].

All species showed a fast response to total precipitation for time lags up to 6 h, in particular during the day. While the contribution of precipitation to stem swelling was highly significant for all species, PIMO was the only species for which precipitation had an immediate effect (in the previous hour) on irreversible growth. This behavior shows that PIMO is able to use even small amounts of precipitation during the growing season [23], likely due to a combination of higher transpiration demands and lower internal stem water reserves. Large trees can store water in the canopy and in the sapwood [17] to compensate for hydraulic limitations imposed by reduced moisture availability [16]. A species like PIPO can rely on internal reserves of stem water thanks to a particularly high proportion of sapwood [63], while species characterized by smaller trees like PIMO might rely more on water stored in the canopy. Since canopy reserves are rapidly depleted by transpiration [50], species with lower stem water reserves (i.e., sapwood) would then be more dependent on external sources (i.e., precipitation) to replenish them. In this sense, small individuals might be favored against larger ones in a drier/warmer climate [62]. The presence of a negative, yet non-significant, effect of precipitation in the logistic regressions describing RG for PILO and PIFL may suggest that climate-growth relationships near treeline are more complex than at lower elevations [64].

4.3. Implications for Species Adaptation to Climate Change

The fate of tree species under future warming scenarios depends upon their plasticity to adapt, in particular to more frequent and severe droughts [65], hence research on drought-induced tree mortality and its causes is essential to predict future vulnerability of forest species [66,67]. Compensatory mechanisms have been observed via increased water use efficiency [68], physiological short-term regulation [69], or phenological avoidance of dry periods [70]. At the same time, more severe and frequent droughts have been identified as main drivers of reduced tree growth [71,72] and extensive forest die-off, often associated with increased fire frequency and insect outbreaks [73,74]. A trade-off between drought resistance and post-drought recovery has been observed in conifers growing under contrasting climatic conditions, with trees from mesic sites showing higher resistance and trees from xeric conditions showing higher resilience [75]. However, analyses of drought response patterns in relation to stem size variations measured by dendrometers are still scarce in the western US [76].

Conifer species included in our study showed a high level of adaptation to moisture and heat stress, which is reflected in their rapid response to even limited water supplies and to cooler temperatures. The ability of conifer species growing at the xeric southern site to utilize episodic precipitation events during an otherwise hyperarid growing season may represent an adaptation to extremely variable conditions. In particular, it points toward opportunistic growth strategies driven by a high phenological plasticity, which allows xeric southern species to adjust the timing of stem water dynamics with improved moisture conditions (i.e., higher soil water content) and reduced evapotranspiration demands (i.e., lower vapor pressure deficit). The fast response of PIMO to episodic rainfall might explain why single-needle pinyon populations were minimally affected by early 2000s drought episodes, and have continued expanding in the Great Basin [77], whereas Colorado pinyon (*Pinus edulis* Engelm.) has shown unusually high mortality rates in recent decades. Ponderosa pine was previously shown to possess high phenological plasticity [43] and greater capacity to recover pre-drought growth rates in old trees compared to young ones [78], pointing again to the importance of internal stem water reserves for this species.

Rather than on phenological plasticity, adaptability of northern, five-needle species (PILO and PIFL) to a changing climate might depend on the presence of favorable microclimates [79], which determine less demanding evapotranspiration conditions. Rugged montane topography might reduce the effect of climate warming by sheltering temperature-sensitive species [80], and some populations of limber pine in the eastern Sierra Nevada have in fact retreated downslope in cool ravines to escape increasingly dry conditions [81]. High-elevation species like PILO, whose growth dynamics are tightly linked to snowpack dynamics, might also develop different crown architectures in response to rising temperatures [82]. Bristlecone pine growing under water limitation regimes at the White Mountains maintains erect growth up to the treeline, whereas at the Snake Range krummholz growth forms are more frequent [83]. In general, higher soil moisture, steep slopes, and shallow rocky soils, all conditions found at the Snake Range site, favor strategies of drought resistance rather than post-drought recovery [65], and the sparse structure of such high-elevation stands might actually promote drought adaptation [81].

5. Conclusions

Dendrometer observations provided valuable information about tree responses to environmental stressors. We found that stem circadian phenology of four foundation conifer species growing in sky-island forests on desert mountains in the western USA were controlled directly by moisture or indirectly by those factors favoring accessibility to available water. In warmer and drier habitats, evapotranspiration demands, defined by a combination of air temperature and water availability, were crucial in determining the possibility for conifer species to mitigate tree water deficit and produce irreversible radial growth. In cooler and moister habitats, warm soil temperature was required to facilitate the uptake of soil moisture, derived from a melting snowpack, by subalpine conifers in order to recharge their internal reserves. The different thermal regimes at the Sheep Range (southern site) and at the Snake Range (northern site) resulted in phase-reversed circadian cycles between cold and warm seasons at the northern site but not at the southern one.

Our findings highlighted drought responses that can help understand future tree growth and survival under a changing climate. At the southern site (Sheep Range), conifer species growing in xeric conditions were able to utilize episodic precipitation events during an otherwise hyperarid growing season. Such pattern was not present in near-treeline five-needle pines from the northern site (Snake Range), where seasonal dynamics of stem radial phenology were linked to spring conditions and their linkages to moisture provided by snowpack melting.

Supplementary Materials: The following are available online at http://www.mdpi.com/1999-4907/9//396/s1, Figure S1: Relationships between duration and magnitude of circadian phases; Figure S2: Discrimination capacity of logistic regression for SS and RG; Figure S3: Stem-cycle phases of contraction, expansion, and increment for the Snake Range and the Sheep Range; Table S1: Statistics for logistic regression models for day hours; Table S2: Statistics for logistic regression models for night hours.

Author Contributions: Conceptualization, F.B. and E.Z.; Methodology, E.Z. and F.B.; Formal Analysis, E.Z.; Writing-Original Draft Preparation, E.Z.; Writing-Review & Editing, F.B.; Visualization, E.Z.; Supervision, F.B.; Project Administration, F.B.; Funding Acquisition, F.B.

Funding: This research was funded, in part, by the US National Science Foundation under grants AGS-P2C2-1401381 to F. Biondi and AGS-P2C2-1502379 to F. Biondi and E. Ziaco. Additional funding was provided to F. Biondi by the Office of the Vice President for Research and Innovation at the University of Nevada, Reno, in relation to NevCAN. The views and conclusions contained in this document are those of the authors and should not be interpreted as representing the opinions or policies of the funding agencies and supporting institutions.

Acknowledgments: We thank S. Strachan for NevCAN maintenance, R. Hill for compiling dendrometer data, E. van der Maaten and M. van der Maaten-Theunissen for helpful discussion of the R package dendrometeR.

Conflicts of Interest: The authors declare no conflict of interest. The funders had no role in the design of the study; in the collection, analyses, or interpretation of data; in the writing of the manuscript, and in the decision to publish the results.

References

1. Mendivelso, H.A.; Camarero, J.J.; Gutiérrez, E.; Zuidema, P.A. Time-dependent effects of climate and drought on tree growth in a Neotropical dry forest: Short-term tolerance vs. long-term sensitivity. *Agr. Forest Meteorol.* **2014**, *188*, 13–23. [CrossRef]
2. George, S.S. An overview of tree-ring width records across the Northern Hemisphere. *Quat. Sci. Rev.* **2014**, *95*, 132–150. [CrossRef]

3. Kocher, P.; Horna, V.; Leuschner, C. Environmental control of daily stem growth patterns in five temperate broad-leaved tree species. *Tree Physiol.* **2012**, *32*, 1021–1032. [CrossRef] [PubMed]
4. Deslauriers, A.; Anfodillo, T.; Rossi, S.; Carraro, V. Using simple causal modeling to understand how water and temperature affect daily stem radial variation in trees. *Tree Physiol.* **2007**, *27*, 1125–1136. [CrossRef] [PubMed]
5. Sevanto, S.; McDowell, N.G.; Dickman, L.T.; Pangle, R.; Pockman, W.T. How do trees die? A test of the hydraulic failure and carbon starvation hypotheses. *Plant Cell Environ.* **2014**, *37*, 153–161. [CrossRef] [PubMed]
6. Steppe, K.; Sterck, F.; Deslauriers, A. Diel growth dynamics in tree stems: Linking anatomy and ecophysiology. *Trends Plant Sci.* **2015**, *20*, 335–343. [CrossRef] [PubMed]
7. Coussement, J.R.; De Swaef, T.; Lootens, P.; Roldán-Ruiz, I.; Steppe, K. Introducing turgor-driven growth dynamics into functional-structural plant models. *Ann. Bot.* **2018**, *121*, 849–861. [CrossRef] [PubMed]
8. van der Maaten, E.; Bouriaud, O.; van der Maaten-Theunissen, M.; Mayer, H.; Spiecker, H. Meteorological forcing of day-to-day stem radius variations of beech is highly synchronic on opposing aspects of a valley. *Agr. Forest Meteorol.* **2013**, *181*, 85–93. [CrossRef]
9. Oberhuber, W. Soil water availability and evaporative demand affect seasonal growth dynamics and use of stored water in co-occurring saplings and mature conifers under drought. *Trees-Struct. Funct.* **2017**, *31*, 467–478. [CrossRef] [PubMed]
10. De Swaef, T.; De Schepper, V.; Vandegehuchte, M.W.; Steppe, K. Stem diameter variations as a versatile research tool in ecophysiology. *Tree Physiol.* **2015**, *35*, 1047–1061. [CrossRef] [PubMed]
11. Herzog, K.M.; Häsler, R.; Thum, R. Diurnal changes in the radius of a subalpine Norway spruce stem: their relation to the sap flow and their use to estimate transpiration. *Trees* **1995**, *10*, 94–101. [CrossRef]
12. Zweifel, R.; Item, H.; Hasler, R. Link between diurnal stem radius changes and tree water relations. *Tree Physiol.* **2001**, *21*, 869–877. [CrossRef] [PubMed]
13. Braekke, F.H.; Kozlowski, T.T. Shrinkage and swelling of stems of Pinus resinosa and Betula papyrifera in northern Wisconsin. *Plant Soil* **1975**, *43*, 387–410. [CrossRef]
14. Ziaco, E.; Biondi, F. Tree growth, cambial phenology, and wood anatomy of limber pine at a Great Basin (USA) mountain observatory. *Trees-Struct. Funct.* **2016**, *30*, 1507–1521. [CrossRef]
15. Zweifel, R.; Haeni, M.; Buchmann, N.; Eugster, W. Are trees able to grow in periods of stem shrinkage? *New Phytol.* **2016**, *211*, 839–849. [CrossRef] [PubMed]
16. Phillips, N.G.; Ryan, M.G.; Bond, B.J.; McDowell, N.G.; Hinckley, T.M.; Čermák, J. Reliance on stored water increases with tree size in three species in the Pacific Northwest. *Tree Physiol.* **2003**, *23*, 237–245. [CrossRef] [PubMed]
17. Čermák, J.; Kučera, J.; Bauerle, W.L.; Phillips, N.; Hinckley, T.M. Tree water storage and its diurnal dynamics related to sap flow and changes in stem volume in old-growth Douglas-fir trees. *Tree Physiol.* **2007**, *27*, 181–198. [CrossRef] [PubMed]
18. Barnard, D.M.; Meinzer, F.C.; Lachenbruch, B.; McCulloh, K.A.; Johnson, D.M.; Woodruff, D.R. Climate-related trends in sapwood biophysical properties in two conifers: avoidance of hydraulic dysfunction through coordinated adjustments in xylem efficiency, safety and capacitance. *Plant Cell Environ.* **2011**, *34*, 643–654. [CrossRef] [PubMed]
19. Meinzer, F.C.; Bond, B.J.; Warren, J.M.; Woodruff, D.R. Does water transport scale universally with tree size? *Funct. Ecol.* **2005**, *19*, 558–565. [CrossRef]
20. King, G.; Fonti, P.; Nievergelt, D.; Büntgen, U.; Frank, D. Climatic drivers of hourly to yearly tree radius variations along a 6 °C natural warming gradient. *Agr. Forest Meteorol.* **2013**, *168*, 36–46. [CrossRef]
21. Turcotte, A.; Rossi, S.; Deslauriers, A.; Krause, C.; Morin, H. Dynamics of depletion and replenishment of water storage in stem and roots of black spruce measured by dendrometers. *Front. Plant Sci.* **2011**, *2*, 21. [CrossRef] [PubMed]
22. Deslauriers, A.; Morin, H.; Urbinati, C.; Carrer, M. Daily weather response of balsam fir (Abies balsamea (L.) Mill.) stem radius increment from dendrometer analysis in the boreal forests of Québec (Canada). *Trees* **2003**, *17*, 477–484. [CrossRef]
23. Biondi, F.; Rossi, S. Plant-water relationships in the Great Basin Desert of North America derived from Pinus monophylla hourly dendrometer records. *Int. J. Biometeorol.* **2015**, *59*, 939–953. [CrossRef] [PubMed]
24. Drew, D.M.; Downes, G.M. The use of precision dendrometers in research on daily stem size and wood property variation: A review. *Dendrochronologia* **2009**, *27*, 159–172. [CrossRef]

25. Grayson, D.K. *The Great Basin: A Natural Prehistory, Revised and Expanded Edition*; University of California Press: Berkeley, CA, USA, 2011.
26. Osmond, C.B.; Pitelka, L.F.; Hidy, G.M. *Plant Biology of the Basin and Range*; Springer-Verlag: Berlin, Germany, 1990; p. 375.
27. Adams, D.K.; Comrie, A.C. The North American Monsoon. *Bull. Am. Meteorol. Soc.* **1997**, *78*, 2197–2213. [CrossRef]
28. Szejner, P.; Wright, W.E.; Babst, F.; Belmecheri, S.; Trouet, V.; Leavitt, S.W.; Ehleringer, J.R.; Monson, R.K. Latitudinal gradients in tree ring stable carbon and oxygen isotopes reveal differential climate influences of the North American Monsoon System. *J. Geophys. Res. Biogeosci.* **2016**, *121*, 1978–1991. [CrossRef]
29. Ellison, A.M.; Bank, M.S.; Clinton, B.D.; Colburn, E.A.; Elliott, K.; Ford, C.R.; Foster, D.R.; Kloeppel, B.D.; Knoepp, J.D.; Lovett, G.M.; et al. Loss of foundation species: Consequences for the structure and dynamics of forested ecosystems. *Front. Ecol. the Environ.* **2005**, *3*, 479–486. [CrossRef]
30. Mensing, S.; Strachan, S.; Arnone, J.; Fenstermaker, L.; Biondi, F.; Devitt, D.; Johnson, B.; Bird, B.; Fritzinger, E. A network for observing Great Basin climate change. *EOS Trans. Amer. Geophys. Union* **2013**, *94*, 105–106. [CrossRef]
31. Johnson, B.G.; Verburg, P.S.J.; Arnone, J.A. Effects of climate and vegetation on soil nutrients and chemistry in the Great Basin studied along a latitudinal-elevational climate gradient. *Plant Soil* **2014**, *382*, 151–163. [CrossRef]
32. Gochis, D.; Schemm, J.; Shi, W.; Long, L.; Higgins, W.; Douglas, A. A Forum for Evaluating Forecasts of the North American Monsoon. *Eos Trans. Amer. Geophys. Union* **2009**, *90*, 249–251. [CrossRef]
33. Biondi, F.; Hartsough, P.C.; Estrada, I.G. Daily weather and tree growth at the tropical treeline of North America. *Arct. Antarctic Alp. Res.* **2005**, *37*, 16–24. [CrossRef]
34. Biondi, F.; Hartsough, P. Using automated point dendrometers to analyze tropical treeline stem growth at Nevado de Colima, Mexico. *Sensors* **2010**, *10*, 5827. [CrossRef] [PubMed]
35. Ziaco, E.; Biondi, F.; Rossi, S.; Deslauriers, A. Environmental drivers of cambial phenology in Great Basin bristlecone pine. *Tree Physiol.* **2016**, *36*, 818–831. [CrossRef] [PubMed]
36. Lawrence, M.G. The Relationship between Relative Humidity and the Dewpoint Temperature in Moist Air: A Simple Conversion and Applications. *Bull. Amer. Meteorol. Soc.* **2005**, *86*, 225–233. [CrossRef]
37. Feld, S.I.; Cristea, N.C.; Lundquist, J.D. Representing atmospheric moisture content along mountain slopes: Examination using distributed sensors in the Sierra Nevada, California. *Water Resour. Res.* **2013**, *49*, 4424–4441. [CrossRef]
38. McEvoy, D.J.; Mejia, J.F.; Huntington, J.L. Use of an Observation Network in the Great Basin to Evaluate Gridded Climate Data. *J. Hydrometeorol.* **2014**, *15*, 1913–1931. [CrossRef]
39. Kunkel, K.E. Simple procedures for extrapolation of humidity variables in the mountainous western United States. *J. Clim.* **1989**, *2*, 656–669. [CrossRef]
40. Maaten, E.V.D.; Maaten-Theunissen, M.V.D.; Smiljanić, M.; Rossi, S.; Simard, S.; Wilmking, M.; Deslauriersb, A.; Fontie, P.; Arxe, G.; Bouriaudf, O. DendrometeR: Analyzing the pulse of trees in R. *Dendrochronologia* **2016**, *40*, 12–16. [CrossRef]
41. R Core Team. R: A Language and Environment for Statistical Computing. R Foundation for Statistical Computing: Vienna, Austria, 2015.
42. Deslauriers, A.; Rossi, S.; Turcotte, A.; Morin, H.; Krause, C. A three-step procedure in SAS to analyze the time series from automatic dendrometers. *Dendrochronologia* **2011**, *29*, 151–161. [CrossRef]
43. Ziaco, E.; Truettner, C.; Biondi, F.; Bullock, S. Moisture-driven xylogenesis in Pinus ponderosa from a Mojave Desert mountain reveals high phenological plasticity. *Plant Cell Environ.* **2018**, *41*, 823–836. [CrossRef] [PubMed]
44. Chan, T.; Berninger, F.; Mencuccini, M.; Nikinmaa, E. Separating water-potential induced swelling and shrinking from measured radial stem variations reveals a cambial growth and osmotic concentration signal. *Plant Cell Environ.* **2016**, *39*, 233–244. [CrossRef] [PubMed]
45. Hosmer, J.D.W.; Lemeshow, S.; Sturdivant, R.X. *The Multiple Logistic Regression Model, in Applied Logistic Regression*; John Wiley & Sons: Hoboken, NJ, USA, 2013; pp. 35–47.
46. Quinn, G.P.; Keough, M.J. *Experimental Design and Data Analysis for Biologists*; Cambridge University Press: Cambridge, UK, 2002.

47. Pearce, J.; Ferrier, S. Evaluating the predictive performance of habitat models developed using logistic regression. *Ecol. Modell.* **2000**, *133*, 225–245. [CrossRef]
48. Robin, X.; Turck, N.; Hainard, A.; Tiberti, N.; Lisacek, F.; Sanchez, J.C.; Müller, M. pROC: an open-source package for R and S+ to analyze and compare ROC curves. *BMC Bioinformatics* **2011**, *12*, 77. [CrossRef] [PubMed]
49. Oberhuber, W.; Hammerle, A.; Kofler, W. Tree water status and growth of saplings and mature Norway spruce (Picea abies) at a dry distribution limit. *Front. Plant Sci.* **2015**, *6*, 703. [CrossRef] [PubMed]
50. Zweifel, R.; Häsler, R. Dynamics of water storage in mature subalpine Picea abies: Temporal and spatial patterns of change in stem radius. *Tree Physiol.* **2001**, *21*, 561–569. [CrossRef] [PubMed]
51. Maherali, H.; DeLucia, E.H. Influence of climate-driven shifts in biomass allocation on water transport and storage in ponderosa pine. *Oecologia* **2001**, *129*, 481–491. [CrossRef] [PubMed]
52. Zweifel, R.; Häsler, R. Frost-induced reversible shrinkage of bark of mature subalpine conifers. *Agr. Forest Meteorol.* **2000**, *102*, 213–222. [CrossRef]
53. Vieira, J.; Rossi, S.; Campelo, F.; Freitas, H.; Nabais, C. Seasonal and daily cycles of stem radial variation of Pinus pinaster in a drought-prone environment. *Agr. Forest Meteorol.* **2013**, *180*, 173–181. [CrossRef]
54. Tardif, J.; Flannigan, M.; Bergeron, Y. An Analysis of the Daily Radial Activity of 7 Boreal Tree Species, Northwestern Quebec. *Environ. Monit. Assess.* **2001**, *67*, 141–160. [CrossRef] [PubMed]
55. Güney, A.; Küppers, M.; Rathgeber, C.; Şahin, M.; Zimmermann, R. Intra-annual stem growth dynamics of Lebanon Cedar along climatic gradients. *Trees* **2017**, *31*, 1–20. [CrossRef]
56. Linares, J.C.; Camarero, J.J.; Carreira, J.A. Plastic responses of Abies pinsapo xylogenesis to drought and competition. *Tree Physiol.* **2009**, *29*, 1525–1536. [CrossRef] [PubMed]
57. Rossi, S.; Morin, H.; Deslauriers, A. Multi-scale influence of snowmelt on xylogenesis of black spruce. *Arct. Antarct. Alp. Res.* **2011**, *43*, 457–464. [CrossRef]
58. Gaylord, M.L.G.L.; Kolb, T.E.K.E.; Wallin, K.F.W.F.; Wagner, M.R.W.R. Seasonal dynamics of tree growth, physiology, and resin defenses in a northern Arizona ponderosa pine forest. *Can. J. Forest Res.* **2007**, *37*, 1173–1183. [CrossRef]
59. Kolb, T.E.; Stone, J.E. Differences in leaf gas exchange and water relations among species and tree sizes in an Arizona pine-oak forest. *Tree Physiol.* **2000**, *20*, 1–12. [CrossRef] [PubMed]
60. Kozlowski, T.T. Soil moisture and absorption of water by tree roots. *J. Arboric.* **1987**, *13*, 39–46.
61. Gambetta, G.A.; Knipfer, T.; Fricke, W.; Mcelrone, A.J. Aquaporins and Root Water Uptake. In *Plant Aquaporins: From Transport to Signaling*; Chaumont, F., Tyerman, S.D., Eds.; Springer International Publishing: New York, NY, USA, 2017; pp. 133–153.
62. Bennett, A.C.; Mcdowell, N.G.; Allen, C.D.; Andersonteixeira, K.J. Larger trees suffer most during drought in forests worldwide. *Nat. Plants* **2015**, *1*, 15139. [CrossRef] [PubMed]
63. Domec, J.C.; Gartner, B.L. Relationship between growth rates and xylem hydraulic characteristics in young, mature and old-growth ponderosa pine trees. *Plant Cell Environ.* **2003**, *26*, 471–483. [CrossRef]
64. Hughes, M.K.; Funkhouser, G. Frequency-Dependent Climate Signal in Upper and Lower Forest Border Tree Rings in the Mountains of the Great Basin. *Clim. Change* **2003**, *59*, 233–244. [CrossRef]
65. Gazol, A.; Camarero, J.J.; Anderegg, W.R.L.; Vicente-Serrano, S.M. Impacts of droughts on the growth resilience of Northern Hemisphere forests. *Global Ecol. Biogeogr.* **2017**, *26*, 166–176. [CrossRef]
66. Allen, C.D.; Breshears, D.D.; McDowell, N.G. On underestimation of global vulnerability to tree mortality and forest die-off from hotter drought in the Anthropocene. *Ecosphere* **2015**, *6*, art129. [CrossRef]
67. Grossiord, C.; Sevanto, S.; Borrego, I.; Chan, A.M.; Collins, A.D.; Dickman, L.T.; Hudson, P.J.; McBranch, N.; Michaletz, S.T.; Pockman, W.T.; et al. Tree water dynamics in a drying and warming world. *Plant Cell Environ.* **2017**, *40*, 1861–1873. [CrossRef] [PubMed]
68. Peñuelas, J.; Canadell, J.G.; Ogaya, R. Increased water-use efficiency during the 20th century did not translate into enhanced tree growth. *Global Ecol. Biogeogr.* **2011**, *20*, 597–608. [CrossRef]
69. Mencuccini, M. The ecological significance of long-distance water transport: short-term regulation, long-term acclimation and the hydraulic costs of stature across plant life forms. *Plant Cell Environ.* **2003**, *26*, 163–182. [CrossRef]
70. Limousin, J.M.; Rambal, S.; Ourcival, J.M.; Rodríguezcalcerrada, J.; Pérezramos, I.M.; Rodríguezcortina, R.; Misson, L.; Joffre, R. Morphological and phenological shoot plasticity in a Mediterranean evergreen oak facing long-term increased drought. *Oecologia* **2012**, *169*, 565–577. [CrossRef] [PubMed]

71. Griffin, D.; Anchukaitis, K.J. How unusual is the 2012–2014 California drought? *Geophys. Res. Lett.* **2014**, *41*, 9017–9023. [CrossRef]
72. Anderegg, W.R.; Schwalm, C.; Biondi, F.; Camarero, J.J.; Koch, G.; Litvak, M.; Ogle, K.; Shaw, J.D.; Shevliakova, E.; Williams, P.; et al. Pervasive drought legacies in forest ecosystems and their implications for carbon cycle models. *Science* **2015**, *349*, 528–532. [CrossRef] [PubMed]
73. Anderegg, W.R.L.; Kane, J.M.; Anderegg, L.D.L. Consequences of widespread tree mortality triggered by drought and temperature stress. *Nat. Clim. Change* **2013**, *3*, 30–36. [CrossRef]
74. Williams, A.P.; Allen, C.D.; Macalady, A.K.; Griffin, D.; Woodhouse, C.A.; Meko, D.M.; Swetnam, T.W.; Rauscher, S.A.; Seager, R.; Grissino-Mayer, H.D.; et al. Temperature as a potent driver of regional forest drought stress and tree mortality. *Nat. Clim. Change* **2012**, *3*, 292. [CrossRef]
75. Sánchez-Salguero, R.; Camarero, J.J.; Rozas, V.; Génova, M.; Olano, J.M.; Arzac, A.; Gazol, A.; Caminero, L.; Tejedor, E.; Luis, M.; et al. Resist, recover or both? Growth plasticity in response to drought is geographically structured and linked to intraspecific variability in Pinus pinaster. *J. Biogeogr.* **2018**, *45*, 1–14. [CrossRef]
76. Martin, J.; Looker, N.; Hoylman, Z.; Jencso, K.; Hu, J. Hydrometeorology organizes intra-annual patterns of tree growth across time, space and species in a montane watershed. *New Phytol.* **2017**, *215*, 1387–1398. [CrossRef] [PubMed]
77. Biondi, F.; Bradley, M.L. Long-term survivorship of single-needle pinyon (Pinus monophylla) in mixed-conifer ecosystems of the Great Basin, USA. *Ecosphere* **2013**, *4*, art120. [CrossRef]
78. Lloret, F.; Keeling, E.G.; Sala, A. Components of tree resilience: effects of successive low-growth episodes in old ponderosa pine forests. *Oikos* **2011**, *120*, 1909–1920. [CrossRef]
79. Daly, C.; Conklin, D.R.; Unsworth, M.H. Local atmospheric decoupling in complex topography alters climate change impacts. *Int. J. Climatol.* **2010**, *30*, 1857–1864. [CrossRef]
80. Loarie, S.R.; Duffy, P.B.; Hamilton, H.; Asner, G.P.; Field, C.B.; Ackerly, D.D. The velocity of climate change. *Nature* **2009**, *462*, 1052. [CrossRef] [PubMed]
81. Millar, C.I.; Westfall, R.D.; Delany, D.L. Response of high-elevation limber pine (Pinus flexilis) to multiyear droughts and 20th-century warming, Sierra Nevada, California, USA. *Can. J. Forest Res.* **2007**, *37*, 2508–2520. [CrossRef]
82. Millar, C.I.; Westfall, R.D.; Delany, D.L.; King, J.C.; Graumlich, L.J. Response of Subalpine Conifers in the Sierra Nevada, California, U.S.A., to 20th-Century Warming and Decadal Climate Variability. *Arc. Antarct. Alp. Res.* **2004**, *36*, 181–200. [CrossRef]
83. LaMarche, V.C., Jr.; Mooney, H.A. Recent climatic change and development of the bristlecone pine (P. longaeva Bailey) krummholz zone, Mt. Washington Nevada. *Arc. Alp. Res.* **1972**, *4*, 61–72. [CrossRef]

MDPI
St. Alban-Anlage 66
4052 Basel
Switzerland
Tel. +41 61 683 77 34
Fax +41 61 302 89 18
www.mdpi.com

Forests Editorial Office
E-mail: forests@mdpi.com
www.mdpi.com/journal/forests

www.ingramcontent.com/pod-product-compliance
Lightning Source LLC
Chambersburg PA
CBHW041753220926
51847CB00014B/192
* 9 7 8 3 0 3 8 9 7 4 2 5 3 *